Plate Structures

SOLID MECHANICS AND ITS APPLICATIONS
Volume 178

Aims and Scope of the Series

The fundamental questions arising in mechanics are: *Why?*, *How?*, and *How much?* The aim of this series is to provide lucid accounts written by authoritative researchers giving vision and insight in answering these questions on the subject of mechanics as it relates to solids.

The scope of the series covers the entire spectrum of solid mechanics. Thus it includes the foundation of mechanics; variational formulations; computational mechanics; statics, kinematics and dynamics of rigid and elastic bodies: vibrations of solids and structures; dynamical systems and chaos; the theories of elasticity, plasticity and viscoelasticity; composite materials; rods, beams, shells and membranes; structural control and stability; soils, rocks and geomechanics; fracture; tribology; experimental mechanics; biomechanics and machine design.

The median level of presentation is the first year graduate student. Some texts are monographs defining the current state of the field; others are accessible to final year undergraduates; but essentially the emphasis is on readability and clarity.

For further volumes:
http://www.springer.com/series/6557

Victor Birman

Plate Structures

Springer

Victor Birman
Engineering Education Center
Missouri University of Science
and Technology
One University Blvd.
St. Louis, Missouri 63121
USA
vbirman@mst.edu

ISSN 0925-0042
ISBN 978-94-007-1714-5 e-ISBN 978-94-007-1715-2
DOI 10.1007/978-94-007-1715-2
Springer Dordrecht Heidelberg London New York

Library of Congress Control Number: 2011932669

Cover design: SPi Publisher Services

Printed on acid-free paper

Springer is part of Springer Science+Business Media (www.springer.com)

Preface

Plate structures are typically thin structures, working under bending and/or in-plane loads and are extensively used in engineering applications, such as aerospace, mechanical and civil engineering and naval architecture. Plates can be manufactured from such isotropic materials as steel, aluminum or titanium alloys. Notably, the spectrum of material systems available to a designer of plate structures has significantly expanded in recent decades, so that composite and sandwich plates are now being widely used in modern engineering. Advanced plate structures, incorporating piezoelectric and shape memory elements as well as nanotube-reinforced materials are now gathering added attention from scientists and engineers.

While in the majority of books, plates are considered jointly with shells, there are important differences in the response of these types of structures, justifying a separate book exclusively concerned with the plate analysis. Shell structures are characterized by curvature in one or several planes (e.g., cylindrical, spherical, ellipsoidal and conical shells). The curvature may increase the stiffness of the shell, but it also makes it more vulnerable in certain loading scenarios. For example, while a spherical shell or panel possesses high strength when loaded by pressure applied at the inner surface, it may become unstable and buckle, with catastrophic consequences, if pressure is applied at the outer surface. On the other hand, a flat plate subject to pressure does not exhibit buckling tendencies and although its bending response is inferior to that of a comparable spherical shell, it may be preferable. The difference in the qualitative response of plate and shell structures, combined with a broad range of available plate materials and designs, motivated the author to write this book specially dedicated to the behavior and response of plates.

The purpose of the book is to present the foundations, methods of analysis, and guidelines for scientists and engineers working with plate structures. Accordingly, a brief paragraph on design philosophy is included at the end of each chapter; while by no means comprehensive, these paragraphs emphasize some of the most critical aspects of design and provide recommendations to designers. The subject of elastic-plastic response of plates is excluded from consideration since plastic stresses are usually disallowed in applications. The first chapter of the book presents the general theoretical foundations employed in the analysis of plates. It describes the fundamental concepts and mathematical apparatus of mechanics that are applied in the subsequent chapters to address specific plate problems. The second chapter of the book concentrates on widely used rectangular isotropic plates. The problems of boundary conditions, response to various loading classes, including bending and buckling, the presence of an elastic foundation, and the effect of initial imperfections are discussed. In addition, it illustrates the analysis of reinforced plates and discusses nonlinear postbuckling response.

Plates of non-rectangular shapes, including circular plates, are considered in the third chapter of the book. As most of these plates defy a convenient and accurate analytical solution, we restrict ourselves to the case of axisymmetric bending of circular plates and several non-circular classes of plates, where closed form solutions are known. For other cases, such as an asymmetric bending of circular plates, we present only the theoretical approach and mathematical formulation of the problem. The solutions for these cases are usually obtained via numerical methods, which are out of the scope of this book.

Dynamic problems are reviewed in the fourth chapter of the book. Besides the subject of free and forced vibrations, and vibrations of stringer-reinforced plates, the qualitative effect of large amplitudes, i.e. nonlinear motion, is discussed. The response of plates to non-periodic loads is demonstrated via the example of blast loading. In addition, dynamic instability constituted in large-amplitude vibrations, in response to in-plane harmonic-in-time loads, is illustrated.

Plates manufactured from composites are discussed in the fifth chapter. Composite materials have become popular in an amazing spectrum of applications due to their features enabling a designer to tailor and optimize the structure. This chapter provides a formulation of the analysis of thin-walled laminated structures and includes bending and buckling problems of thin composite laminates as well as static and dynamic response of stringer-reinforced composite plates. In addition, the formulation for shear deformable plates exhibiting non-negligible transverse shear deformability is illustrated using the first-order shear deformation theory. The theory of shear deformable plates is followed with the discussion on the analysis of sandwich plates since these structures can seldom be adequately modeled by a classical thin plate theory.

Thermoelastic problems of plate structures are discussed in the sixth chapter. We start with the heat transfer problem, establishing the profile of temperature throughout the plate. Only after the distribution of temperature and its effect on material properties are known, the solution can proceed to thermomechanical stress or stability analysis. The corresponding solutions are shown in the chapter, while a representative example of the response of a composite plate to fire demonstrates a typical modern engineering problem.

The last chapter of the book is concerned with representative advanced applications of plates using modern materials and concepts on the examples of plates with piezoelectric sensors and actuators and functionally graded plates. Governing equations for the former class of plates are formulated, both for thin and for relatively thick plates (the latter requires analysis through first-order shear deformation theory). Active control of plates using composite piezoelectric stringers is discussed. The effect of temperature on the readings from piezoelectric sensors bonded to plates is considered. In this chapter we also address a comprehensive analysis of functionally graded plates subject to thermomechanical loading, with the exception of the solution of the micromechanical problem, i.e. heat transfer and the subsequent thermomechanical stress problem.

The goal of the book is to provide the reader with a vision and an insight into the problems of analysis and design of plate structures. The comprehension of the corresponding problems and the understanding of the limitations and the applicability of various solutions and plate models are particularly emphasized. Many representative solutions are either shown in their entirety or outlined so that the reader can clearly see the subsequent steps. Some available solutions that result in a very time-consuming analysis and/or approximate results are omitted since industry relies on numerical analyses of the corresponding problems. For example, while the solution of the problem of axisymmetric bending of circular plates is exact, so that using a numerical approach is not warranted (it is important to realize that finite element or finite difference methods only provide approximate solutions), asymmetric bending of circular plates can be quicker and more accurately analyzed using a numerical procedure.

The book can be used as a textbook for a one-semester or one-quarter course. It can also be useful to engineers whose work involves design of thin-walled structures, including plates, and to researchers working in this area.

Acknowledgements

I considered writing this book in 2008 and discussed it with my friend and colleague, Dr. George A. Kardomateas, from Georgia Institute of Technology. George encouraged me to write the book on the subject of mechanics and analysis of plates and recommended to contact Springer with the project. Subsequently, I worked with Publishing Editor for Engineering in the Amsterdam Office of Springer, Mrs. Nathalie Jacobs, and Editorial Assistant, Ms. Anneke Pot, whose support I warmly appreciate.

I was fortunate to collaborate during my career with a number of outstanding scientists whom I owe gratitude both for the knowledge they shared with me as well as for their friendship. In particular, I would like to acknowledge Drs. Charles W. Bert who worked at the University of Oklahoma prior to his retirement, George J. Simitses who retired from Georgia Institute of Technology, Liviu Librescu who died heroically protecting his students from a deranged gunman on campus of Virginia Polytechnic Institute and State University and Larry W. Byrd from Wright-Patterson Air Force Base. I would like to express my deepest appreciation to my assistant at Missouri University of Science and Technology, Mrs. Debbie Benenati, who prepared most of figures appearing in the book. Drs. Guy M. Genin (Washington University), Larry W. Byrd (Wright-Patterson Air Force Base) and Dimitris A. Saravanos (University of Patras, Greece) kindly agreed to review several chapters and provided me with invaluable advice. Finally, last but not least, I am grateful to my wife Anna, my children Michael and Shirley, their spouses Mari and Ajey and my granddaughter Sasha for warm and loving family environment that made my work, including writing this book, so pleasant and enjoyable.

Contents

Chapter 1
Introduction and Basic Concepts

This chapter represents a collection of concepts and mathematical formulations that are employed to develop the theory of plates. In the subsequent chapters the equations introduced here will be reduced to the form used in the relevant version of the theory. The material outlined in this chapter refers to derivations and concepts that can be found in relevant references concerned with solid mechanics or theory of elasticity (some of these sources are referred to below). Accordingly, the goal is to both illustrate that the background of the theory of plates can be traced to the fundamental concepts of mechanics as well as to outline details of this background so that we can refer to them in the subsequent chapters, without the need in further justification or elucidation.

1.1 Theoretical Foundations of the Theory of Plates

Four classes of equations in any solid mechanics model, including the theory of plates, involve equations of motion, constitutive law, strain-displacement equations and compatibility equations. Equations of motion guarantee equilibrium of stresses and their moments acting on an infinitesimal element detached from the structure, including dynamic effects. Constitutive law provides a link between stresses and strains in the material. Strain-displacement equations specify relationships between strains and displacements (and rotations). Compatibility equations establish conditions that must be satisfied to ensure single-valuedness of displacements found by integration of strain-displacement equations.

We begin with well known concepts of tensors of stress and strain and the vector of displacements. These tensors and vector characterize the state of stress and deformation within an infinitesimal element encompassing the point of interest within the body.

Consider a particle in an arbitrary Cartesian (rectangular) coordinate system in a three-dimensional space (Fig. 1.1). Two sets of coordinates can be used to characterize the position of the particle. The coordinates of the particle in the

V. Birman, *Plate Structures*, Solid Mechanics and Its Applications 178,
DOI 10.1007/978-94-007-1715-2_1, © Springer Science+Business Media B.V. 2011

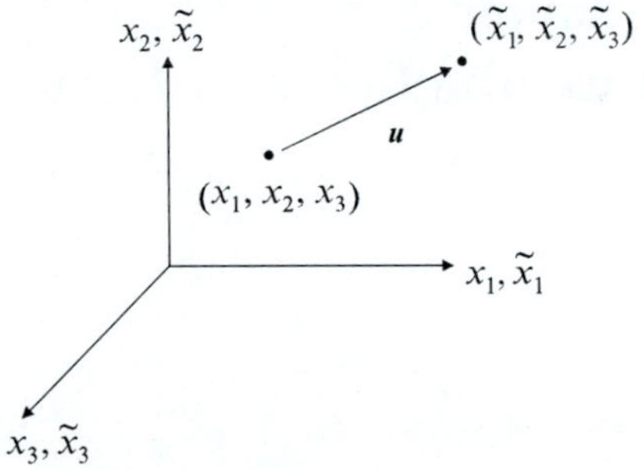

Fig. 1.1 Displacements of a point in the Cartesian coordinate system (x_i = undeformed coordinates, $\tilde{x}_i$= deformed coordinates, u= displacement vector)

undeformed (original) position of the body are (x_1, x_2, x_3), while the coordinates moving with the particle to a new location as the body experiences deformations and referred to as deformed coordinates are $(\tilde{x}_1, \tilde{x}_2, \tilde{x}_3)$. Accordingly, one can refer to "undeformed" (x_1, x_2, x_3) and "deformed" $(\tilde{x}_1, \tilde{x}_2, \tilde{x}_3)$ coordinate systems, the former defining the position of the particle before deformations and the latter referring to the position of the same particle after deformations took place. The displacements of the point produced by external loads applied to the body form the elements of the vector

$$\mathbf{u} = \begin{Bmatrix} u_1(x_1, x_2, x_3) \\ u_2(x_1, x_2, x_3) \\ u_3(x_1, x_2, x_3) \end{Bmatrix} = \begin{Bmatrix} \tilde{x}_1(x_1, x_2, x_3) - x_1 \\ \tilde{x}_2(x_1, x_2, x_3) - x_2 \\ \tilde{x}_3(x_1, x_2, x_3) - x_3 \end{Bmatrix} \tag{1.1}$$

where u_i identifies a displacement along the x_i-axis.

While (1.1) specifies displacements in the undeformed coordinate system, it is also possible to define displacements using deformed coordinates:

$$\mathbf{u} = \begin{Bmatrix} u_1(\tilde{x}_1, \tilde{x}_2, \tilde{x}_3) \\ u_2(\tilde{x}_1, \tilde{x}_2, \tilde{x}_3) \\ u_3(\tilde{x}_1, \tilde{x}_2, \tilde{x}_3) \end{Bmatrix} = \begin{Bmatrix} \tilde{x}_1 - x_1(\tilde{x}_1, \tilde{x}_2, \tilde{x}_3) \\ \tilde{x}_2 - x_2(\tilde{x}_1, \tilde{x}_2, \tilde{x}_3) \\ \tilde{x}_3 - x_3(\tilde{x}_1, \tilde{x}_2, \tilde{x}_3) \end{Bmatrix} \tag{1.2}$$

Equations 1.1 and 1.2 imply that the displacement vector can be determined if the transformations from the deformed to undeformed position or vice versa are known, i.e.

$$\tilde{x}_i = \tilde{x}_i\left(x_j\right), \quad x_i = x_i\left(\tilde{x}_j\right), \quad i, j = 1, 2, 3 \tag{1.3}$$

The strains at the point represent both the relative stretching or contraction along the corresponding coordinate direction and shearing in the plane formed by a couple of coordinate axes (there are three such mutually perpendicular planes). While the concept of linear (normal) strain is well understood, it is useful to indicate that the shear strain is simply the change of the right angle between coordinate directions that are mutually perpendicular in the undeformed state but change their orientation as a result of deformation. The strains are derived according to either Lagrange's or Euler's approach. The Lagrangian tensor of strain refers to strains in the undeformed coordinate system (this tensor is also called Green's strain tensor). In the contrary, the Eulerian tensor of strain is introduced in the system of coordinates of the deformed body (this tensor is sometimes called Cauchy's tensor of strains). In terms of displacements the Green and Cauchy tensors are given by (Fung 1994):

$$\varepsilon_{ii} \equiv \varepsilon_{x_i x_i} = \frac{\partial u_i}{\partial x_i} + \frac{1}{2}\sum_{n=1}^{3}\left(\frac{\partial u_n}{\partial x_i}\right)^2$$

$$\varepsilon_{ij} \equiv \varepsilon_{x_i x_j} = \frac{\gamma_{ij}}{2} = \frac{1}{2}\left[\frac{\partial u_j}{\partial x_i} + \frac{\partial u_i}{\partial x_j} + \sum_{n=1}^{3}\left(\frac{\partial u_n}{\partial x_i}\frac{\partial u_n}{\partial x_j}\right)\right] \quad n = 1, 2, 3 \tag{1.4}$$

and

$$\tilde{\varepsilon}_{ii} \equiv \tilde{\varepsilon}_{\tilde{x}_i \tilde{x}_i} = \frac{\partial u_i}{\partial \tilde{x}_i} - \frac{1}{2}\sum_{n=1}^{3}\left(\frac{\partial u_n}{\partial \tilde{x}_i}\right)^2$$

$$\tilde{\varepsilon}_{ij} \equiv \tilde{\varepsilon}_{\tilde{x}_i \tilde{x}_j} = \frac{\tilde{\gamma}_{ij}}{2} = \frac{1}{2}\left[\frac{\partial u_j}{\partial \tilde{x}_i} + \frac{\partial u_i}{\partial \tilde{x}_j} - \sum_{n=1}^{3}\left(\frac{\partial u_n}{\partial \tilde{x}_i}\frac{\partial u_n}{\partial \tilde{x}_j}\right)\right] \quad n = 1, 2, 3 \tag{1.5}$$

respectively, where as above, $\tilde{x}_i$ are the coordinates in the deformed position and x_i are undeformed coordinates. The first equations in (1.4) and (1.5) define linear strains along the corresponding axes, while the second equations introduce shear strains in the plane formed by the corresponding couple of axes. For example, ε_{11} is the Lagrange (Green) strain along the x_1-axis, and $\tilde{\varepsilon}_{23}$ is the Euler (Cauchy) shear strain in the $\tilde{x}_2\tilde{x}_3$ plane. It is evident that the strains defined by (1.4) or (1.5) are symmetric, i.e. $\varepsilon_{ij} = \varepsilon_{ji}$, $\tilde{\varepsilon}_{ij} = \tilde{\varepsilon}_{ji}$. The Cauchy strain tensor is also sometimes called the Almansi tensor.

Note that in the case of a geometrically linear formulation the difference between the Lagrangian and Eulerian strain tensors disappears and the strains are defined as the following functions of displacements:

$$\varepsilon_{ii} = \frac{\partial u_i}{\partial x_i}$$

$$\varepsilon_{ij} = \frac{\gamma_{ij}}{2} = \frac{1}{2}\left(\frac{\partial u_j}{\partial x_i} + \frac{\partial u_i}{\partial x_j}\right) \tag{1.6}$$

In cases where environmental effects, such as temperature or moisture are present, the strain-displacement relationships are modified to reflect the corresponding contributions as shown in Sect. 1.9.

The strain components that are defined by (1.4), (1.5) or (1.6) form the tensor of strain. The tensor represents an object whose properties are independent of the reference system of coordinates. Furthermore, this object is characterized by its components that change from one system of coordinates to another according to a certain transformation law (see for example, Sokolnikoff 1964). A vector represents a tensor of rank 1 that is characterized by three components, while the tensor of strain characterized by nine components is called a tensor of rank 2.

It should be emphasized that while there are six independent components of the tensor of strain, they depend on three displacements only. Therefore, additional constraints should be superimposed on strains to ensure that the integration process yields single-valued solutions for displacements. For example, in the case of infinitesimal strains (geometrically linear formulation where the difference between Lagrangian and Eulerian strains disappears), the 81 compatibility equations in a Cartesian coordinate system are (Fung 1994):

$$\frac{\partial^2 \varepsilon_{ij}}{\partial x_k \partial x_l} + \frac{\partial^2 \varepsilon_{kl}}{\partial x_i \partial x_j} - \frac{\partial^2 \varepsilon_{ik}}{\partial x_j \partial x_l} - \frac{\partial^2 \varepsilon_{jl}}{\partial x_i \partial x_k} = 0, \quad i, j, k, l = 1, 2, 3 \tag{1.7}$$

Equations 1.7 can be verified by differentiating the second equation (1.6) with respect to x_k and x_l yielding $\frac{\partial^2 \varepsilon_{ij}}{\partial x_k \partial x_l}$. Subsequently one can obtain the other strain derivatives in (1.7) by interchanging the subscripts. Finally, the substitution of the corresponding expressions for strains in terms of displacements into (1.7) yields an identity.

As a result of symmetry of strains, only six equations (1.7) are independent. They are

$$\begin{aligned}
\frac{\partial^2 \varepsilon_{11}}{\partial x_2 \partial x_3} &= \frac{\partial^2 \varepsilon_{12}}{\partial x_1 \partial x_3} - \frac{\partial^2 \varepsilon_{23}}{\partial x_1^2} + \frac{\partial^2 \varepsilon_{13}}{\partial x_1 \partial x_2} \\
\frac{\partial^2 \varepsilon_{22}}{\partial x_1 \partial x_3} &= \frac{\partial^2 \varepsilon_{23}}{\partial x_1 \partial x_2} - \frac{\partial^2 \varepsilon_{13}}{\partial x_2^2} + \frac{\partial^2 \varepsilon_{12}}{\partial x_2 \partial x_3} \\
\frac{\partial^2 \varepsilon_{33}}{\partial x_1 \partial x_2} &= \frac{\partial^2 \varepsilon_{13}}{\partial x_2 \partial x_3} - \frac{\partial^2 \varepsilon_{12}}{\partial x_3^2} + \frac{\partial^2 \varepsilon_{23}}{\partial x_1 \partial x_3} \\
2\frac{\partial^2 \varepsilon_{12}}{\partial x_1 \partial x_2} &= \frac{\partial^2 \varepsilon_{11}}{\partial x_2^2} + \frac{\partial^2 \varepsilon_{22}}{\partial x_1^2} \\
2\frac{\partial^2 \varepsilon_{23}}{\partial x_2 \partial x_3} &= \frac{\partial^2 \varepsilon_{22}}{\partial x_3^2} + \frac{\partial^2 \varepsilon_{33}}{\partial x_2^2} \\
2\frac{\partial^2 \varepsilon_{13}}{\partial x_1 \partial x_3} &= \frac{\partial^2 \varepsilon_{33}}{\partial x_1^2} + \frac{\partial^2 \varepsilon_{11}}{\partial x_3^2}
\end{aligned} \tag{1.8}$$

The compatibility equations where strains are replaced by stresses through the linear constitutive relations are called the Beltrami-Mitchell equations.

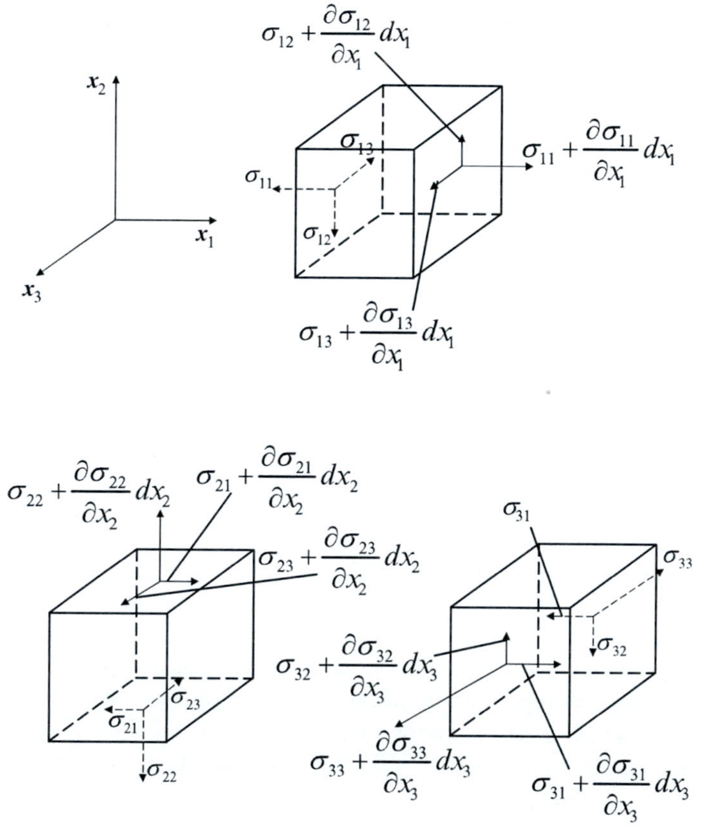

Fig. 1.2 Stresses acting on an infinitesimal element with dimensions $dx_1,\ dx_2,\ dx_3$. Each figure shows the system of stresses acting on a couple of opposite faces of the element

The analysis of stresses within the domain occupied by the body requires us to formulate the equations of motion (or equations of equilibrium, in the static problem). We begin by considering the state of stress at a point. Naturally, it is impossible to identify the stresses at a mathematical point, so we draw an infinitesimal parallelepiped element encompassing the point in question. The stresses acting on the faces of this element replace the effect of the cut-off part of the body (in other words, we analyze the equilibrium of the element considering its free-body diagram). As is shown in Fig. 1.2, there are three stress components applied at each face, including one component perpendicular to the face and two mutually perpendicular shear components. The latter components represent the effect of the shear stress (the vector of the shear stress does not have to be oriented along one of the coordinate axes, but it can be resolved into two components that are shown in the Fig. 1.2).

Equations of motion can be derived from the principle of conservation of the linear momentum of the body that is actually Newton's second law. According to this law, the time rate of change of the linear momentum of the body is equal to the applied force. Mathematically, this law can be expressed in the three-dimensional system of coordinates (x_1, x_2, x_3) as follows:

$$\frac{\partial \sigma_{ij}}{\partial x_j} + X_i = \rho \frac{D v_i}{D t} \tag{1.9}$$

where X_i are the projections of the body force on the x_i coordinate axis, ρ is the mass density, v_i is the component of the velocity vector, t is time, and $\frac{Dv_i}{Dt}$ the material derivative of the velocity representing the rate of its change with respect to time t, i.e. acceleration. In the Lagrangian formulation, i.e. if the motion is analyzed using the undeformed system of coordinates,

$$\frac{D v_i}{D t} = \frac{\partial v_i}{\partial t} = \frac{\partial^2 u_i}{\partial t^2} \tag{1.10}$$

In the Eulerian description, one has to account for the nonhomogeneous velocity field and the expression for the acceleration includes additional terms, introducing the convective component (e.g., Reddy 2008).

A further discussion along the lines of continuum mechanics should refer to the Navier equations that represent the linear version of equations of motion (1.9) in terms of displacements. A derivation of such equations requires us to substitute stresses in terms of strains (see Sect. 1.2) into (1.9) and subsequently use the strain-displacement relationships (1.6).

The principle of conservation of the angular momentum states that the rate of change of the angular momentum of the body is equal to the sum of applied external moments. This principle results in the conclusion that the tensor of stress that is work conjugate to the linearized strain tensor (1.6) is symmetric, so that

$$\sigma_{ij} = \sigma_{ji} \tag{1.11}$$

The principle of conservation of mass implies that the mass within a certain domain occupied by the material remains without change as this domain is transformed to a new position as a result of motion of the body. Equations of continuity can be derived from this principle (Reddy 2008). The first and second laws of thermodynamics factor into limitations of the constitutive law. Material can dissipate but cannot create energy. The existence of a strain energy function places certain restrictions on a linear elastic constitutive law including symmetry that is discussed in the next paragraph. The second law also serves as a foundation of the energy methods that are discussed in Sect. 1.5. A detailed discussion of these principles that can be found in books on continuum mechanics is outside the scope of this monograph.

1.2 Constitutive Relations for Composite, Isotropic and Piezoelectric Materials

As follows from the previous paragraph, the relationships between the stresses and strains "tie together" the equations of motion in terms of components of the tensor of stress with the kinematic and strain-displacement equations. Stress-strain relationships that are also referred to as constitutive relations describe the actual material that may be isotropic or anisotropic, physically linear or nonlinear, etc. In this book we concentrate on materials operating within the physically linear range.

We begin with a linear anisotropic material whose stress-strain relationships, excluding the effects of moisture or temperature, can be written as

$$\{\sigma_{ij}\} = \left[C_{ijkl}\right]\{\varepsilon_{kl}\}, \quad i, j, k, l = 1, 2, 3 \tag{1.12}$$

where repeated indices imply summation (according to Einstein's notation). There are nine components in tensors of stress and strain and the order of the square matrix of stiffness coefficients that are also called elastic constants $\left[C_{ijkl}\right]$ is also equal to 9.

As a result of the principle of conservation of angular momentum discussed in the previous Sect. (see Eq. 1.11), the elements of the tensor of stress are symmetric. The tensor of strain is also symmetric. Accordingly, $\sigma_{ij} = \sigma_{ji}$, $\varepsilon_{kl} = \varepsilon_{lk}$ and the number of independent elements of the tensors of stress and strain is reduced to 6. The order of the matrix of stiffness of an anisotropic material is also reduced to 6. Furthermore, the symmetry that is observed with respect to the first two and last two subscripts dictates that $C_{ijkl} = C_{jikl}$, $C_{ijkl} = C_{ijlk}$ (Gibson 2007) that also follows from the energy conservation condition.

A conventional contracted notation for the stresses, strains and stiffness components is

$$\begin{aligned}
&\sigma_1 = \sigma_{11}, && \sigma_2 = \sigma_{22}, && \sigma_3 = \sigma_{33}, \\
&\sigma_4 = \tau_{23} = \sigma_{23}, && \sigma_5 = \tau_{13} = \sigma_{13}, && \sigma_6 = \tau_{12} = \sigma_{12}, \\
&\varepsilon_1 = \varepsilon_{11}, && \varepsilon_2 = \varepsilon_{22}, && \varepsilon_3 = \varepsilon_{33}, \\
&\varepsilon_4 = \gamma_{23} = 2\varepsilon_{23}, && \varepsilon_5 = \gamma_{13} = 2\varepsilon_{13}, && \varepsilon_6 = \gamma_{12} = 2\varepsilon_{12}
\end{aligned} \tag{1.13}$$

where we acknowledge a popular notation for shear stresses in the ij plane, i.e. τ_{ij}.

The elements of the contracted stiffness matrix are defined as follows

$$\begin{aligned}
C_{ijkl} \to C_{\alpha\beta}: \quad & 11 \to 1, \quad 22 \to 2, \quad 33 \to 3, \quad 23 \to 4, \\
& 13 \to 5, \quad 12 \to 6
\end{aligned} \tag{1.14}$$

Accordingly, the contracted form of the constitutive relations is

$$\{\sigma_\alpha\} = \left[C_{\alpha\beta}\right]\{\varepsilon_\beta\} \tag{1.15}$$

where repeated indices imply summation. Note that the elements of the matrices of stiffness that are multiplied by shear strains in (1.12) and (1.15) differ by a factor of 2.

As indicated above, the number of equations (1.15) is equal to 6. The matrix of stiffness coefficients includes 36 components, but only 21 of them are independent since $C_{\alpha\beta} = C_{\beta\alpha}$. Furthermore, in practical situations many components of the matrix of stiffness are equal to zero. Consider for example, the case of a monoclinic material with one plane of material properties symmetry (1–2 plane). Then the constitutive relations are (Jones 1999):

$$\begin{Bmatrix} \sigma_1 \\ \sigma_2 \\ \sigma_3 \\ \tau_{23} \\ \tau_{13} \\ \tau_{12} \end{Bmatrix} = \begin{bmatrix} C_{11} & C_{12} & C_{13} & 0 & 0 & C_{16} \\ C_{12} & C_{22} & C_{23} & 0 & 0 & C_{26} \\ C_{13} & C_{23} & C_{33} & 0 & 0 & C_{36} \\ 0 & 0 & 0 & C_{44} & C_{45} & 0 \\ 0 & 0 & 0 & C_{45} & C_{55} & 0 \\ C_{16} & C_{26} & C_{36} & 0 & 0 & C_{66} \end{bmatrix} \begin{Bmatrix} \varepsilon_1 \\ \varepsilon_2 \\ \varepsilon_3 \\ \gamma_{23} \\ \gamma_{13} \\ \gamma_{12} \end{Bmatrix} \tag{1.16}$$

The number of independent stiffness coefficients (elastic constants) in (1.16) is equal to 13.

The case where the material has three planes of property symmetry is found in orthotropic materials, such as composite laminae (layers) with unidirectional fibers oriented along one of the axes 1, 2 or 3 discussed in Chap. 5. In such case, Eq. 1.16 include only nine independent elastic constants (the planes of symmetry are 1–2, 1–3 and 2–3):

$$\begin{Bmatrix} \sigma_1 \\ \sigma_2 \\ \sigma_3 \\ \tau_{23} \\ \tau_{13} \\ \tau_{12} \end{Bmatrix} = \begin{bmatrix} C_{11} & C_{12} & C_{13} & 0 & 0 & 0 \\ C_{12} & C_{22} & C_{23} & 0 & 0 & 0 \\ C_{13} & C_{23} & C_{33} & 0 & 0 & 0 \\ 0 & 0 & 0 & C_{44} & 0 & 0 \\ 0 & 0 & 0 & 0 & C_{55} & 0 \\ 0 & 0 & 0 & 0 & 0 & C_{66} \end{bmatrix} \begin{Bmatrix} \varepsilon_1 \\ \varepsilon_2 \\ \varepsilon_3 \\ \gamma_{23} \\ \gamma_{13} \\ \gamma_{12} \end{Bmatrix} \tag{1.17}$$

A further simplification is possible in the material with one plane where the properties are isotropic, i.e. independent of the direction. If this plane is 1–2 and the planes 1–3 and 2–3 are planes of material property symmetry, the constitutive relations become

$$\begin{Bmatrix} \sigma_1 \\ \sigma_2 \\ \sigma_3 \\ \tau_{23} \\ \tau_{13} \\ \tau_{12} \end{Bmatrix} = \begin{bmatrix} C_{11} & C_{12} & C_{13} & 0 & 0 & 0 \\ C_{12} & C_{11} & C_{13} & 0 & 0 & 0 \\ C_{13} & C_{13} & C_{33} & 0 & 0 & 0 \\ 0 & 0 & 0 & C_{44} & 0 & 0 \\ 0 & 0 & 0 & 0 & C_{44} & 0 \\ 0 & 0 & 0 & 0 & 0 & \frac{C_{11}-C_{12}}{2} \end{bmatrix} \begin{Bmatrix} \varepsilon_1 \\ \varepsilon_2 \\ \varepsilon_3 \\ \gamma_{23} \\ \gamma_{13} \\ \gamma_{12} \end{Bmatrix} \tag{1.18}$$

The number of independent elastic constants is now reduced to 5. The material characterized by (1.18) is called transversely isotropic. An example of such material is a composite lamina (layer) with fibers of a circular cross section that are oriented in the direction of axis 3 and uniformly distributed in cross sections perpendicular to this axis. Accordingly, the properties of such lamina are direction-independent in the plane 1–2, while 1–3 and 2–3 are planes of property symmetry. Note that we will refer to composite materials characterized by equations shown above in Chap. 5.

Finally, an isotropic material has the same properties in all directions and possesses an infinite number of planes of symmetry. Such a material has only two independent elastic constants:

$$\begin{Bmatrix} \sigma_1 \\ \sigma_2 \\ \sigma_3 \\ \tau_{23} \\ \tau_{13} \\ \tau_{12} \end{Bmatrix} = \begin{bmatrix} C_{11} & C_{12} & C_{12} & 0 & 0 & 0 \\ C_{12} & C_{11} & C_{12} & 0 & 0 & 0 \\ C_{12} & C_{12} & C_{11} & 0 & 0 & 0 \\ 0 & 0 & 0 & \frac{C_{11}-C_{12}}{2} & 0 & 0 \\ 0 & 0 & 0 & 0 & \frac{C_{11}-C_{12}}{2} & 0 \\ 0 & 0 & 0 & 0 & 0 & \frac{C_{11}-C_{12}}{2} \end{bmatrix} \begin{Bmatrix} \varepsilon_1 \\ \varepsilon_2 \\ \varepsilon_3 \\ \gamma_{23} \\ \gamma_{13} \\ \gamma_{12} \end{Bmatrix} \tag{1.19}$$

While the constitutive relations and elastic constants for composite materials are discussed in Chap. 5, in the case of isotropic materials the elastic constants are immediately available in terms of Lame constants λ, G the latter being the shear modulus of the material or alternatively, in terms of the modulus of elasticity E and the Poisson ratio ν:

$$C_{11} = \lambda + 2G, \qquad C_{12} = \lambda,$$
$$\lambda = \frac{G\,(2G - E)}{E - 3G}, \qquad G = \frac{E}{2\,(1+\nu)} \tag{1.20}$$

The stress-strain constitutive relations shown above could be inverted yielding the strain-stress relations. In particular, the inverted form of (1.15) is

$$\{\varepsilon_\alpha\} = \left[S_{\alpha\beta}\right]\{\sigma_\beta\} \tag{1.21}$$

where $\left[S_{\alpha\beta}\right] = \left[C_{\alpha\beta}\right]^{-1}$ is the matrix of compliance coefficients. The number of independent compliance coefficients is equal to the number of independent elastic constants in the same class of materials.

Piezoelectric materials have attracted the interest of engineers using them both as sensors as well as actuators. The attractive feature of these materials is related to their ability to acquire an electric potential as a result of applied mechanical strains (direct piezoelectric effect). These materials also exhibit the converse piezoelectric effect generating mechanical strains in response to the applied electric field. Sensory applications utilize the direct effect, while piezoelectric actuators are based on the application of the converse effect.

The constitutive relations for a piezoelectric medium that is not affected by environmental effects are (Berlincourt et al. 1964; Patron and Kudryavtsev 1993):

$$\{\sigma_\alpha\} = \left[C_{\alpha\beta}\right]\left\{\varepsilon_\beta\right\} - [e_{\alpha m}]\{E_m\}$$
$$\{D_m\} = [e_{m\alpha}]\{\varepsilon_\alpha\} + [\varepsilon_{mk}]\{E_k\} \tag{1.22}$$

where $\{E_m\}$ is the vector of electric field defined as the gradient of the electric potential in the corresponding direction, $e_{\alpha m} = -\left(\frac{\partial\sigma_\alpha}{\partial E_m}\right)_\varepsilon$ and $\varepsilon_{mk} = \left(\frac{\partial D_m}{\partial E_k}\right)_\varepsilon$ are piezoelectric constants and dielectric coefficients respectively, evaluated at a constant strain, and the "electric displacement" $D_m = -\left(\frac{\partial G_e}{\partial E_m}\right)_\varepsilon$ is a derivative of the Gibbs free energy determined at a constant strain. Note that Eqs. 1.22 could be written for a more general thermo-electro-mechanical case, including temperature, pyroelectric constants and entropy (e.g., Patron and Kudryavtsev 1993).

In the case of a transversely isotropic piezoelectric material, such as a polarized ceramic of a mm6 symmetry class (Patron and Kudryavtsev 1993) with the 3-axis oriented along the polarization direction, the matrix of elastic constants corresponds to that in (1.18), while the matrices of piezoelectric and dielectric coefficients are

$$[e_{m\alpha}] = \begin{bmatrix} 0 & 0 & 0 & 0 & e_{15} & 0 \\ 0 & 0 & 0 & e_{15} & 0 & 0 \\ e_{31} & e_{31} & e_{33} & 0 & 0 & 0 \end{bmatrix}, \quad [\varepsilon_{mk}] = \begin{bmatrix} \varepsilon_{11} & 0 & 0 \\ 0 & \varepsilon_{11} & 0 \\ 0 & 0 & \varepsilon_{33} \end{bmatrix} \tag{1.23}$$

An alternative form of equations (1.22) presents strains and electric displacements in terms of stresses and the applied electric field:

$$\{\varepsilon_\alpha\} = \left[S_{\alpha\beta}\right]\left\{\sigma_\beta\right\} + [d_{\alpha m}]\{E_m\}$$
$$\{D_m\} = [d_{m\alpha}]\{\sigma_\alpha\} + [\varepsilon_{mk}]\{E_k\} \tag{1.24}$$

where $d_{\alpha m} = \left(\frac{\partial\varepsilon_\alpha}{\partial E_m}\right)_\sigma$ are piezoelectric coefficients determined at a constant stress.

The constitutive or stress-strain relations outlined in this paragraph can be simplified in the analysis of thin plates using the assumption of plane stress. According to this assumption, in-plane stresses are much higher than the stresses acting in the thickness direction. Accordingly, in the classical thin plate theory of plates $\sigma_3,\ \tau_{13},\ \tau_{23}$ are neglected as compared to $\sigma_1,\ \sigma_2,\ \tau_{12}$.

1.3 Strain-Displacement Relations for Plates and Relevant Kinematic Assumptions

In this paragraph we outline kinematic assumptions used in the analysis of plates by various theories. Specific cases, such as sandwich plates, are discussed in relevant sections of the book. We consider a rectangular coordinate system shown in Fig. 1.3

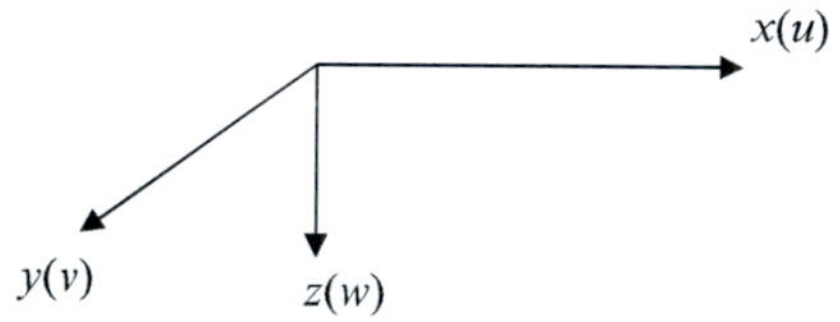

Fig. 1.3 System of rectangular (Cartesian) coordinates that is often used in the analyses of plates and notation for displacements along the corresponding axes

where x and y are in-plane coordinate (i.e., these coordinate axes are in the middle plane that is equidistant from the two surfaces of the plate). The corresponding displacements along these axes are denoted by u and v, respectively. The coordinate z is counted from the middle plane and the deflection of the plate in the z-direction is denoted by w. Note that following a typical orientation of axes in the theory of plates we rotated the coordinate system as compared to that shown in Figs. 1.1 and 1.2. Naturally, this does not alter the following analysis.

Kinematic and strain-displacement relations do not explicitly restrict the choice of the material of the plate, but rather reflect its geometric features, i.e. a relatively small thickness. However, the choice of these relations may be influenced by material properties, such as low transverse shear modulus. These relations are paramount to the development of the theory of plates specifying the range of validity of the particular theory.

Kinematic assumptions of the theory of plates determine the accuracy adopted to describe its deformations throughout the thickness. The classical (thin) plate theory is the simplest, assuming in-plane displacements to be linear functions of the thickness coordinate. Additionally, the thickness of the plate is assumed unaffected by its deformations. Such assumptions referred to as Kirchhoff-Love assumptions represent an extension of the slender or Euler-Bernoulli beam theory to plates. The classical plate theory will be considered in this paragraph. So-called first-order and higher-order theories reflect either a larger thickness of the plate or a low transverse shear stiffness of its material by modifying the assumption of the classical theory. For example, the third-order theory represents in-plane displacements as cubic functions of the thickness coordinate. Some higher-order theories discard the assumption that the thickness remains constant during deformations. An example where such assumption becomes invalid is found in sandwich structures with a "soft" core. Plates that are either thick or have a low transverse shear stiffness are often referred to as "shear deformable."

In addition to the assumptions related to the effect of thickness and material shear deformability, the choice of the plate theory requires consideration of the relative magnitude of deformations as compared to the plate dimensions. In the case where deflections in the thickness direction become comparable with the thickness of the plate, geometric nonlinearity should be taken into account. In industry, the boundary between the applicability of geometrically linear and nonlinear theories is sometimes associated with the deflections being smaller or larger than half-thickness, respectively. Of course, this boundary has no theoretical justification, i.e.

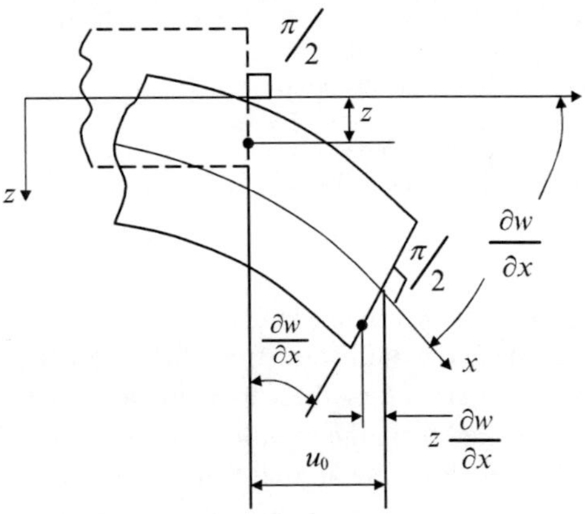

Fig. 1.4 Deformations in the xz-plane according to the Kirchhoff-Love hypothesis

it is simply an arbitrary limit indicating when the engineer can disregard nonlinear terms in the strain-displacement relationships without making a large error. It will be illustrated later in this book that neglecting geometric nonlinearity in the analysis of plate structures leads to a conservative estimate of stresses.

Before proceedings to the formulation of the strain-displacement relations for thin plates, it is worth mentioning that geometrically nonlinear effects are essential in relatively thin and flexible structures. Such structures are often adequately characterized by the classical thin plate theory, without the need to account for transverse shear deformations and to use first-order or higher-order theories of plates. On the other hand, "thick" structures that experience significant transverse shear deformations and require the application of first-order of higher-order theories are relatively rigid, so that failure occurs at small deformations that do not require a geometrically nonlinear formulation. Accordingly, solutions for typical isotropic material plates including both shear deformability as well as geometric nonlinearity are often unnecessary since the overlap of two phenomena is negligible. The situation may sometimes be different in composite and sandwich structures where the material exhibits low transverse shear stiffness. If this is the case, the plate may experience large nonlinear deformations, accompanied by noticeable transverse shear strains.

The analysis of thin plates is conducted by the Kirchhoff-Love assumption that cross sections of the plate perpendicular to the middle plane prior to deformation remain plane and perpendicular to the deformed middle plane after the deformation (Fig. 1.4). Of course, this implies that in-plane displacements are linear functions of curvature and the thickness coordinate as was already indicated above. In addition, the thickness of the plate being assumed constant, the normal strain in the direction perpendicular to the middle plane is equal to zero.

Following the Kirchhoff-Love assumption, displacements of an arbitrary point of the plate are

$$\mathbf{u} = \begin{Bmatrix} u(x_1, x_2, x_3) \\ v(x_1, x_2, x_3) \\ w(x_1, x_2, x_3) \end{Bmatrix} \equiv \begin{Bmatrix} u(x, y, z) \\ v(x, y, z) \\ w(x, y, z) \end{Bmatrix} = \begin{Bmatrix} u_0(x, y) - z\frac{\partial w(x,y)}{\partial x} \\ v_0(x, y) - z\frac{\partial w(x,y)}{\partial y} \\ w(x, y) \end{Bmatrix} \tag{1.25}$$

where $u_0(x, y)$ and $v_0(x, y)$ are displacements of the middle plane. The first equation (1.25) immediately follows from the consideration of Fig. 1.4.

The strains in the middle plane can be obtained as linear functions of the displacements from (1.6) or as nonlinear functions following the Lagrange or Euler definitions (1.4) and (1.5), respectively. In particular, combining (1.4) with (1.25) yields

$$\begin{aligned}
\varepsilon_x &= \frac{\partial u_0}{\partial x} - z\frac{\partial^2 w}{\partial x^2} + \frac{1}{2}\left[\underline{\left(\frac{\partial u}{\partial x}\right)^2 + \left(\frac{\partial v}{\partial x}\right)^2} + \left(\frac{\partial w}{\partial x}\right)^2\right] \\
\varepsilon_y &= \frac{\partial v_0}{\partial y} - z\frac{\partial^2 w}{\partial y^2} + \frac{1}{2}\left[\underline{\left(\frac{\partial u}{\partial y}\right)^2 + \left(\frac{\partial v}{\partial y}\right)^2} + \left(\frac{\partial w}{\partial y}\right)^2\right] \\
\varepsilon_{xy} &= \frac{\gamma_{xy}}{2} = \frac{1}{2}\left(\frac{\partial u_0}{\partial y} + \frac{\partial v_0}{\partial x} - 2z\frac{\partial^2 w}{\partial x \partial y} + \underline{\frac{\partial u}{\partial x}\frac{\partial u}{\partial y} + \frac{\partial v}{\partial x}\frac{\partial v}{\partial y}} + \frac{\partial w}{\partial x}\frac{\partial w}{\partial y}\right)
\end{aligned} \tag{1.26}$$

where u and v in the underlined nonlinear terms denote in-plane displacements at the point. The strains in the left side of (1.26) represent normal strains in the x and y directions and the in-plane shear strain, respectively.

The nonlinear expressions for the strains can be simplified since the underlined terms are usually negligible, so that the geometrically nonlinear equations that are customarily adopted in the theory of thin plates are

$$\begin{aligned}
\varepsilon_x &= \frac{\partial u_0}{\partial x} - z\frac{\partial^2 w}{\partial x^2} + \frac{1}{2}\left(\frac{\partial w}{\partial x}\right)^2 \\
\varepsilon_y &= \frac{\partial v_0}{\partial y} - z\frac{\partial^2 w}{\partial y^2} + \frac{1}{2}\left(\frac{\partial w}{\partial y}\right)^2 \\
\varepsilon_{xy} &= \frac{\gamma_{xy}}{2} = \frac{1}{2}\left(\frac{\partial u_0}{\partial y} + \frac{\partial v_0}{\partial x} - 2z\frac{\partial^2 w}{\partial x \partial y} + \frac{\partial w}{\partial x}\frac{\partial w}{\partial y}\right)
\end{aligned} \tag{1.27}$$

The strains given by (1.27) can further be separated into two components: those occurring at the middle plane of the plate and those dependent on the distance z between the point where the strains are evaluated and the middle plane. The middle plane strains are

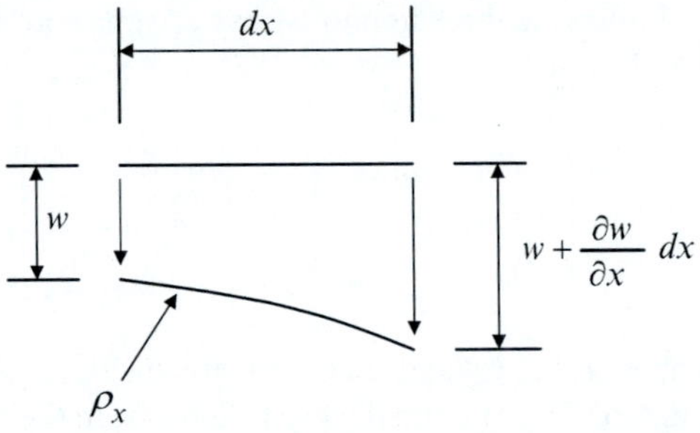

Fig. 1.5 Element of the middle plane in the xz-plane before and after deformation

$$\varepsilon_x^0 = \frac{\partial u_0}{\partial x} + \frac{1}{2}\left(\frac{\partial w}{\partial x}\right)^2$$

$$\varepsilon_y^0 = \frac{\partial v_0}{\partial y} + \frac{1}{2}\left(\frac{\partial w}{\partial y}\right)^2$$

$$\varepsilon_{xy}^0 = \frac{\gamma_{xy}^0}{2} = \frac{1}{2}\left(\frac{\partial u_0}{\partial y} + \frac{\partial v_0}{\partial x} + \frac{\partial w}{\partial x}\frac{\partial w}{\partial y}\right) \tag{1.28}$$

The components of the strains dependent on the z-coordinate are given by

$$\varepsilon_x(z) = z\kappa_x = -z\frac{\partial^2 w}{\partial x^2}, \quad \varepsilon_y(z) = z\kappa_y = -z\frac{\partial^2 w}{\partial y^2}, \quad \varepsilon_{xy}(z) = \frac{z\kappa_{xy}}{2} = -z\frac{\partial^2 w}{\partial x \partial y} \tag{1.29}$$

Physically, κ_x, κ_y and κ_{xy} are the changes of curvature in the planes xz and yz and the twist of the middle plane, respectively, with respect to a stress-free reference state. This is illustrated below for the change of curvature in the xz plane.

Consider an element of the middle plane of the plate in the xz plane (Fig. 1.5). Prior to deformation, the length of this element was equal to dx. As a result of deformation, the plate acquired a deflection $w(x, y)$, so that in the xz plane ($y = const$) this deflection is simply $w(x)$. We can characterize the deformed position of the element by $w(x)$ and by the local curvature acquired by the middle plane. The local radius of curvature in the xz plane is denoted by ρ_x. This radius is determined using the definition of curvature of the curve $w = w(x)$:

$$\frac{1}{\rho_x} = \frac{-\frac{d^2 w}{dx^2}}{\left[1 + \left(\frac{dw}{dx}\right)^2\right]^{\frac{3}{2}}} \approx -\frac{d^2 w}{dx^2} \tag{1.30}$$

where the nonlinear contribution in the denominator is neglected compared to unity. The curvature in the yz plane is specified by analogy. The illustration of the concept of twist of the middle plane omitted here for brevity can be found in many books (e.g., Timoshenko and Woinowsky-Krieger 1959).

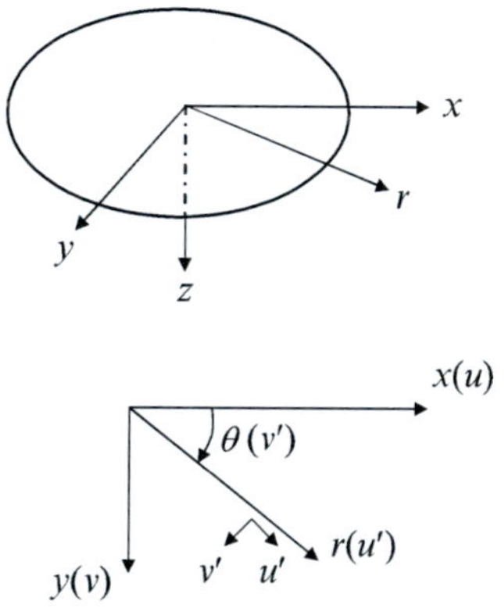

Fig. 1.6 Polar coordinate system (axes and displacements)

The strains evaluated above can be written in the form where the middle plane strains and the strains associated with the change of the curvature and twist of the middle plane are decomposed:

$$\begin{aligned}\varepsilon_x &= \varepsilon_x^0 + z\kappa_x \\ \varepsilon_y &= \varepsilon_y^0 + z\kappa_y \\ \gamma_{xy} &= \gamma_{xy}^0 + z\kappa_{xy}\end{aligned} \tag{1.31}$$

Note that in this theory geometric nonlinearity affects only the strains in the middle plane, while the changes in curvature and twist appear unaffected by the magnitude of deflections. This reflects the degree of accuracy of the theory employed to develop the strain-displacement equations. Such accuracy is sufficient for the analysis of most plate structures.

In case of different coordinate systems, the strain-displacement relations are modified accordingly. In particular, in the polar coordinate system shown in Fig. 1.6 where r, θ and z are radial, tangential (circumferential) and thickness coordinates, respectively, kinematic relations (1.25) become

$$\mathbf{u} = \begin{Bmatrix} u'(r,\theta,z) \\ v'(r,\theta,z) \\ w'(r,\theta,z) \end{Bmatrix} = \begin{Bmatrix} u'_0(r,\theta) - z\frac{\partial w(r,\theta)}{\partial r} \\ v'_0(r,\theta) - z\frac{\partial w(r,\theta)}{r\partial\theta} \\ w(r,\theta) \end{Bmatrix} \tag{1.32}$$

where the prime is introduced to distinguish in-plane displacement components in the polar system from those in the Cartesian system of coordinates.

The total strains at the distance z from the middle plane can be represented similarly to (1.31) in terms of middle plane strains and the changes of curvature and twist:

$$\varepsilon_r = \varepsilon_r^0 + z\kappa_r$$

$$\varepsilon_\theta = \varepsilon_\theta^0 + z\kappa_\theta$$
$$\gamma_{r\theta} = \gamma_{r\theta}^0 + z\kappa_{r\theta} \tag{1.33}$$

The strains in (1.33) can be evaluated using the transformation relations between rectangular and polar coordinates in the case where both coordinate systems share the origin (naturally, the z-coordinate is not affected by this change of the coordinate system):

$$x = r\cos\theta, \qquad y = r\sin\theta$$
$$r = \sqrt{x^2 + y^2}, \quad \theta = \tan^{-1}\frac{y}{x} \tag{1.34}$$

It immediately follows that

$$\frac{\partial r}{\partial x} = \frac{x}{r} = \cos\theta, \qquad \frac{\partial r}{\partial y} = \frac{y}{r} = \sin\theta$$
$$\frac{\partial \theta}{\partial x} = -\frac{y}{r^2} = -\frac{\sin\theta}{r}, \quad \frac{\partial \theta}{\partial y} = \frac{x}{r^2} = \frac{\cos\theta}{r} \tag{1.35}$$

The derivatives with respect to the Cartesian coordinates can now be expressed in terms of the derivatives in the polar coordinate system by (Fung 1994):

$$\frac{\partial(\ldots)}{\partial x} = \cos\theta\frac{\partial(\ldots)}{\partial r} - \frac{\sin\theta}{r}\frac{\partial(\ldots)}{\partial\theta}$$
$$\frac{\partial(\ldots)}{\partial y} = \sin\theta\frac{\partial(\ldots)}{\partial r} + \frac{\cos\theta}{r}\frac{\partial(\ldots)}{\partial\theta} \tag{1.36}$$

The inspection of Fig. 1.6 immediately yields simple relationships between in-plane displacements in rectangular and polar coordinate systems:

$$u = u'\cos\theta - v'\sin\theta$$
$$v = u'\sin\theta + v'\cos\theta \tag{1.37}$$

As indicated above, the tensor of strain is transformed according to the tensor transformation law. Accordingly, the strains in the Cartesian coordinate system given by (1.4), (1.5) or (1.6) are transformed to the components of the vector of strains in the polar coordinate system through the transformation equations

$$\begin{Bmatrix} \varepsilon_r \\ \varepsilon_\theta \\ \varepsilon_{r\theta} \end{Bmatrix} = \begin{bmatrix} \cos^2\theta & \sin^2\theta & \sin 2\theta \\ \sin^2\theta & \cos^2\theta & -\sin 2\theta \\ -\frac{1}{2}\sin 2\theta & \frac{1}{2}\sin 2\theta & \cos^2\theta - \sin^2\theta \end{bmatrix} \begin{Bmatrix} \varepsilon_x \\ \varepsilon_y \\ \varepsilon_{xy} \end{Bmatrix} \tag{1.38}$$

Now one can substitute the strains in the Cartesian coordinate system into (1.38) simultaneously replacing displacements and derivatives with their expressions in the polar coordinate system according to (1.37) and (1.36). For example, in the geometrically linear formulation we obtain

$$\varepsilon_r = \frac{\partial u}{\partial r}, \quad \varepsilon_\theta = \frac{u}{r} + \frac{1}{r}\frac{\partial v}{\partial \theta}, \quad \varepsilon_{r\theta} = \frac{1}{2}\left(\frac{1}{r}\frac{\partial u}{\partial \theta} + \frac{\partial v}{\partial r} - \frac{v}{r}\right) \tag{1.39}$$

where the prime is eliminated in understanding that in-plane displacements are in the polar coordinate system.

This expression could be generalized to include geometrically nonlinear terms. The corresponding relations are (Reddy 2007):

$$\begin{aligned} \varepsilon_r &= \frac{\partial u}{\partial r} + \frac{1}{2}\left(\frac{\partial w}{\partial r}\right)^2 \\ \varepsilon_\theta &= \frac{u}{r} + \frac{1}{r}\frac{\partial v}{\partial \theta} + \frac{1}{2r^2}\left(\frac{\partial w}{\partial \theta}\right)^2 \\ \varepsilon_{r\theta} &= \frac{\gamma_{r\theta}}{2} = \frac{1}{2}\left(\frac{1}{r}\frac{\partial u}{\partial \theta} + \frac{\partial v}{\partial r} - \frac{v}{r} + \frac{1}{r}\frac{\partial w}{\partial r}\frac{\partial w}{\partial \theta}\right) \end{aligned} \tag{1.40}$$

Decomposing the strains into the middle plane components and the changes of curvature in the rz plane (radial plane), θz plane (tangential plane) and twist of the middle plane we obtain

$$\begin{aligned} \varepsilon_r^0 &= \frac{\partial u_0}{\partial r} + \underline{\frac{1}{2}\left(\frac{\partial w}{\partial r}\right)^2} \\ \varepsilon_\theta^0 &= \frac{u_0}{r} + \frac{1}{r}\frac{\partial v_0}{\partial \theta} + \underline{\frac{1}{2r^2}\left(\frac{\partial w}{\partial \theta}\right)^2} \\ \varepsilon_{r\theta}^0 &= \frac{\gamma_{r\theta}^0}{2} = \frac{1}{2}\left(\frac{1}{r}\frac{\partial u_0}{\partial \theta} + \frac{\partial v_0}{\partial r} - \frac{v_0}{r} + \underline{\frac{1}{r}\frac{\partial w}{\partial r}\frac{\partial w}{\partial \theta}}\right) \end{aligned} \tag{1.41}$$

where nonlinear terms are underlined and

$$\begin{aligned} \kappa_r &= -\frac{\partial^2 w}{\partial r^2} \\ \kappa_\theta &= -\frac{1}{r}\left(\frac{\partial w}{\partial r} + \frac{1}{r}\frac{\partial^2 w}{\partial \theta^2}\right) \\ \kappa_{r\theta} &= -\frac{2}{r}\left(\frac{\partial^2 w}{\partial r \partial \theta} - \frac{1}{r}\frac{\partial w}{\partial \theta}\right) \end{aligned} \tag{1.42}$$

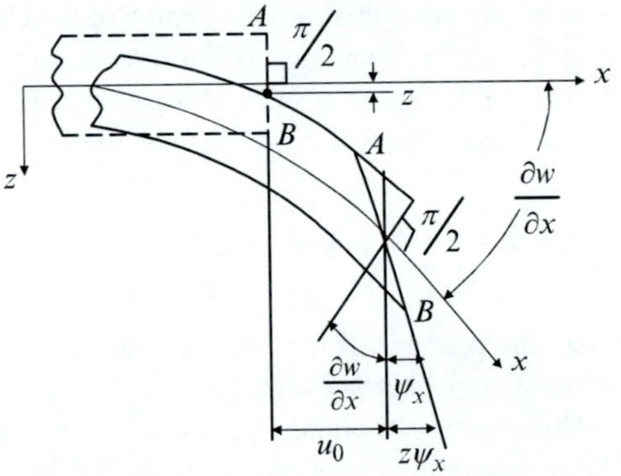

Fig. 1.7 Deformations in the xz-plane according to the first-order shear deformation theory

In the case of axisymmetric deformations the derivatives with respect to the tangential coordinate are equal to zero and the displacement in this direction $v_0 = 0$. Accordingly, (1.41) and (1.42) simplify to

$$\varepsilon_r^0 = \frac{\partial u_0}{\partial r} + \frac{1}{2}\left(\frac{\partial w}{\partial r}\right)^2$$
$$\varepsilon_\theta^0 = \frac{u_0}{r}$$
$$\varepsilon_{r\theta}^0 = 0 \tag{1.43}$$

and

$$\kappa_x = -\frac{\partial^2 w}{\partial r^2}$$
$$\kappa_y = -\frac{1}{r}\frac{\partial w}{\partial r}$$
$$\kappa_{xy} = 0 \tag{1.44}$$

In this book we will also consider shear-deformable composite plates (Chap. 5). As we have already indicated, although transverse shear deformability is seldom essential in isotropic plate structures, it is often important in the analysis of composite and particularly sandwich plates. We limit the illustration to the first-order theory that is based on the assumption that a plane perpendicular to the middle plane remains flat, without warping, after deformation but rotates about the deformed middle plane (Fig. 1.7). Accordingly, referring to deformations in the xz plane we consider the slope of the middle plane $\frac{\partial w}{\partial x}$ and the rotation of the plane AB from the original orientation that was perpendicular to the undeformed middle plane ψ_x. The combination of these two angles represents the transverse shear strain. Extrapolating the same approach to strains in the yz plane we obtain

$$\gamma_{xz} = \frac{\partial w}{\partial x} + \psi_x$$
$$\gamma_{yz} = \frac{\partial w}{\partial y} + \psi_y \tag{1.45}$$

As is obvious from Fig. 1.7 and a counterpart of this figure that could be shown for the yz plane, the kinematic relations of the first-order shear deformation theory are

$$u = \left\{ \begin{array}{c} u_0(x,y) + z\psi_x(x,y) \\ v_0(x,y) + z\psi_y(x,y) \\ w(x,y) \end{array} \right\} \tag{1.46}$$

It is easily observed that in the case where transverse shear deformations are negligible the classical plate theory is recovered by setting

$$\psi_x = -\frac{\partial w}{\partial x}, \quad \psi_y = -\frac{\partial w}{\partial y} \tag{1.47}$$

The in-plane strains produced by the changes of curvature and twist according to the first order shear deformation theory are obtained by substituting (1.46) into (1.4). The middle plane strains are not affected by shear deformability of the material; they are given by (1.28). However, the strain components associated with the changes of curvature and twist according to the first-order shear deformation theory are

$$\varepsilon_x(z) = z\kappa_x = z\frac{\partial \psi_x}{\partial x}, \quad \varepsilon_y(z) = z\kappa_y = z\frac{\partial \psi_y}{\partial y},$$
$$\varepsilon_{xy}(z) = \frac{z\kappa_{xy}}{2} = \frac{z}{2}\left(\frac{\partial \psi_x}{\partial y} + \frac{\partial \psi_y}{\partial x}\right) \tag{1.48}$$

1.4 Stress Resultants and Stress Couples

The three-dimensional state of stresses in a structure can be analyzed solving the equations of equilibrium (or equations of motion) at each point. However, in the theory of plates (as well as in the theory of shells) it was found more convenient to analyze the equilibrium replacing the actual three-dimensional stress distribution in a relatively thin structure with functions of stresses that depends only on the middle plane coordinates. These functions of stress are chosen in such manner that they are statically equivalent to the actual system of stresses. This enables us to reduce the equations of equilibrium from three to two dimensions (the thickness coordinate is eliminated). Once the equations of equilibrium are solved, we evaluate the strains

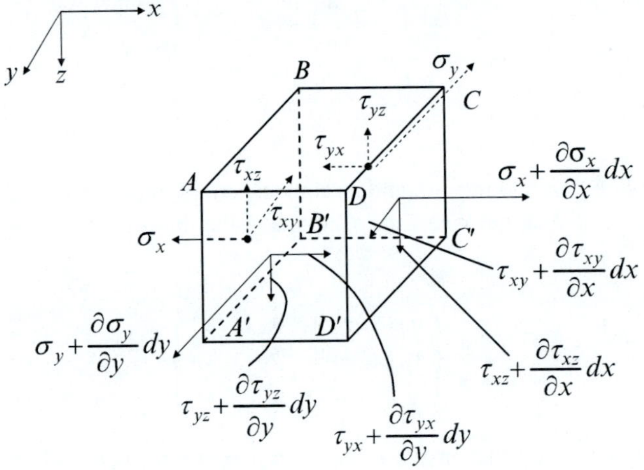

Fig. 1.8 Stresses acting on an infinitesimal element with dimensions $dx,\ dy,\ dz$. Following the assumptions of the classical and first-order shear deformation theories stresses on the planes ABCD and $A'B'C'D'$ are disregarded

in the three-dimensional domain occupied by the plate and subsequently, determine the stresses throughout the entire plate.

Consider the domain occupied by a plate where the in-plane stress vector at each point in the Cartesian coordinate system is given by

$$\boldsymbol{\sigma} \equiv \{\sigma_i\} = \begin{Bmatrix} \sigma_x(x, y, z) \\ \sigma_y(x, y, z) \\ \tau_{xy}(x, y, z) \end{Bmatrix} \tag{1.49}$$

Note that following a typical notation of the theory of plates, the system of coordinates 1-2-3 referred to above is replaced with the system x-y-z (Fig. 1.3). As a result of symmetry, in-plane shear stresses $\tau_{xy} = \tau_{yx}$. The stresses referred to in (1.49) and applied to an infinitesimal plate element are depicted in Fig. 1.8.

We replace the actual system of stresses with so-called stress resultants and stress couples that represent integrals of the stresses and of the moments of the stresses with respect to the middle plane, respectively. These integrals are taken through the thickness of the plate. Accordingly, the vector of in-plane stress resultants is

$$\{N\} \equiv \{N_i\} = \begin{Bmatrix} N_x(x, y) \\ N_y(x, y) \\ N_{xy}(x, y) \end{Bmatrix} = \int_{-\frac{h}{2}}^{\frac{h}{2}} \begin{Bmatrix} \sigma_x(x, y, z) \\ \sigma_y(x, y, z) \\ \tau_{xy}(x, y, z) \end{Bmatrix} dz \tag{1.50}$$

h being the thickness of the plate.

The vector of stress couples is

$$\{M\} \equiv \{M_i\} = \left\{ \begin{array}{l} M_x(x,y) \\ M_y(x,y) \\ M_{xy}(x,y) \end{array} \right\} = \int_{-\frac{h}{2}}^{\frac{h}{2}} \left\{ \begin{array}{l} \sigma_x(x,y,z) \\ \sigma_y(x,y,z) \\ \tau_{xy}(x,y,z) \end{array} \right\} zdz \tag{1.51}$$

Note that in-plane shear stresses τ_{xy} and τ_{yx} produce equal stress resultants and stress couples.

Transverse shear stresses $\tau_{xz},\ \tau_{yz}$ can be replaced with the vector of transverse shear stress resultants:

$$\{Q\} \equiv \{Q_i\} = \left\{ \begin{array}{l} Q_x(x,y) \\ Q_y(x,y) \end{array} \right\} = \int_{-\frac{h}{2}}^{\frac{h}{2}} \left\{ \begin{array}{l} \tau_{xz}(x,y,z) \\ \tau_{yz}(x,y,z) \end{array} \right\} dz \tag{1.52}$$

Obviously, the moment of transverse shear stress with respect to the middle plane is equal to zero. The normal transverse stress σ_z is usually much smaller than transverse shear stresses, differing from them by an order of magnitude in typical plate geometries. Accordingly, the effect of this stress on the equations of equilibrium is disregarded.

A similar approach can be applied to derive the vectors of in-plane and transverse shear stress resultants and stress couples in other coordinate systems. For example, in a polar coordinate system characterized by the radial (r), circumferential (θ) and thickness (z) coordinates the corresponding vectors expressed in terms of in-plane and transverse shear stresses are:

$$\{N\} = \left\{ \begin{array}{l} N_r(r,\theta) \\ N_\theta(r,\theta) \\ N_{r\theta}(r,\theta) \end{array} \right\} = \int_{-\frac{h}{2}}^{\frac{h}{2}} \left\{ \begin{array}{l} \sigma_r(r,\theta,z) \\ \sigma_\theta(r,\theta,z) \\ \tau_{r\theta}(r,\theta,z) \end{array} \right\} dz$$

$$\{M\} = \left\{ \begin{array}{l} M_r(r,\theta) \\ M_\theta(r,\theta) \\ M_{r\theta}(r,\theta) \end{array} \right\} = \int_{-\frac{h}{2}}^{\frac{h}{2}} \left\{ \begin{array}{l} \sigma_r(r,\theta,z) \\ \sigma_\theta(r,\theta,z) \\ \tau_{r\theta}(r,\theta,z) \end{array} \right\} zdz \tag{1.53}$$

$$\{Q\} = \left\{ \begin{array}{l} Q_r(r,\theta) \\ Q_\theta(r,\theta) \end{array} \right\} = \int_{-\frac{h}{2}}^{\frac{h}{2}} \left\{ \begin{array}{l} \tau_{rz}(r,\theta,z) \\ \tau_{\theta z}(r,\theta,z) \end{array} \right\} dz$$

where the notation of the stresses is evident (also, see Chap. 3).

The stress resultants and stress couples in the rectangular system of coordinates are shown in Fig. 1.9. It is obvious that stress resultants $N_x,\ N_{xy},\ Q_x$ and stress

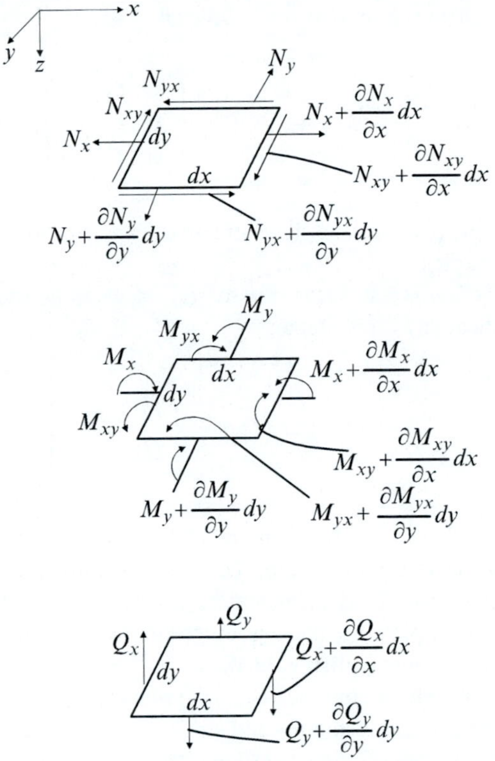

Fig. 1.9 Stress resultants and stress couples acting on the element of the plate. *Top view*: in-plane stress resultants, *middle view*: stress couples, *bottom view*; transverse shear stress resultants

couples $M_x,\ M_{xy}$ are applied at the cross sections $x = const$, while $N_y,\ N_{yx},\ Q_y$ and $M_y,\ M_{yx}$ are applied at $y = const$. The stress resultant N_{yx} and stress couple M_{yx} are produced by the in-plane shear stress τ_{yx} acting at the front and back edges of the element. However, as follows from (1.11), $N_{yx} = N_{xy}$ and $M_{yx} = M_{xy}$.

Considering that stress resultants represent integrals of the corresponding stresses through the thickness, while stress couples are integrals of the moments of these stresses, we observe that stress resultants and stress couples are the forces and moments per unit length of the corresponding cross section, respectively. Accordingly, the units of stress resultants and those of stress couples are force per unit length (e.g., N/m or lbf/in) and moment per unit length (e.g., N or lbf), respectively.

It is obvious that at the macromechanical scale stress resultants and couples are equivalent to the actual system of stresses in the cross section. However, even if stress resultants and couples are found, we still need to know stresses at each point through the thickness of the plate to analyze the strength. This problem can be addressed as follows.

Consider a thin plate where the vector of in-plane stresses is related to the vector of in-plane strains by the system of equations similar to (1.12) or (1.15). This system is reduced to three in-plane equations that are written here in the form

$$\{\sigma\} = [C]\{\varepsilon\} \tag{1.54}$$

where $[C]$ is a matrix of stiffness coefficients.

According to the previous discussion, the in-plane stress resultants and stress couples are

$$\{N\} = \int_{-\frac{h}{2}}^{\frac{h}{2}} [C]\{\varepsilon\}\,dz, \quad \{M\} = \int_{-\frac{h}{2}}^{\frac{h}{2}} [C]\{\varepsilon\}\,zdz \tag{1.55}$$

As was shown above, in the theory of thin and first-order shear deformable plates the vector of in-plane strains is a linear function of the thickness coordinate, so that it can be represented by (1.31), i.e.

$$\{\varepsilon\} = \left\{\varepsilon^0\right\} + z\{\kappa\} \tag{1.56}$$

where the first vector in the right side represents the strains in the middle plane, and the second vector refers to the changes of curvature and twist of the plate.

The substitution of in-plane strains given by (1.56), into (1.55), yields the in-plane stress resultants and stress couples:

$$\begin{Bmatrix} N \\ M \end{Bmatrix} = \begin{bmatrix} \int_{-\frac{h}{2}}^{\frac{h}{2}} [C]\,dz & \int_{-\frac{h}{2}}^{\frac{h}{2}} [C]\,zdz \\ \int_{-\frac{h}{2}}^{\frac{h}{2}} [C]\,zdz & \int_{-\frac{h}{2}}^{\frac{h}{2}} [C]\,z^2dz \end{bmatrix} \begin{Bmatrix} \varepsilon^0 \\ \kappa \end{Bmatrix} \tag{1.57}$$

Given the terms in the left side of the above equation, we can evaluate the components of the strain vector ε^0, κ. Subsequently, in-plane stresses can be determined from the constitutive equations at every point. Such approach to the solution involves the reduction of the actual equilibrium problem to a simplified but macromechanically equivalent two-dimensional formulation in terms of stress resultants and stress couples that depend on the coordinates x and y and subsequently, the return to in-plane strain and stress fields that depend on all three coordinates x, y and z.

In the case of a thin isotropic plate experiencing geometrically nonlinear deformations the substitution of (1.19), (1.20), (1.28) and (1.29) into (1.57) yields

$$N_x = \frac{Eh}{1-\nu^2}\left\{\frac{\partial u_0}{\partial x} + \frac{1}{2}\left(\frac{\partial w}{\partial x}\right)^2 + \nu\left[\frac{\partial v_0}{\partial y} + \frac{1}{2}\left(\frac{\partial w}{\partial y}\right)^2\right]\right\}$$

$$
\begin{aligned}
N_y &= \frac{Eh}{1-\nu^2}\left\{\frac{\partial v_0}{\partial y}+\frac{1}{2}\left(\frac{\partial w}{\partial y}\right)^2+\nu\left[\frac{\partial u_0}{\partial x}+\frac{1}{2}\left(\frac{\partial w}{\partial x}\right)^2\right]\right\}\\
N_{xy} &= \frac{Eh}{2(1+\nu)}\left(\frac{\partial u_0}{\partial y}+\frac{\partial v_0}{\partial x}+\frac{\partial w}{\partial x}\frac{\partial w}{\partial y}\right)\\
M_x &= -D\left(\frac{\partial^2 w}{\partial x^2}+\nu\frac{\partial^2 w}{\partial y^2}\right)\\
M_y &= -D\left(\frac{\partial^2 w}{\partial y^2}+\nu\frac{\partial^2 w}{\partial x^2}\right)\\
M_{xy} &= -D(1-\nu)\frac{\partial^2 w}{\partial x\partial y}
\end{aligned}
\tag{1.58}
$$

where

$$
D=\frac{Eh^3}{12(1-\nu^2)}
\tag{1.59}
$$

is the so-called cylindrical or bending stiffness of the plate. Similar relationships can be derived for the polar coordinate system and for non-isotropic materials.

1.5 Introduction to the Rayleigh-Ritz and Galerkin Methods

As indicated in Sect. 1.1, the first law of thermodynamics or the principle of conservation of energy serves as the foundation for energy-based methods employed in the analysis of structures, including plates. In the absence of energy dissipation and other non-conservative forces, i.e. if the forces acting on the system are conservative, this principle is reduced to the principle of stationery total energy (e.g. Thomson 1993):

$$
\Pi + K = const \quad or \quad \delta(\Pi + K) = 0
\tag{1.60}
$$

where Π and K are the potential and kinetic energies of the system, respectively and δ is a variational operator. In static problems the principle of stationary total energy reduces to the principle of minimum total potential energy implying that the virtual work of forces acting on the system in equilibrium is equal to zero, so that $\delta\Pi = 0$ or $\Pi = const.$

The Lagrange equation that can be derived for conservative systems from the principle of stationery total energy (e.g., Thomson 1993) is:

$$
\frac{d}{dt}\left(\frac{\partial L}{\partial \dot{q}_i}\right)-\frac{\partial L}{\partial q_i}=0
\tag{1.61}
$$

where the functional $L = K - \Pi$ (it is also called the Lagrangian), and q_i and $\dot{q}_i$ are generalized coordinates and generalized velocities, respectively. Generalized coordinates referred to above represent independent coordinates that characterize the instantaneous position of the system and equal in number to that of its degrees of freedom.

The Hamilton principle for conservative systems can be derived from the Lagrange equation (e.g., Goldstein 1950):

$$\int_{t_0}^{t_1} \delta(K - \Pi)dt = \int_{t_0}^{t_1} \delta(K - U - V)dt = 0 \tag{1.62}$$

where U is the strain energy and V is the energy of applied loads. The Hamilton principle implies that a true motion path of the system between states specified at arbitrary time instances is such that the Lagrangian has an extreme (and constant) value.

Using appropriate constitutive and strain-displacement relations and applying Green's theorem one can derive equations of motion and the boundary conditions for the system from Hamilton's principle (e.g., Reddy 2002, and Whitney 1987). This procedure is demonstrated in Sect. 1.6 for plate structures.

Consider the case where the motion of the system can be represented by the approximation

$$\mathbf{u} = \begin{Bmatrix} u_1(x_1, x_2, x_3, t) \\ u_2(x_1, x_2, x_3, t) \\ u_3(x_1, x_2, x_3, t) \end{Bmatrix} = \begin{Bmatrix} \sum_{i=1}^{n_1} k_{1i}(t) f_{1i}(x_1, x_2, x_3) \\ \sum_{j=1}^{n_2} k_{2j}(t) f_{2j}(x_1, x_2, x_3) \\ \sum_{s=1}^{n_3} k_{3s}(t) f_{3s}(x_1, x_2, x_3) \end{Bmatrix} \tag{1.63}$$

In (1.63), $k_{1i}(t)$, $k_{2j}(t)$, $k_{3s}(t)$ are unknown functions of time and $f_{1i}(x_1, x_2, x_3)$, $f_{2j}(x_1, x_2, x_3)$, $f_{3s}(x_1, x_2, x_3)$ are assumed functions of the coordinates that satisfy kinematic boundary conditions, i.e. the conditions for displacements and slopes along the boundaries of the domain occupied by the system. Note that the number of terms in each of series (1.63) can vary, though in many applications it is convenient to use $n_1 = n_2 = n_3$.

The requirement that the Lagrangian of the system remains stationary implies that the variation of this functional is equal to zero. Accordingly, assuming motion in the form (1.63) one obtains

$$\delta(\Pi - K) = \sum_{i=1}^{n_1} \frac{\partial(\Pi - K)}{\partial k_{1i}} \delta k_{1i} + \sum_{j=1}^{n_2} \frac{\partial(\Pi - K)}{\partial k_{2j}} \delta k_{2j} + \sum_{s=1}^{n_3} \frac{\partial(\Pi - K)}{\partial k_{3s}} \delta k_{3s} = 0 \tag{1.64}$$

However, the variations of the functions $k_{1i}(t)$, $k_{2j}(t)$, $k_{3s}(t)$ are arbitrary. Therefore, the condition (1.64) can be satisfied only if

$$\frac{\partial(\Pi - K)}{\partial k_{1i}} = \frac{\partial(\Pi - K)}{\partial k_{2j}} = \frac{\partial(\Pi - K)}{\partial k_{3s}} = 0$$

$$i = 1, 2, ..., n_1, \quad j = 1, 2, ..., n_2, \quad s = 1, 2, ..., n_3, \tag{1.65}$$

Equations 1.65 represent the Rayleigh-Ritz method. Historically, Rayleigh was the first to propose this approach for the study of vibrations of an undamped structure using a single-degree approximation (Rayleigh 1877). This approach was extended by Ritz (1909) who suggested using a multi-degree of freedom approach, i.e. more than one term approximation in the series representing the motion.

Equations 1.65 enable us to avoid solving the partial differential equations of motion where displacements depend on the spatial coordinates and time. Instead, the Rayleigh-Ritz method reduces the problem to a system of ordinary differential equations with only one independent variable, i.e. time. Furthermore, in static problems, this method results in a system of algebraic equations since in such problems k_{1i}, k_{2j}, k_{3s} are unknown constant coefficients in the series approximation (1.63).

The Rayleigh-Ritz method formulated by (1.65) could also be derived from the principle of stationary total energy (1.60) by indicating that the constant value of the total energy of a conservative system should be equal to the maximum value of potential energy achieved when the system reached an extreme deviation from static equilibrium and stopped so that its kinetic energy is equal to zero. On the other hand, the same constant value corresponds to the maximum kinetic energy that is reached when the system passes through the static equilibrium position and its potential energy is equal to zero. Then we can require that the series for displacements (1.63) should be such that the difference between the maximum potential and kinetic energies is minimized. Of course, this difference depends on the coefficients of series (1.63). The result of the error minimization is the system of equations (1.65).

Note that the static boundary conditions referring to boundary values of stress resultants and stress couples do not have to be addressed in the Rayleigh-Ritz method. However, if the series (1.63) accurately reflect these conditions, the accuracy of the solution may be improved.

The Galerkin method (Galerkin 1915) is considered an energy method, although its implementation does not directly employ the energy of the system. Bypassing the theoretical foundation of this method, we refer to the equations of motion and boundary conditions that can be derived from the Hamilton principle. The solution is sought in the form (1.63), except for the requirement that the functions of coordinates in these series must satisfy both kinematic and static boundary conditions. The procedure associated with the Galerkin method belongs to a class of weighted residual methods (Reddy 2002). According to these methods we require that the residual of the equations of motion must be orthogonal to a set of independent weight functions. In the Galerkin method the weight functions are identical with the functions $f_{1i}(x_1, x_2, x_3)$, $f_{2j}(x_1, x_2, x_3)$, $f_{3s}(x_1, x_2, x_3)$. Accordingly, the method results in the mathematical formulation:

$$\iiint\limits_{\Omega} L_q[\mathbf{u}(x_1, x_2, x_3)]f_{1i}(x_1, x_2, x_3)\,dx_1dx_2dx_3 = 0$$

$$\iiint\limits_{\Omega} L_q[\mathbf{u}(x_1, x_2, x_3)]f_{2j}(x_1, x_2, x_3)\,dx_1dx_2dx_3 = 0$$

$$\iiint\limits_{\Omega} L_q[\mathbf{u}(x_1, x_2, x_3)]f_{3s}(x_1, x_2, x_3)\,dx_1dx_2dx_3 = 0 \tag{1.66}$$

where $L_q[\mathbf{u}(x_1, x_2, x_3)] = 0$ are equations of motion reflecting the equilibrium of forces in the q-th direction ($q = x_1, x_2, x_3$) and Ω is the volume occupied by the structure. Although the Galerkin method is applied directly to equations of equilibrium or motion, it follows from the Hamilton principle (e.g. Whitney 1987).

The Galerkin procedure requires that series (1.63) satisfy all kinematic and static boundary conditions. If some of these conditions are violated, the Generalized Galerkin procedure enables us to incorporate them in the formulation resulting in a more accurate solution. Although this procedure is relatively little known, contrary to the standard Galerkin procedure, solutions utilizing it have been published (Houbolt and Brooks 1958; Simitses 1986; Birman and Suhir 2007). It can be shown that the Rayleigh-Ritz and Galerkin methods using the same series (1.63) result in identical results if the boundary conditions are kinematic or if the functions of coordinates satisfy both kinematic and static boundary conditions.

There are numerous other energy-based methods that are applied to the analysis of structures, including plates. While the review of these methods is outside the scope of the book, we refer readers to extensive literature on the subject, such as the book of Wunderlich and Pilkey (2003) or the monograph of Reddy (2002).

Considering the fact that equations of motion of plate structures and their boundary conditions are derived in this chapter, it is useful to understand the place of energy methods in the analysis. The combination of equations of motion and boundary conditions for plate structures often denies an exact integration. This may make attractive an alternative approach based on the spectrum of available energy methods, such as the Rayleigh-Ritz and Galerkin methods. Numerical procedures, including finite element, finite difference or boundary element methods, can be developed, using either the equations of motion or energy methods.

1.6 Equations of Motion and Boundary Conditions: Derivation from the Hamilton Principle

Equations of motion and boundary conditions of a plate can be derived from the Hamilton principle. Alternatively, it is possible to obtain these equations by analyzing the equilibrium of an infinitesimal element detached from the plate where the effect of adjacent parts of the plate separated from the element is represented by stress resultants and stress couples. The advantage of the former approach is related to a simultaneous derivation of both the equations of motion as well as the boundary conditions. In this paragraph, we derive such equations for the case of a shear deformable plate. The following derivation is based on the solutions presented by Vol'mir (1972), Whitney (1987) and Reddy (2007).

The strain energy of an elastic plate in a Cartesian coordinate system is given by

$$U = \frac{1}{2} \iiint_{\Omega} \left(\sigma_x \varepsilon_x + \sigma_y \varepsilon_y + \sigma_z \varepsilon_z + \tau_{yz} \gamma_{yz} + \tau_{xz} \gamma_{xz} + \tau_{xy} \gamma_{xy}\right) d\Omega \qquad (1.67)$$

The energy of transverse pressure $p\,(x, y, t)$ applied to the surface of the plate is

$$V = -\iint_{A} pw\,dA \qquad (1.68)$$

where A is the surface area.

The kinetic energy of the plate is given by

$$K = \frac{1}{2} \iiint_{\Omega} \rho\,(x, y, z) \left[\left(\frac{\partial u}{\partial t}\right)^2 + \left(\frac{\partial v}{\partial t}\right)^2 + \left(\frac{\partial w}{\partial t}\right)^2\right] d\Omega \qquad (1.69)$$

where $\rho\,(x, y, z)$ is a mass density of the material.

The expression for the strain energy can be transformed using the constitutive relations. Substituting (1.15) into (1.67), using the definition of stress resultants and stress couples according to (1.50)–(1.52) and accounting for the previously introduced assumption that $\varepsilon_z = 0$ we obtain

$$U = \frac{1}{2} \iint_{A} \left(N_x \varepsilon_x^0 + N_y \varepsilon_y^0 + N_{xy} \gamma_{xy}^0 + M_x \kappa_x + M_y \kappa_y + M_{xy} \kappa_{xy} \right.$$
$$\left. + Q_x \gamma_{xz} + Q_y \gamma_{yz}\right) dA \qquad (1.70)$$

Now we can substitute the expressions for the strains in (1.70). In particular, considering geometrically nonlinear plates and adopting the first-order shear deformation theory, i.e. combining (1.31), (1.28) and (1.48) we obtain

$$U = \frac{1}{2}\iint_A \left[\begin{array}{l} N_x\left(\frac{\partial u_0}{\partial x} + \frac{1}{2}\left(\frac{\partial w}{\partial x}\right)^2\right) + N_y\left(\frac{\partial v_0}{\partial y} + \frac{1}{2}\left(\frac{\partial w}{\partial y}\right)^2\right) \\ + N_{xy}\left(\frac{\partial u_0}{\partial y} + \frac{\partial v_0}{\partial x} + \frac{\partial w}{\partial x}\frac{\partial w}{\partial y}\right) + M_x\frac{\partial \psi_x}{\partial x} + M_y\frac{\partial \psi_y}{\partial y} \\ + M_{xy}\left(\frac{\partial \psi_x}{\partial y} + \frac{\partial \psi_y}{\partial x}\right) + Q_x\left(\psi_x + \frac{\partial w}{\partial x}\right) + Q_y\left(\psi_y + \frac{\partial w}{\partial y}\right) \end{array} \right] dA \tag{1.71}$$

The kinetic energy is expressed in terms of displacements and rotations by substituting (1.46) into (1.69):

$$K = \frac{1}{2}\hat{m}\iint_A \left[\left(\frac{\partial u_0}{\partial t}\right)^2 + \left(\frac{\partial v_0}{\partial t}\right)^2 + \left(\frac{\partial w}{\partial t}\right)^2\right] dA + \frac{1}{2}I\iint_A \left[\left(\frac{\partial \psi_x}{\partial t}\right)^2 + \left(\frac{\partial \psi_y}{\partial t}\right)^2\right] dA \tag{1.72}$$

where $\hat{m} = \int_z \rho(x, y, z)\, dz$ is the mass of the plate per unit surface area and

$$I = \int_z \rho(x, y, z)\, z^2 dz \tag{1.73}$$

Equation 1.72 is obtained by assumption that the mass density is symmetric about the middle plane. The first integral in the right side of (1.72) represents in-plane and transverse inertias, while the second integral includes rotational inertia terms.

The kinetic and strain energy and the energy of applied pressure can now be substituted into the Hamiltonian. Consider the integral of the variation of the kinetic energy. As follows from (1.72),

$$\int_{t_0}^{t_1} \delta K dt = \int_{t_0}^{t_1}\iint_A \left\{\hat{m}\left[\frac{\partial u_0}{\partial t}\delta\left(\frac{\partial u_0}{\partial t}\right) + \frac{\partial v_0}{\partial t}\delta\left(\frac{\partial v_0}{\partial t}\right) + \frac{\partial w}{\partial t}\delta\left(\frac{\partial w}{\partial t}\right)\right] + I\left[\frac{\partial \psi_x}{\partial t}\delta\left(\frac{\partial \psi_x}{\partial t}\right) + \frac{\partial \psi_y}{\partial t}\delta\left(\frac{\partial \psi_y}{\partial t}\right)\right]\right\} dA dt \tag{1.74}$$

Integrating (1.74) by parts yields

$$\int_{t_0}^{t_1} \delta K dt = \iint_A \left\{ \hat{m} \left[\frac{\partial u_0}{\partial t} \delta u_0 + \frac{\partial v_0}{\partial t} \delta v_0 + \frac{\partial w}{\partial t} \delta w \right] + I \left[\frac{\partial \psi_x}{\partial t} \delta \psi_x + \frac{\partial \psi_y}{\partial t} \delta \psi_y \right] \right\}_{t_0}^{t_1} dA$$
$$- \int_{t_0}^{t_1} \iint_A \left\{ \hat{m} \left[\frac{\partial^2 u_0}{\partial t^2} \delta u_0 + \frac{\partial^2 v_0}{\partial t^2} \delta v_0 + \frac{\partial^2 w}{\partial t^2} \delta w \right] \right.$$
$$\left. + I \left[\frac{\partial^2 \psi_x}{\partial t^2} \delta \psi_x + \frac{\partial^2 \psi_y}{\partial t^2} \delta \psi_y \right] \right\} dA dt \tag{1.75}$$

The variations of displacements and rotations are arbitrary. Therefore, it is possible to assume that they are different from the actual motion at any time except for the initial and final instants so that

$$\delta u_0 (t_0) = \delta u_0 (t_1) = \delta v_0 (t_0) = \delta v_0 (t_1) = \delta w (t_0) = \delta w (t_1) = \delta \psi_x (t_0)$$
$$= \delta \psi_x (t_1) = \delta \psi_y (t_0) = \delta \psi_y (t_1) = 0 \tag{1.76}$$

Then the first integral in the right side of (1.75) yields zero since every term in this integral is multiplied by one of initial or final values of displacement or rotation variations. Accordingly, we do not consider this integral in the subsequent derivations.

The terms that constitute potential energy can also be integrated by parts. Let us consider a part of the plate limited by the contour formed by lines x = const that will be denoted by S_x and by lines y = const will be denoted by S_y. Consider for example the variation of the first term in (1.71). The integration by parts yields

$$\iint_A N_x \frac{\partial \delta u_0}{\partial x} dA = \int_{S_x} N_x \delta u_0 dy - \iint_A \frac{\partial N_x}{\partial x} \delta u_0 dA \tag{1.77}$$

Similar procedure can be applied to all terms in (1.71). While the terms related to variations of in-plane displacements u_0 and v_0 and those due to variations of rotations ψ_x and ψ_y are easily evaluated, it is useful to show terms related to the variation of the transverse deflection since they include geometrically nonlinear contributions:

$$
\begin{aligned}
\delta U(w) &= \iint\limits_A \left\{ \begin{matrix} N_x \dfrac{\partial w}{\partial x}\dfrac{\partial \delta w}{\partial x} + N_y \dfrac{\partial w}{\partial y}\dfrac{\partial \delta w}{\partial y} + N_{xy}\left(\dfrac{\partial w}{\partial x}\dfrac{\partial \delta w}{\partial y} + \dfrac{\partial w}{\partial y}\dfrac{\partial \delta w}{\partial x}\right) + \\ Q_x \dfrac{\partial \delta w}{\partial x} + Q_y \dfrac{\partial \delta w}{\partial y} \end{matrix} \right\} dA \\
&= \int\limits_{S_x} \left(N_x \frac{\partial w}{\partial x} + N_{xy}\frac{\partial w}{\partial y}\right)\delta w dy + \int\limits_{S_y}\left(N_{xy}\frac{\partial w}{\partial x} + N_y \frac{\partial w}{\partial y}\right)\delta w dx \\
&\quad + \int\limits_{S_x} Q_x \delta w dy + \int\limits_{S_y} Q_y \delta w dx - \iint\limits_A \left[\frac{\partial}{\partial x}\left(N_x \frac{\partial w}{\partial x} + N_{xy}\frac{\partial w}{\partial y}\right)\right. \\
&\quad \left. + \frac{\partial}{\partial y}\left(N_{xy}\frac{\partial w}{\partial x} + N_y \frac{\partial w}{\partial y}\right) + \frac{\partial Q_x}{\partial x} + \frac{\partial Q_y}{\partial y}\right]\delta w dA
\end{aligned} \tag{1.78}
$$

The variation of the energy of applied pressure given by (1.68) is

$$
\delta V = -\iint\limits_A p\delta w dA \tag{1.79}
$$

Collecting all terms in Hamiltonian (1.62) we obtain the following variational equation:

$$
\begin{aligned}
&\int\limits_{t_0}^{t_1}\iint\limits_A \left\{ \begin{matrix} \left(\dfrac{\partial N_x}{\partial x} + \dfrac{\partial N_{xy}}{\partial y} - \hat{m}\dfrac{\partial^2 u_o}{\partial t^2}\right)\delta u_0 + \left\{ \dfrac{\partial N_{xy}}{\partial x} + \dfrac{\partial N_y}{\partial y} - \hat{m}\dfrac{\partial^2 v_o}{\partial t^2}\right\}\delta v_0 \\ + \left[\dfrac{\partial Q_x}{\partial x} + \dfrac{\partial Q_y}{\partial y} + \dfrac{\partial}{\partial x}\left(N_x \dfrac{\partial w}{\partial x} + N_{xy}\dfrac{\partial w}{\partial y}\right)\right. \\ \left. + \dfrac{\partial}{\partial y}\left(N_{xy}\dfrac{\partial w}{\partial x} + N_y\dfrac{\partial w}{\partial y}\right) + p - \hat{m}\dfrac{\partial^2 w}{\partial t^2}\right]\delta w \\ + \left(\dfrac{\partial M_x}{\partial x} + \dfrac{\partial M_{xy}}{\partial y} - Q_x - I\dfrac{\partial^2 \psi_x}{\partial t^2}\right)\delta\psi_x \\ + \left(\dfrac{\partial M_{xy}}{\partial x} + \dfrac{\partial M_y}{\partial y} - Q_y - I\dfrac{\partial^2 \psi_y}{\partial t^2}\right)\delta\psi_y \end{matrix} \right\} dAdt \\
&- \int\limits_{t_0}^{t_1}\int\limits_{S_x}\left[N_x\delta u_0 + N_{xy}\delta v_0 + \left(Q_x + N_x\frac{\partial w}{\partial x} + N_{xy}\frac{\partial w}{\partial y}\right)\delta w + M_x\delta\psi_x + M_{xy}\delta\psi_y\right] \\
&\times dydt - \int\limits_{t_0}^{t_1}\int\limits_{S_x}\left[N_{xy}\delta u_0 + N_y\delta v_0 + \left(Q_y + N_{xy}\frac{\partial w}{\partial x} + N_y\frac{\partial w}{\partial y}\right)\delta w \right. \\
&\quad \left. + M_{xy}\delta\psi_x + M_y\delta\psi_y\right] dydt
\end{aligned} \tag{1.80}
$$

The equations of motion are obtained by requiring that the coefficients at the variations of displacements and rotations in the area integrals must be equal to zero:

$$\begin{aligned}
&\frac{\partial N_x}{\partial x} + \frac{\partial N_{xy}}{\partial y} = \hat{m}\frac{\partial^2 u_o}{\partial t^2} \\
&\frac{\partial N_{xy}}{\partial x} + \frac{\partial N_y}{\partial y} = \hat{m}\frac{\partial^2 v_o}{\partial t^2} \\
&\frac{\partial Q_x}{\partial x} + \frac{\partial Q_y}{\partial y} + \frac{\partial}{\partial x}\left(N_x\frac{\partial w}{\partial x} + N_{xy}\frac{\partial w}{\partial y}\right) + \frac{\partial}{\partial y}\left(N_{xy}\frac{\partial w}{\partial x} + N_y\frac{\partial w}{\partial y}\right) \\
&\quad + p = \hat{m}\frac{\partial^2 w}{\partial t^2} \\
&\frac{\partial M_x}{\partial x} + \frac{\partial M_{xy}}{\partial y} - Q_x = I\frac{\partial^2 \psi_x}{\partial t^2} \\
&\frac{\partial M_{xy}}{\partial x} + \frac{\partial M_y}{\partial y} - Q_y = I\frac{\partial^2 \psi_y}{\partial t^2}
\end{aligned} \tag{1.81}$$

Consider the case where besides transverse pressure $p(x, y, t)$ the plate is subject to in-plane stress resultants $\bar{N}_x$, $\bar{N}_y$ and $\bar{N}_{xy}$ applied along the edges (throughout this book, an overbar identifies applied loads). For example, it is possible to apply along the edge x = const stress resultants $\bar{N}_x$ and $\bar{N}_{xy}$.

In a geometrically linear formulation in-plane stress resultants are constant over the surface area of the plate. Furthermore, in most practical problems in-plane inertia terms in the first two equations (1.81) are negligible. Accordingly, in such case equations (1.81) become

$$\begin{aligned}
&\frac{\partial N_x}{\partial x} + \frac{\partial N_{xy}}{\partial y} = 0 \\
&\frac{\partial N_{xy}}{\partial x} + \frac{\partial N_y}{\partial y} = 0 \\
&\frac{\partial Q_x}{\partial x} + \frac{\partial Q_y}{\partial y} + \bar{N}_x\frac{\partial^2 w}{\partial x^2} + 2\bar{N}_{xy}\frac{\partial^2 w}{\partial x \partial y} + \bar{N}_y\frac{\partial^2 w}{\partial y^2} + p = \hat{m}\frac{\partial^2 w}{\partial t^2} \\
&\frac{\partial M_x}{\partial x} + \frac{\partial M_{xy}}{\partial y} - Q_x = I\frac{\partial^2 \psi_x}{\partial t^2} \\
&\frac{\partial M_{xy}}{\partial x} + \frac{\partial M_y}{\partial y} - Q_y = I\frac{\partial^2 \psi_y}{\partial t^2}
\end{aligned} \tag{1.82}$$

Another form of the equations of motion applicable to both linear and nonlinear problems is obtained if rotational inertia terms in the right side of the last two equations of motion (1.81) are negligible. Accordingly, transverse shear stress resultants expressed in terms of stress couples can be substituted in the third equation of motion yielding

$$\frac{\partial^2 M_x}{\partial x^2} + 2\frac{\partial^2 M}{\partial x \partial y} + \frac{\partial^2 M_y}{\partial y^2} + \frac{\partial}{\partial x}\left(N_x \frac{\partial w}{\partial x} + N_{xy}\frac{\partial w}{\partial y}\right) + \frac{\partial}{\partial y}\left(N_{xy}\frac{\partial w}{\partial x} + N_y \frac{\partial w}{\partial y}\right)$$
$$+ p = \hat{m}\frac{\partial^2 w}{\partial t^2} \tag{1.83}$$

Neglecting in-plane inertia terms, the equations of motion become

$$\begin{aligned}
&\frac{\partial N_x}{\partial x} + \frac{\partial N_{xy}}{\partial y} = 0\\
&\frac{\partial N_{xy}}{\partial x} + \frac{\partial N_y}{\partial y} = 0\\
&\frac{\partial^2 M_x}{\partial x^2} + 2\frac{\partial^2 M_{xy}}{\partial x \partial y} + \frac{\partial^2 M_y}{\partial y^2} + p + \underline{\underline{N_x \frac{\partial^2 w}{\partial x^2} + 2N_{xy}\frac{\partial^2 w}{\partial x \partial y} + N_y \frac{\partial^2 w}{\partial y^2}}} = \hat{m}\frac{\partial^2 w}{\partial t^2}
\end{aligned} \tag{1.84}$$

The doubly-underlined terms in the last equation (1.84) account for the bending-stretching coupling in geometrically nonlinear problems. In such problems, these terms incorporating stress resultants given by (1.58) are nonlinear functions of displacements. However, in a geometrically linear formulation bending and twisting are uncoupled from in-plane stretching and shear. Accordingly, in linear problems if the plate is not subject to external in-plane loads, $N_x = N_y = N_{xy} = 0$ and the doubly-underlined terms in the last equation (1.84) are equal to zero.

Equations 1.81 or 1.82 are applicable to the analysis of plates characterized by the first-order shear deformation theory. This is because this theory involves five unknown displacements and rotations, i.e. u_0, v_0, w, ψ_x, ψ_y. Accordingly, upon the substitution of the constitutive relations, the number of equations is equal to the number of equations of motion (or equilibrium), i.e. the problem is statically determinate. On the other hand, three equations (1.84) are suitable for the analysis of the classical theory of plates since they involve three displacements only.

The boundary conditions are formulated from the requirement that the integrals along the contours S_x and S_y must be equal to zero. Accordingly, we list these conditions that directly follow from (1.80). For example, for the former contour they are:

$$\begin{aligned}
&N_x \delta u_0 = 0\\
&N_{xy} \delta v_0 = 0\\
&\left(Q_x + N_x \frac{\partial w}{\partial x} + N_{xy}\frac{\partial w}{\partial y} + \underline{\frac{\partial M_{xy}}{\partial y}}\right)\delta w = 0
\end{aligned}$$

$$M_x \delta \psi_x = 0$$
$$M_{xy} \delta \psi_y = 0 \tag{1.85}$$

where the underlined term in the third equation is added to account for the energy of the twisting stress couple acting along the contour S_x. The presence of this term becomes clear from the discussion on the boundary conditions in Sect. 2.1 (see discussion related to Fig. 2.2). This term is directly derived using the Hamilton principle in the case of thin plates characterized by the classical theory (e.g., Whitney 1987, and Reddy 2007).

The above requirements are satisfied if one of the terms forming each of the five products in (1.85) is equal to a prescribed value (such prescribed values are identified in the following equations by a "hat"):

$$\begin{aligned}
&N_x = \bar{N}_x = \hat{N}_x && or && u_0 = \hat{u}_0\\
&N_{xy} = \bar{N}_{xy} = \hat{N}_{xy} && or && v_0 = \hat{v}_0\\
&Q_x + N_x \frac{\partial w}{\partial x} + N_{xy} \frac{\partial w}{\partial y} + \frac{\partial M_{xy}}{\partial y} = \hat{V}_x && or && w = \hat{w}_0\\
&M_x = \bar{M}_x = \hat{M}_x && or && \psi_x = \hat{\psi}_x\\
&M_{xy} = \bar{M}_{xy} = \hat{M}_{xy} && or && \psi_y = \hat{\psi}_y
\end{aligned} \tag{1.86}$$

Similar boundary conditions can be formulated along S_y:

$$\begin{aligned}
&N_y = \bar{N}_y = \hat{N}_y && or && v_0 = \hat{v}_0\\
&N_{xy} = \bar{N}_{xy} = \hat{N}_{xy} && or && u_0 = \hat{u}_0\\
&Q_y + N_{xy} \frac{\partial w}{\partial x} + N_y \frac{\partial w}{\partial y} + \frac{\partial M_{xy}}{\partial x} = \hat{V}_y && or && w = \hat{w}_0\\
&M_y = \bar{M}_y = \hat{M}_y && or && \psi_y = \hat{\psi}_y\\
&M_{xy} = \bar{M}_{xy} = \hat{M}_{xy} && or && \psi_x = \hat{\psi}_x
\end{aligned} \tag{1.87}$$

The boundary conditions have an identical form in static and dynamic problems. However, in the case of a thin plate the number of boundary conditions is reduced to four. For example, for the contour S_x these conditions are:

$$\begin{aligned}
&N_x = \bar{N}_x = \hat{N}_x && or && u_0 = \hat{u}_0\\
&N_{xy} = \bar{N}_y = \hat{N}_{xy} && or && v_0 = \hat{v}_0\\
&\frac{\partial M_x}{\partial x} + 2\frac{\partial M_{xy}}{\partial y} + N_x \frac{\partial w}{\partial x} + N_{xy} \frac{\partial w}{\partial y} = \hat{V}_x && or && w = \hat{w}_0\\
&M_x = \bar{M}_x = \hat{M}_x && or && \frac{\partial w}{\partial x} = \frac{\partial \hat{w}}{\partial x}
\end{aligned} \tag{1.88}$$

In the conclusion of this paragraph, it is useful to show the expression for the energy of applied in-plane stress resultants that may be useful in the application of the Rayleigh-Ritz method. In the linear problem, neglecting stretching of the middle plane of the plate, this energy:

$$\bar{V} = -\frac{1}{2}\iint_A \left[\bar{N}_x\left(\frac{\partial w}{\partial x}\right)^2 + \bar{N}_y\left(\frac{\partial w}{\partial y}\right)^2 + 2\bar{N}_{xy}\frac{\partial w}{\partial x}\frac{\partial w}{\partial y}\right]dA \quad (1.89)$$

1.7 Equations of Motion and Boundary Conditions: Derivation from the Analysis of an Infinitesimal Plate Element

Equations of motion derived in the previous paragraph from Hamilton's principle could also be obtained by the analysis of equilibrium of forces and moments acting on an infinitesimal element detached from the plate (Fig. 1.9). According to this approach, the action of adjacent removed parts of the plate is represented by stress resultants and stress couples applied to the edges of the element. Additionally, external loads acting on the surface of the element are included in the consideration.

Consider an infinitesimal element of a plate shown in Fig. 1.9. The inertia contribution will be included in the equations of equilibrium using d'Alambert's principle. We should derive three equations of dynamic equilibrium of forces acting in the direction of the coordinate axes and three equations of equilibrium of moments. However, it would be easy to show that the equilibrium of moments about the z-axis is identically satisfied since $N_{xy} = N_{yx}$ and $M_{xy} = M_{yx}$.

The dynamic equilibrium of forces acting in the x-direction is obtained taking the projections of in-plane forces on this axis. The forces in the x-direction produced by in-plane stress resultants N_x and N_{xy} are

$$F(N_x) = \left(N_x + \frac{\partial N_x}{\partial x}dx\right)dy - N_x dy = \frac{\partial N_x}{\partial x}dxdy$$

$$F(N_{xy}) = \left(N_{xy} + \frac{\partial N_{xy}}{\partial y}dy\right)dx - N_{xy}dx = \frac{\partial N_{xy}}{\partial y}dxdy \quad (1.90)$$

Transverse shear stress resultants also produce a projection force in the x-direction. However, the effect of this contribution is small in plates and shallow shells and can be neglected (e.g., Vol'mir 1972). Combining the contributions of the in-plane stress resultants given by (1.90), including the in-plane inertia of the infinitesimal element $\hat{m}\frac{\partial^2 u_0}{\partial t^2}dxdy$ and canceling out the surface area of the element $dxdy$ in all terms we obtain the first equation of motion that is predictably

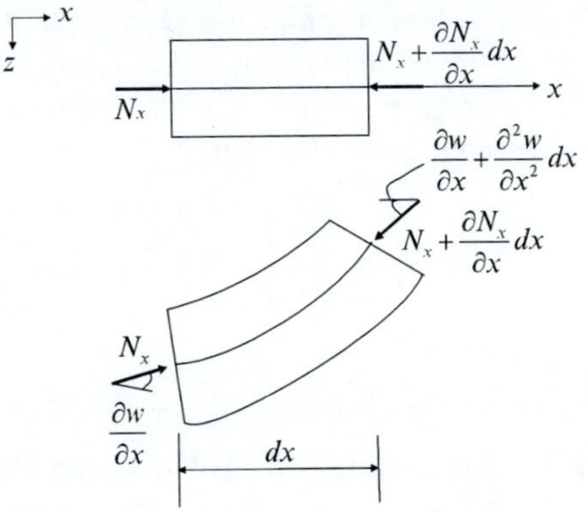

Fig. 1.10 Illustration to the evaluation of the projection of stress resultant N_x on the z-axis

identical with the first equation (1.81). The second equation (1.81) that represents the dynamic equilibrium of forces in the y-direction is obtained in the same manner.

The equilibrium of forces in the z-direction includes the contributions of transverse shear and in-plane stress resultants. The latter stress resultants produce the projection equal to the product of the stress resultant and the sine of the angle formed by the middle plane with the original undeformed position (Fig. 1.10). The rotation of the cross section about the middle plane that is considered in shear deformation theories does not affect this projection.

The contributions of the stress resultants to the dynamic equilibrium in the z-direction are now shown in detail:

$$F(Q_x) = \left(Q_x + \frac{\partial Q_x}{\partial x} dx \right) dy - Q_x dy = \frac{\partial Q_x}{\partial x} dxdy$$

$$F(Q_y) = \left(Q_y + \frac{\partial Q_y}{\partial y} dy \right) dx - Q_y dx = \frac{\partial Q_y}{\partial y} dxdy$$

$$F(N_x) = \left(N_x + \frac{\partial N_x}{\partial x} dx \right) \left(\frac{\partial w}{\partial x} + \frac{\partial^2 w}{\partial x^2} dx \right) dy - N_x \frac{\partial w}{\partial x} dy$$

$$= \frac{\partial N_x}{\partial x} \frac{\partial w}{\partial x} dxdy + N_x \frac{\partial^2 w}{\partial x^2} dxdy + \underline{\frac{\partial N_x}{\partial x} \frac{\partial^2 w}{\partial x^2} dx^2 dy}$$

$$F(N_y) = \frac{\partial N_y}{\partial y} \frac{\partial w}{\partial y} dxdy + N_y \frac{\partial^2 w}{\partial y^2} dxdy + \underline{\frac{\partial N_y}{\partial y} \frac{\partial^2 w}{\partial y^2} dxdy^2} \tag{1.91}$$

The terms produced by in-plane shear stress resultant are:

$$F\left(N_{xy}\right)=\left(N_{xy}+\frac{\partial N_{xy}}{\partial x}dx\right)\left(\frac{\partial w}{\partial y}+\frac{\partial^2 w}{\partial y^2}dx\right)dy-N_{xy}\frac{\partial w}{\partial y}dy$$
$$+\left(N_{xy}+\frac{\partial N_{xy}}{\partial y}dy\right)\left(\frac{\partial w}{\partial x}+\frac{\partial^2 w}{\partial x^2}dy\right)dx-N_{xy}\frac{\partial w}{\partial x}dx$$
$$=\frac{\partial N_{xy}}{\partial x}\frac{\partial w}{\partial y}dxdy+N_{xy}\frac{\partial^2 w}{\partial y^2}dxdy+\underline{\frac{\partial N_{xy}}{\partial x}\frac{\partial^2 w}{\partial y^2}dx^2dy}+\frac{\partial N_{xy}}{\partial y}\frac{\partial w}{\partial x}dxdy$$
$$+N_{xy}\frac{\partial^2 w}{\partial x^2}dxdy+\underline{\frac{\partial N_{xy}}{\partial y}\frac{\partial^2 w}{\partial x^2}dx^2dy} \tag{1.92}$$

The underlined terms in the above equations are of higher order and can be neglected in many cases. Additional terms are contributed by the transverse pressure acting on the surface of the element, i.e. $pdxdy$ and by transverse inertia, i.e. $\hat{m}\frac{\partial^2 w}{\partial t^2}dxdy$. Combining all above-mentioned terms and canceling the surface area of the element that is present in all terms we obtain the third equation (1.81).

The equations of dynamic equilibrium of moments about the x- and y-axes can be obtained without difficulty. The rotational inertia terms are derived by considering the inertial moment of an element. For example, for a plate where the material is symmetrically distributed about the middle plane, the inertial moment of the element $dxdy$ in the xz plane is

$$\int_{-\frac{h}{2}}^{\frac{h}{2}}\rho\frac{\partial^2\left(u_0+z\psi_x\right)}{\partial t^2}zdzdxdy=I\frac{\partial^2\psi_x}{\partial t^2}dxdy \tag{1.93}$$

The contributions of the stress couples and transverse shear stress resultants to the moment in the xz plane about the right edge of the element shown in Fig. 1.9 are (in-plane stress resultants produce a negligible effect on the equilibrium of moments):

$$\left(M_x+\frac{\partial M_x}{\partial x}dx\right)dy-M_xdy+\left(M_{xy}+\frac{\partial M_{xy}}{\partial y}dy\right)dx-M_{xy}dx-Q_xdxdy$$
$$+\left(Q_y+\frac{\partial Q_y}{\partial y}dy\right)\frac{dxdy}{2}-Q_y\frac{dxdy}{2} \tag{1.94}$$

Neglecting higher-order terms and accounting for (1.93) we obtain the fourth equation (1.81). The remaining fifth equation is obtained by analogy.

1.8 An Alternative Formulation of Equations of Equilibrium and Boundary Conditions of Thin Plates in Terms of a Stress Function

The approach considered in this paragraph is applicable to plates that can be analyzed by the classical (thin plate) theory. Consider vibrations of a thin plate neglecting the contribution of in-plane inertia (static problem represents a particular case of the dynamic problem so that the present formulation can be applied to the static case as well). Then equations of motion are given by equations (1.84). The approach considered here is based on reducing these three equations to a single equation through the introduction of the stress function defined by

$$\sigma_x = \frac{N_x}{h} = \frac{\partial^2 \varphi}{\partial y^2}, \quad \sigma_y = \frac{N_y}{h} = \frac{\partial^2 \varphi}{\partial x^2}, \quad \tau_{xy} = \frac{N_{xy}}{h} = -\frac{\partial^2 \varphi}{\partial x \partial y} \tag{1.95}$$

It is immediately observed that the first two equations (1.84) where in-plane inertia is neglected are identically satisfied. The third equation is represented in terms of the stress function while the stress couples are expressed in terms of deflection w according to (1.58). This yields

$$\frac{D}{h}\nabla^4 w = \frac{\partial^2 \varphi}{\partial x^2}\frac{\partial^2 w}{\partial y^2} + \frac{\partial^2 \varphi}{\partial y^2}\frac{\partial^2 w}{\partial x^2} - 2\frac{\partial^2 \varphi}{\partial x \partial y}\frac{\partial^2 w}{\partial x \partial y} + \frac{p}{h} - \frac{\hat{m}}{h}\frac{\partial^2 w}{\partial t^2} \tag{1.96}$$

In the above equation the Laplacian $\nabla^4(\ldots) = \frac{\partial^4(\ldots)}{\partial x^4} + 2\frac{\partial^4(\ldots)}{\partial x^2 \partial y^2} + \frac{\partial^4(\ldots)}{\partial y^4}$ in Cartesian coordinates.

Now equations of motion are reduced to a single equation that contains two unknowns, i.e. the deflection and the stress function. The second equation is the equation of compatibility of deformations. It is obtained from conditions (1.8). In the case of plane stress the only applicable condition is the fourth equation (1.8) relating in-plane strains. Substituting the strains expressed in terms of displacements we obtain

$$\frac{\partial^2 \varepsilon_x^0}{\partial y^2} + \frac{\partial^2 \varepsilon_y^0}{\partial x^2} - \frac{\partial^2 \gamma_{xy}^0}{\partial x \partial y} = \left(\frac{\partial^2 w}{\partial x \partial y}\right)^2 - \frac{\partial^2 w}{\partial x^2}\frac{\partial^2 w}{\partial y^2} \tag{1.97}$$

The middle-plane strains can be expressed in terms of the stress function using the inverted form of the constitutive relations for the case of plane stress that are available from (1.19):

$$\varepsilon_x^0 = \frac{\sigma_x - \nu\sigma_y}{E} = \frac{N_x - \nu N_y}{Eh} = \frac{1}{E}\left(\frac{\partial^2 \varphi}{\partial y^2} - \nu\frac{\partial^2 \varphi}{\partial x^2}\right)$$

$$\varepsilon_y^0 = \frac{\sigma_y - \nu\sigma_x}{E} = \frac{N_y - \nu N_x}{Eh} = \frac{1}{E}\left(\frac{\partial^2 \varphi}{\partial x^2} - \nu\frac{\partial \varphi}{\partial y^2}\right)$$

$$\gamma_{xy}^{0} = \frac{\tau_{xy}}{G} = \frac{N_{xy}}{Gh} = -\frac{1}{G}\frac{\partial^2\varphi}{\partial x\partial y} \tag{1.98}$$

Now we can substitute the strains in terms of the stress function from (1.98) into the left side of the compatibility equation (1.97) yielding the second equation necessary for the analysis:

$$\frac{1}{E}\nabla^4\varphi = \frac{\partial^2 w}{\partial x^2}\frac{\partial^2 w}{\partial y^2} - \left(\frac{\partial^2 w}{\partial x\partial y}\right)^2 \tag{1.99}$$

Equations 1.96 and 1.99 that pose a statically determinate problem (two unknowns and two equations) are often used in the analysis of plates, particularly in geometrically nonlinear cases.

1.9 Effect of Temperature on Constitutive Relations and Material Constants

Environmental conditions, such as temperature and moisture, affect the analysis of structures in two ways. Firstly, the constitutive stress-strain relations include thermally-induced (or moisture-induced) contributions to the strains. Accordingly, the equations of motion when written in terms of displacements and therefore depend on the constitutive relations are affected, by the corresponding terms. The second effect is sometimes disregarded or forgotten but it may be as important as explicit environment-dependent terms in the constitutive relations. This is the effect of the environment on the material constants, such as the moduli of elasticity and shear, the Poisson ratio, the coefficient of thermal expansion and the strength of the material (of course, this effect raises the question whether a reference to these properties as "constants" is misleading). Moreover, the problem of heat transfer that predicts the distribution of temperature throughout the structure is dependent on the effect of temperature on thermal conductivity and thermal boundary conditions (similar considerations apply to the problem of moisture distribution). In this book we limit the analysis to thermal problems, although moisture can significantly affect the behavior of polymer-matrix composite plates. Furthermore, we restrict the discussion to plate structures that can be treated by the classical or first-order shear deformation theories.

The strain-displacement relations are modified by including thermally-induced components of strain. For example, geometrically nonlinear relationships (1.27) in an orthotropic or isotropic material that composes a structure analyzed by the classical theory (i.e. neglecting transverse shear deformations) become

$$\varepsilon'_x = \varepsilon_x - \int_{T_0}^{T} \alpha_x dT = \frac{\partial u_0}{\partial x} - z\frac{\partial^2 w}{\partial x^2} + \frac{1}{2}\left(\frac{\partial w}{\partial x}\right)^2 - \int_{T_0}^{T} \alpha_x dT$$

$$\varepsilon'_y = \varepsilon_y - \int_{T_0}^{T} \alpha_y dT = \frac{\partial v_0}{\partial y} - z\frac{\partial^2 w}{\partial y^2} + \frac{1}{2}\left(\frac{\partial w}{\partial y}\right)^2 - \int_{T_0}^{T} \alpha_y dT$$

$$\varepsilon'_{xy} = \frac{\gamma'_{xy}}{2} = \frac{\gamma_{xy}}{2} - \frac{1}{2}\int_{T_0}^{T} \alpha_{xy} dT = \frac{1}{2}\left(\frac{\partial u_0}{\partial y} + \frac{\partial v_0}{\partial x} - 2z\frac{\partial^2 w}{\partial x \partial y} + \frac{\partial w}{\partial x}\frac{\partial w}{\partial y}\right)$$
$$- \frac{1}{2}\int_{T_0}^{T} \alpha_{xy} dT \qquad (1.100)$$

where the quantities with prime in the left side represent mechanical strains, ε_x, ε_y and γ_{xy} are total strains, α_i $(i = x, y, xy)$ are coefficients of thermal expansion and the integration in the right side is performed over the range of temperature (T) from the reference to the current value.

The counterpart of the above equations for the case of a shear-deformable problem is

$$\varepsilon'_x = \varepsilon_x - \int_{T_0}^{T} \alpha_x dT = \frac{\partial u_0}{\partial x} + z\frac{\partial \psi_x}{\partial x} + \frac{1}{2}\left(\frac{\partial w}{\partial x}\right)^2 - \int_{T_0}^{T} \alpha_x dT$$

$$\varepsilon'_y = \varepsilon_y - \int_{T_0}^{T} \alpha_y dT = \frac{\partial v_0}{\partial y} + z\frac{\partial \psi_y}{\partial y} + \frac{1}{2}\left(\frac{\partial w}{\partial y}\right)^2 - \int_{T_0}^{T} \alpha_y dT$$

$$\varepsilon'_{xy} = \frac{\gamma'_{xy}}{2} = \frac{\gamma_{xy}}{2} - \frac{1}{2}\int_{T_0}^{T} \alpha_{xy} dT$$

$$= \frac{1}{2}\left(\frac{\partial u_0}{\partial y} + \frac{\partial v_0}{\partial x} + z\left(\frac{\partial \psi_x}{\partial y} + \frac{\partial \psi_y}{\partial x}\right) + \frac{\partial w}{\partial x}\frac{\partial w}{\partial y}\right) - \frac{1}{2}\int_{T_0}^{T} \alpha_{xy} dT \qquad (1.101)$$

In addition, transverse shear strains are present in the first-order shear deformation theory (see Eq. 1.45). However, they are not affected by temperature as long as the plane of the plate contains two principal material axes.

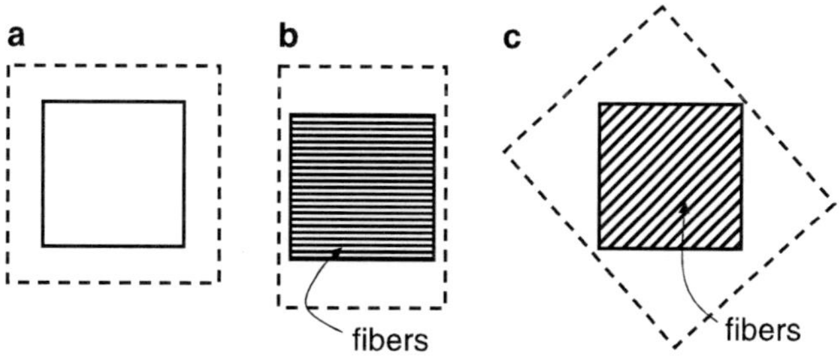

Fig. 1.11 Response to an elevated temperature: (**a**) isotropic plate, (**b**) fiber-reinforced layer with the fibers oriented in the x-direction, (**c**) layer with the fibers oriented at an angle to the x- and y-axes

In the case of an isotropic material the coefficient of thermal expansion is identical in all directions. Moreover, an isotropic plate does not experience shearing deformations due to temperature. Accordingly, in such plate $\alpha_x = \alpha_y,\ \alpha_{xy} = 0$. In a fiber-reinforced layer the coefficients of thermal expansion in the direction of fibers and the direction perpendicular to the fibers differ. However, the layer does not experience in-plane shear due to temperature in the coordinate system with one of the axes oriented along the fibers. Accordingly, if the fibers are oriented along one of the axes (say, the x-axis), $\alpha_x \neq \alpha_y,\ \alpha_{xy} = 0$. Finally, if the fibers are oriented at an angle to the x and y axes, the layer exhibits in-plane shear as well as axial deformations in response to temperature changes. Accordingly, $\alpha_x \neq \alpha_y,\ \alpha_{xy} \neq 0$. The difference in the response of an isotropic plate, a fiber-reinforced layer with the fibers oriented in the x-direction, and a layer with the fibers oriented at an angle to the x- and y-axes is depicted in Fig. 1.11. As is shown in this figure, isotropic materials experience identical expansion in all directions as a result of an elevated temperature. Specially orthotropic materials (materials with the fibers oriented along the x-axis) experience a larger expansion in the direction perpendicular to the fibers than in the fiber direction, reflecting a much smaller thermal expansion coefficient in the latter direction. In the case of materials where the fibers are inclined relative to the coordinate axes (generally orthotropic materials), an elevated temperature causes both expansion in the coordinate directions as well as in-plane shear. If temperature decreases, the pattern of deformations remains similar to that described above, but expansion is replaced with contraction.

It should also be noted that as long as the fibers remain in the plane of the layer, temperature does not affect transverse shear deformations. However, this effect would be present if the fibers were inclined relative to the plane $z = const$. This explains the necessity to carefully analyze local thermal stresses in woven plates, accounting for their contribution to transverse shear. On the other hand, the coefficient of thermal expansion in the z-direction is always present, i.e. $\alpha_z \neq 0$. However, thermal strains and stresses associated with this coefficient often affect the solution of plate problems negligibly, except for high local thermal stresses in the vicinity of such discontinuities as rivets and bolts.

Constitutive relations for in-plane stresses in the case where fibers are arbitrary oriented in the plane can be obtained from (1.16) adopting the assumption of plane stress:

$$\begin{Bmatrix} \sigma_x \\ \sigma_y \\ \tau_{xy} \end{Bmatrix} = \begin{bmatrix} C_{11}(T) & C_{12}(T) & C_{16}(T) \\ C_{12}(T) & C_{22}(T) & C_{26}(T) \\ C_{16}(T) & C_{26}(T) & C_{66}(T) \end{bmatrix} \begin{Bmatrix} \varepsilon_x - \int_{T_0}^{T} \alpha_x(T)dT \\ \varepsilon_y - \int_{T_0}^{T} \alpha_y(T)dT \\ \gamma_{xy} - \int_{T_0}^{T} \alpha_{xy}(T)dT \end{Bmatrix} \tag{1.102}$$

where we account for the effect of temperature on both the stiffness as well as the coefficients of thermal expansion, and the strains ε_i $(i = x, y, xy)$ can include geometrically nonlinear or/and transverse shear effects.

Linear elastic constitutive equations (1.102) can be sufficiently accurate in numerous applications (e.g., Tauchert 1995). In problems where it is essential to reflect the effect of the history of temperature on the stiffness an alternative form of these equations is

$$\begin{Bmatrix} \sigma_x \\ \sigma_y \\ \tau_{xy} \end{Bmatrix} = \int_{T_0}^{T} \begin{bmatrix} C_{11}(T) & C_{12}(T) & C_{16}(T) \\ C_{12}(T) & C_{22}(T) & C_{26}(T) \\ C_{16}(T) & C_{26}(T) & C_{66}(T) \end{bmatrix} \begin{Bmatrix} \varepsilon_x \\ \varepsilon_y \\ \gamma_{xy} \end{Bmatrix} dT - \int_{T_0}^{T} \begin{bmatrix} C_{11}(T) & C_{12}(T) & C_{16}(T) \\ C_{12}(T) & C_{22}(T) & C_{26}(T) \\ C_{16}(T) & C_{26}(T) & C_{66}(T) \end{bmatrix} \begin{Bmatrix} \alpha_x(T) \\ \alpha_y(T) \\ \alpha_{xy}(T) \end{Bmatrix} dT \tag{1.103}$$

In numerous solutions published in literature it is considered acceptable to integrate the terms in the right side of (1.103) effectively neglecting the history of thermal loading. This results in the constitutive relations

$$\begin{Bmatrix} \sigma_x \\ \sigma_y \\ \tau_{xy} \end{Bmatrix} = \begin{bmatrix} C_{11}(T) & C_{12}(T) & C_{16}(T) \\ C_{12}(T) & C_{22}(T) & C_{26}(T) \\ C_{16}(T) & C_{26}(T) & C_{66}(T) \end{bmatrix} \begin{Bmatrix} \varepsilon_x - \alpha_x(T)\Delta T \\ \varepsilon_y - \alpha_y(T)\Delta T \\ \gamma_{xy} - \alpha_{xy}(T)\Delta T \end{Bmatrix} \tag{1.104}$$

where ΔT is the local change of temperature from the stress-free reference value and T is the current temperature. Equations 1.104 yield approximate values of stresses that are not equal to those obtained from more accurate equations (1.103).

The integration of the constitutive equations and their moments with respect to the middle plane through the thickness yields stress resultants and stress couples. Accordingly, using (1.104) we obtain the following counterpart of (1.57):

$$\begin{Bmatrix} N \\ M \end{Bmatrix} = \begin{bmatrix} \int\limits_{-\frac{h}{2}}^{\frac{h}{2}} [C(T)]\,dz & \int\limits_{-\frac{h}{2}}^{\frac{h}{2}} [C(T)]\,zdz \\ \int\limits_{-\frac{h}{2}}^{\frac{h}{2}} [C(T)]\,zdz & \int\limits_{-\frac{h}{2}}^{\frac{h}{2}} [C(T)]\,z^2dz \end{bmatrix} \begin{Bmatrix} \varepsilon^0 - \alpha(T)\Delta T \\ \kappa \end{Bmatrix} \tag{1.105}$$

where the vector $\alpha = \left\{\alpha_x\ \alpha_y\ \alpha_{xy}\right\}^T$.

The extended form of equations (1.105) depends on the material, temperature distribution and the theory of plates adopted for the analysis. For example, for an isotropic geometrically nonlinear plate treated by the classical theory and subject to a uniform temperature these equations become

$$\begin{aligned}
N_x &= \frac{Eh}{1-\nu^2}\left\{\frac{\partial u_0}{\partial x} + \frac{1}{2}\left(\frac{\partial w}{\partial x}\right)^2 + \nu\left[\frac{\partial v_0}{\partial y} + \frac{1}{2}\left(\frac{\partial w}{\partial y}\right)^2\right]\right\} - N_x^T \\
N_y &= \frac{Eh}{1-\nu^2}\left\{\frac{\partial v_0}{\partial y} + \frac{1}{2}\left(\frac{\partial w}{\partial y}\right)^2 + \nu\left[\frac{\partial u_0}{\partial x} + \frac{1}{2}\left(\frac{\partial w}{\partial x}\right)^2\right]\right\} - N_y^T \\
N_{xy} &= \frac{Eh}{2(1+\nu)}\left(\frac{\partial u_0}{\partial y} + \frac{\partial v_0}{\partial x} + \frac{\partial w}{\partial x}\frac{\partial w}{\partial y}\right) \\
M_x &= -D\left(\frac{\partial^2 w}{\partial x^2} + \nu\frac{\partial^2 w}{\partial y^2}\right) \\
M_y &= -D\left(\frac{\partial^2 w}{\partial y^2} + \nu\frac{\partial^2 w}{\partial x^2}\right) \\
M_{xy} &= -D(1-\nu)\frac{\partial^2 w}{\partial x \partial y}
\end{aligned} \tag{1.106}$$

where $E = E(T)$, $\nu = \nu(T)$ and accordingly $D = D(T)$, and thermally-induced terms are$N_x^T = N_y^T = \frac{Eh\alpha}{1-\nu^2}(T - T_0)$, the coefficient of thermal expansion being a function of the current temperature, i.e. $\alpha = \alpha(T)$.

The relations are more complicated if temperature varies through the thickness of the plate, i.e. $T = T(z)$. An example of such situation is found in plates subject to heat flux on one surface and active cooling on the opposite surface. Then material properties vary through the thickness. In such general case the constitutive relations (1.105) can be written as

$$\begin{Bmatrix} N \\ M \end{Bmatrix} = \begin{bmatrix} \int\limits_{-\frac{h}{2}}^{\frac{h}{2}} [C(T(z))]\,dz & \int\limits_{-\frac{h}{2}}^{\frac{h}{2}} [C(T(z))]\,zdz \\ \int\limits_{-\frac{h}{2}}^{\frac{h}{2}} [C(T(z))]\,zdz & \int\limits_{-\frac{h}{2}}^{\frac{h}{2}} [C(T(z))]\,z^2dz \end{bmatrix} \begin{Bmatrix} \varepsilon_0 \\ \kappa \end{Bmatrix} - \begin{Bmatrix} N^T \\ M^T \end{Bmatrix} \tag{1.107}$$

Table 1.1 The modulus of elasticity of selected metals (Msi) as a function of temperature (°F) (From: http://www.engineeringtoolbox.com)

Material	−200	−100	70	200	300	400	500	600	700	800	900
Carbon steel (C<=0.3%)	30.8	30.2	29.5	28.8	28.3	27.7	27.3	26.7	25.5	24.2	22.4
Nickel alloy (Monel 400)	27.3	26.8	26.0	25.4	25.0	24.7	24.3	24.1	23.7	23.1	22.6
Titanium (grades 1,2,3,7)			15.5	15.0	14.6	14.0	13.3	12.6	11.9	11.2	
Aluminum and aluminum alloys	10.8	10.5	10.0	9.6	9.2	8.7					

where thermally-induced terms are

$$\begin{Bmatrix} N^T \\ M^T \end{Bmatrix} = \int_{-\frac{h}{2}}^{\frac{h}{2}} [C\,(T(z))] \begin{Bmatrix} 1 \\ z \end{Bmatrix} \alpha\,(T(z))\,T(z)dz \tag{1.108}$$

It is important to briefly review the effect of temperature on the material properties that has already been reflected in the equations presented in this paragraph. It is impossible to reflect such effects in a universal mathematical form due to variety amongst materials and their different sensitivity to the environment. The relevant information is sometimes provided by manufacturers, but it is often difficult to find a reliable data. An example of empirical data for the modulus of elasticity of a number of metals is presented in Table 1.1.

One of the methods enabling an analytical approach to thermal problems accounting for the effect of temperature on material properties is based on approximating experimental property-temperature relationships with analytical functions. In particular, truncated polynomial functions were proposed by Touloukian (1967) in the form:

$$P = P_0\left(P_{-1}T^{-1} + 1 + P_1T + P_2T^2 + P_3T^3\right) \tag{1.109}$$

where P is a property (modulus of elasticity, Poisson's ratio, etc.), P_i are coefficients, and T is temperature counted from the reference value. The first term in the brackets introduces a singularity at the reference temperature and it is often eliminated by setting $P_{-1}=0$.

Examples of relevant relationships are presented here for silicon nitride and nickel (Yang et al. 2006):

Silicon nitride:

$$E = 348.49 \times 10^9\left(1 - 3.070 \times 10^{-4}T + 2.160 \times 10^{-7}T^2 - 8.946 \times 10^{-11}T^3\right)$$

$$\nu = 0.24$$

$$\alpha = 5.8723 \times 10^{-6} \left(1 + 9.095 \times 10^{-4} T\right)$$

Nickel:

$$E = 223.95 \times 10^{9} \left(1 - 2.794 \times 10^{-4} T + 3.998 \times 10^{-9} T^{2}\right)$$

$$\nu = 0.31$$

$$\alpha = 9.9209 \times 10^{-6} \left(1 + 8.705 \times 10^{-4} T\right) \tag{1.110}$$

where the modulus of elasticity is measured in Pa, and temperature and the coefficients of thermal expansion refer to the Kelvin scale. Note that the Poisson ratio remains constant reflecting a rather general trend of a weak dependence of the Poisson ratio on temperature. The coefficient of thermal expansion in (1.110) is a linear function increasing with temperature, although in other materials this relationship exhibits nonlinearity. The nonlinear relationship between the modulus of elasticity and temperature that could also be observed in Table 1.1 is evident in the above equations.

1.10 Strength Theories

The first experiments concerned with the strength of materials were conducted by Leonardo Da Vinci. While in the case of a uniaxial loading the stress reaching a yield limit of the material can be associated with failure, the situation is more complex in the case of a multiaxial state of stress. Strength theories address the problem providing analytical expressions for combinations of multiaxial stresses or strains that cause failure.

1.10.1 The Maximum Principal Stress Criterion

The first criterion suggested by Rankin is called the maximum principal stress criterion and it predicts failure reflecting its name, i.e. at the maximum principal stress reaching the strength (a yield or failure stress) of the material. Accordingly, the loss of strength is avoided if

$$Max\left(|\sigma_1|,\ |\sigma_2|,\ |\sigma_3|\right) < Y \tag{1.111}$$

where Y is the yield stress (yield limit) and σ_i $(i = 1, 2, 3)$ are the principal stresses. In the case where the yield limits in tension and compression differ, i.e. $Y_t \neq |Y_c|$, the corresponding criterion is

$$Max\,(\sigma_1,\ \sigma_2,\ \sigma_3) < Y_t$$
$$Min\,(|\sigma_1|\,,\ |\sigma_2|\,,\ |\sigma_3|) < |Y_c| \tag{1.112}$$

where compressive stresses and the yield stress in compression are negative. If inequalities (1.112) are satisfied, the loss of strength does not occur.

1.10.2 The Maximum Principal Strain Criterion

This criterion can be written in the form somewhat similar to that for the previous criterion, but the principal stresses are replaced with the principal strains, while the yield stress is replaced with the strain corresponding to yielding of the material (ε_t in tension and ε_c in compression):

$$Max\,(\varepsilon_1,\ \varepsilon_2,\ \varepsilon_3) < \varepsilon_t$$
$$Min\,(|\varepsilon_1|\,,\ |\varepsilon_2|\,,\ |\varepsilon_3|) < |\varepsilon_c| \tag{1.113}$$

It may be more convenient to express this criterion in terms of principal stresses. For example, if the yield stresses in tension and compression are equal ($Y_t = |Y_c| = Y$), the criterion becomes (Boresi and Schmidt 2003):

$$\max_{i \neq j \neq k} \left|\sigma_i - \nu\left(\sigma_j + \sigma_k\right)\right| < Y \tag{1.114}$$

1.10.3 The Maximum Shear Stress (Tresca's) Criterion

This criterion reflects tests first conducted by Coulomb that illustrated that yielding is often associated with the slippage (shearing). Accordingly, the condition of failure is formulated as the requirement that the maximum shear stress at the point must equal the maximum shear stress in yielding under uniaxial loading. These stresses being equal to one-half the maximum difference between the principal stresses and one-half axial yield stress, respectively, the criterion states that the loss of strength does not occur if

$$|\sigma_1 - \sigma_3| < Y \tag{1.115}$$

where σ_1 and σ_3 are the algebraically maximum and minimum principal stresses, respectively.

1.10.4 Maximum Distortional Energy Density Criterion (Von Mises Criterion)

This criterion should be discussed in light of a possible subdivision of the strain energy density into two components, i.e. the contribution associated with the change in volume and the so-called distortional contribution related to the change of shape. The effect of the former contribution on yielding is disregarded. Accordingly, yielding is associated with the distortional strain energy density reaching a critical value equal to its value corresponding to the onset of yielding in the uniaxial state of stress. This occurs when the second deviatoric stress invariant becomes equal to the squared shear yield stress of the material. Mathematical transformations necessary to develop this criterion are straightforward (e.g., Boresi and Schmidt 2003). The convenient form of the criterion in terms of the effective stress is

$$\sigma_e < Y \tag{1.116}$$

where

$$\begin{aligned}\sigma_e &= \sqrt{\frac{1}{2}\left[(\sigma_1-\sigma_2)^2+(\sigma_1-\sigma_3)^2+(\sigma_2-\sigma_3)^2\right]}\\ &= \sqrt{\frac{1}{2}\left[\left(\sigma_x-\sigma_y\right)^2+(\sigma_x-\sigma_z)^2+\left(\sigma_y-\sigma_z\right)^2\right]+3\left(\tau_{xy}^2+\tau_{xz}^2+\tau_{yz}^2\right)}\end{aligned} \tag{1.117}$$

Inequality (1.116) predicts the condition that has to be satisfied to avoid the loss of strength that occurs if the effective stress reaches the yield limit of the material.

1.10.5 Christensen's Yield and Failure Criterion

The four criteria outlined above are typically used in the analysis of ductile materials, such as metals. However, there is a need in a comprehensive theory predicting conditions of failure encompassing both the loss of strength in ductile materials as well as fracture in more brittle materials. Such theory has been proposed by Christensen (2004, 2007). The approach adopted in these papers utilizes the governing yield function in the form of a quadratic polynomial of the invariants of the stress tensor. The resulting criterion consists of two equations, one of them more suitable for ductile materials and the second reflecting fracture in brittle materials. Using the absolute (positive) value of the compressive yield stress, the strength criterion is

$$\left(\frac{1}{Y_t}-\frac{1}{Y_c}\right)(\sigma_x+\sigma_y+\sigma_z)$$
$$+\frac{1}{Y_tY_c}\left\{\frac{1}{2}\left[(\sigma_x-\sigma_y)^2+(\sigma_y-\sigma_z)^2+(\sigma_z-\sigma_x)^2\right]+3\left(\tau_{xz}^2+\tau_{yz}^2+\tau_{xy}^2\right)\right\}\leqslant 1 \tag{1.118}$$

It is evident that if the yield strengths in tension and compression are equal equation (1.118) reduces to the von Mises strength criterion given by (1.116) and (1.117).

For relatively brittle materials where $Y_t \leqslant \frac{1}{2}Y_c$ the fracture condition also has to be considered. As was shown by Christensen, this condition can be reduced to the maximum principal stress criterion referred to above.

The strength criteria for orthotropic materials (such as layers in composite laminates) are discussed in Chap. 5.

1.11 Outline of a Comprehensive Plate Analysis

The material presented in this chapter should be considered from the point of view of the analysis of plate structures. The mathematical formulation necessary for the analysis of such structures includes:

- Kinematic assumptions that specify a distribution of displacements throughout the plate.
- Strain-displacement relationships that may be linear or geometrically nonlinear (in the latter case, the strains are nonlinear functions of displacements).
- Constitutive relationships representing stresses in terms of strains. These relationships can be linear (elastic case) or nonlinear (e.g., elastic-plastic case). Other factors, such as viscosity or environmental effects can also be incorporated in these relationships.
- Integral functions of stress and stress moments about the middle plane, i.e. stress resultants and stress couples, respectively.
- Equations of equilibrium or motion.
- Boundary conditions.
- Initial conditions that have to be considered in dynamic problems (not discussed in this chapter).

Besides solution based on the exact or approximate integration of the equations of equilibrium (or motion), an alternative approach is based on using one of the energy methods. In such cases, equations of equilibrium are replaced with the corresponding energy formulation but the boundary conditions still have to be satisfied. While in the Rayleigh-Ritz method it is sufficient to satisfy only kinematic boundary conditions, static conditions must also be satisfied using other methods.

The solution of the problem can be sought in terms of displacements or alternatively, in terms of the deflection of the plate and the stress function. In both cases, the boundary conditions can be represented in terms of the corresponding unknowns.

The environment, i.e. temperature (and/or moisture), has a dual effect on the solution. Environment-induced terms appear in the strain-displacement relationships and subsequently, in the equations of equilibrium and energy expressions. Boundary conditions formulated in terms of stress resultants and stress couples are also affected by such terms. In addition, environmental factors influence material "constants," such as stiffness, coefficients of thermal expansion, strength, etc. These effects affect the solution of heat transfer problems and the distribution of temperature throughout the plate. As a result of a nonuniform temperature, stiffness terms in the equations of equilibrium and in energy formulations may include variable coefficients dependent on coordinates.

Once the solution for displacements (and for the stress function, if it is included in the analysis) is generated, the stresses can be evaluated throughout the plate using the strain-displacement and constitutive relations. Subsequently, strength criteria yield the prediction of the strength of the plate. In relevant situations, i.e. if the plate material is brittle, the possibility of fracture also has to be considered.

At the closure of this chapter we should discuss the relationship between analytical and numerical solutions. The latter solutions are usually based on finite element, finite difference or boundary element methods. These solutions are only as accurate as the assumptions and theories employed to develop the corresponding numerical procedure. In general, exact solutions, if available, are always preferred. Although the number of plate problems where exact solutions are available is limited, they may be used as benchmarks to check the accuracy of numerical solutions. As will be shown in subsequent chapters, approximate solutions representing deformations of rectangular or circular plates in double or single Fourier series (independent variables being in-plane coordinates) are available for a number of boundary conditions. These solutions are very accurate as long as the series exhibit a satisfactory convergence. Such convergence is usually obtained for deformations but it is worse for stresses that involve derivatives of the Fourier series.

The number of available Fourier-series solutions for plates being limited, numerical methods should be applied for the cases where the plate has a complicated geometry or boundary conditions differ from those suitable for an analytical solution. In all situations, it is important to accurately reflect the actual boundary conditions (see for example, the discussion in Sect. 2.2).

The choice of the theory of plates employed in the analysis depends on the plate geometry and anticipated deformations. If transverse deflections remain small (less than about one-half the thickness of the plate), geometric nonlinearity can often be neglected. On the other hand, if the plate is relatively thick or exhibits high transverse shear deformability, a first order or higher order shear deformation theory has to be employed. The need for such a theory is seldom encountered in the analysis of metallic plates because of their relatively large side-to-thickness ratios. The situation for composite and particularly for sandwich plates is different and shear-deformable theories are more often applied to their analysis as is discussed in Chap. 5 which includes an example of the first order theory. A possible alternative to using a first order or higher order theory in the analysis of plates is using a three-dimensional finite element procedure. Such modeling can be conducted using a commercial software finite element package.

References

Berlincourt, D. A., Curran, D. R., & Jaffe, H. (1964). Piezoelectric and piezomagnetic materials and their function as transducers. In W. P. Mason (Ed.), *Physical acoustics* (Vol. 1A, pp. 169–270). New York: Academic Press.

Birman, V., & Suhir, E. (2007). Effect of material nonlinearity on the mechanical response of some piezo-electric and photonic systems. In E. Suhir, Y. C. Lee, & C.-F. Wong (Eds.), *Micro and opto electronic materials: Physics, mechanics, materials, reliability and packaging* (Vol. 1, pp. 667–700). New York: Springer.

Boresi, A. P., & Schmidt, R. J. (2003). *Advanced mechanics of materials*. Hoboken, NJ: Wiley.

Christensen, R. M. (2004). A two-property yield, failure (fracture) criterion for homogeneous isotropic materials. *Journal of Engineering Materials and Technology, 126*, 45–52.

Christensen, R. M. (2007). A comprehensive theory of yielding and failure for isotropic materials. *Journal of Engineering Materials and Technology, 129*, 173–181.

Fung, Y. C. (1994). *A first course in continuum mechanics* (3rd ed.). Englewood Cliffs, NJ: Prentice Hall.

Galerkin, B. G. (1915). On electrical circuits for the approximate solution of the Laplace equations. *Vestnik Inzhenerov, 19*, 897–908. In Russian.

Gibson, R. F. (2007). *Principles of composite material mechanics*. Boca Raton, FL: CRC Press.

Goldstein, H. (1950). *Classical mechanics*. Cambridge, MA: Addison-Wesley Press, Inc. (later editions are also available).

Houbolt, J. C., & Brooks, G. W., 1958. *Differential equations of motion for combined flapwise bending, chordwise bending, and torsion of twisted nonuniform rotor blades* (NACA Report 1346).

Jones, R. M. (1999). *Mechanics of composite materials* (2nd ed.). Philadelphia, PA: Taylor & Francis.

Patron, V. Z., & Kudryavtsev, B. A. (1993). *Engineering mechanics of composite materials*. Boca Raton, FL: CRC Press.

Rayleigh, J. W. S. (1877). *The theory of sound*. New York: Dover (re-published in 1945 from the original edition in 1877).

Reddy, J. N. (2002). *Energy principles and variational methods* (2nd ed.). Hoboken, NJ: Wiley.

Reddy, J. N. (2007). *Theory and analysis of elastic plates and shells* (2nd ed.). Boca Raton, FL: CRC Press/Taylor & Francis Group.

Reddy, J. N. (2008). *An introduction to continuum mechanics*. Cambridge, NY: Cambridge University Press.

Ritz, W. (1909). Über eine neue Methode zur Lösung gewisser Variationsprobleme der mathematischen Physik. *Journal für Reine und Angewandte Mathematik, 135*, 1–61.

Simitses, G. J. (1986). *An introduction to the elastic stability of structures*. Malabar, FL: Robert Krieger Publishing Company.

Sokolnikoff, I. S. (1964). *Tensor analysis. Theory and applications to geometry and mechanics of continua* (2nd ed.). New York: Wiley.

Tauchert, T. R. (1995). Temperature and absorbed moisture. In G. J. Turvey & I. H. Marshall (Eds.), *Buckling and postbuckling of composite plates*. London: Chapman & Hall.

Thomson, W. T. (1993). *Theory of vibration with applications* (4th ed.). Englewood Cliffs, NJ: Prentice Hall.

Timoshenko, S., & Woinowsky-Krieger, S. (1959). *Theory of plates and shells.* New York: McGraw-Hill.

Touloukian, Y. S. (1967). *Thermophysical properties of high temperature solid materials.* New York: Macmillan.

Vol'mir, A. S. (1972). *Nonlinear dynamics of plates and shells.* Moscow: Nauka Publishing House (in Russian).

Whitney, J. M. (1987). *Structural analysis of laminated anisotropic plates.* Lancaster/Basel: Technomic.

Wunderlich, W., & Pilkey, W. D. (2003). *Mechanics of structures: Variational and computational methods* (2nd ed.). Boca Raton, FL: CRC Press.

Yang, J., Liew, K. M., Wu, Y. F., & Kitipornchai, S. (2006). Thermo-mechanical post-buckling of FGM cylindrical panels with temperature-dependent properties. *International Journal of Solids and Structures, 43*, 307–324.

Chapter 2
Static Problems in Isotropic Rectangular Plates

Plate structures are often subject to static loads that can cause the loss of strength or instability. The solution of relevant problems can be undertaken either employing equations of equilibrium or applying energy methods. This chapter illustrates the formulation and solution of representative problems for isotropic rectangular plates. While mathematical formulations presented in the chapter can be employed in a finite element or finite difference analysis, solved problems are useful either as benchmark solutions or in cases where the plate can accurately be described by the corresponding model. The chapter illustrates various boundary conditions encountered in applications and presents a discussion on their relevance. The effect of initial imperfections that are present in numerous situations is discussed in detail. The approach to the analysis of plates on an elastic foundation that is often necessary in civil engineering applications is elucidated. The analysis of stringer-reinforced plates is also included since stringers are often employed to increase the strength and stability of plates. The discussion of stability includes the peculiarities of the postbuckling response of plates that is important in numerous design applications. Besides solutions employing the integration of the equation of equilibrium, the energy (Rayleigh-Ritz) approach to the analysis is demonstrated. Although problems considered in this chapter have been studied for a long time, their value and practical applicability have not diminished. Besides the introduction to the analysis of rectangular isotropic plates, we elucidate a number of application aspects and limitations to the present solutions that are seldom discussed in textbooks on the theory of plates. This discussion may be valuable to engineers working on design and development of isotropic plate structures.

2.1 Classical Navier's Problem

The first solution of the stress problem for plate structures was presented by Navier. The problem that he considered dealt with a rectangular isotropic plate simply supported along all edges and subject to a lateral load that was represented in double

V. Birman, *Plate Structures*, Solid Mechanics and Its Applications 178,
DOI 10.1007/978-94-007-1715-2_2, © Springer Science+Business Media B.V. 2011

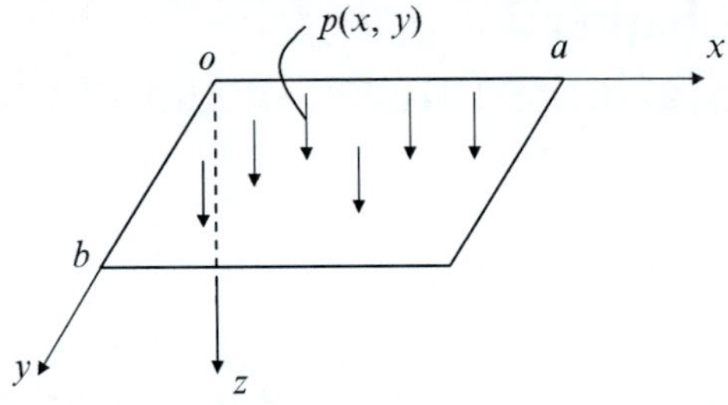

Fig. 2.1 Rectangular plate subject to transverse pressure

Fourier series. As we will see in this paragraph, such problem is applicable to an amazing variety of structures found in diverse applications.

The analysis is conducted by the following assumptions:

1. The problem is physically linear, i.e. the stresses remain in the linear elastic range;
2. The problem is geometrically linear, implying linear strain-displacement relationships;
3. Transverse shear deformability of the plate can be disregarded, i.e. the classical thin plate theory is applicable.

The second and third assumptions are usually applicable to metallic plates that have the ratio of the thickness to the length of the shorter edge smaller than 1/20 and undergo maximum deflections that are smaller than half-thickness of the plate.

The plate of length and width, a and b, respectively, that is subject to an arbitrary distributed transverse pressure p(x,y) is shown in Fig. 2.1. The equation of equilibrium used in the analysis is simplified since there are no applied in-plane loads. In addition, bending deformations of the plate being small, they do not cause stretching of the middle plane. Accordingly, Eq. 1.96 becomes

$$D\nabla^4 w = p(x, y) \tag{2.1}$$

The boundary conditions corresponding to simply supported edges where deflections and bending stress couples are equal to zero are available from (1.88) and (1.58):

$$\begin{aligned} x = 0,\ x = a:\quad w = 0,\ M_x = -D\left(\frac{\partial^2 w}{\partial x^2} + \nu\frac{\partial^2 w}{\partial y^2}\right) = 0 \rightarrow \frac{\partial^2 w}{\partial x^2} = 0 \\ y = 0,\ y = b:\quad w = 0,\ M_y = -D\left(\frac{\partial^2 w}{\partial y^2} + \nu\frac{\partial^2 w}{\partial x^2}\right) = 0 \rightarrow \frac{\partial^2 w}{\partial y^2} = 0 \end{aligned} \tag{2.2}$$

The simplification of conditions for stress couples in (2.2) is possible because the deflections of the plate do not vary along the corresponding edges. For example, at $x = 0$ $w = 0$, so that $\frac{\partial^2 w}{\partial y^2} = 0$, etc.

Note that we used two out of four boundary conditions at each edge of the plate. There is no need to address "in-plane" boundary conditions since they refer to in-plane displacements at the middle plane of the plate and in-plane stress resultants. In the present geometrically linear problem, such displacements and stress resultants are equal to zero (however, they would not necessarily be negligible, if the problem involved geometrically nonlinear terms).

The load that can be an arbitrary function of the coordinates is represented in double Fourier series

$$p(x, y) = \sum_{m=1}^{\infty}\sum_{n=1}^{\infty} p_{mn} \sin\alpha_m x \sin\beta_n y \tag{2.3}$$

where m and n are integer numbers, $\alpha_m = \frac{m\pi}{a}$, $\beta_n = \frac{n\pi}{b}$, and the coefficients of series are available from

$$p_{mn} = \frac{4}{ab}\int_0^b\int_0^a p(x, y)\sin\alpha_m x \sin\beta_n y dx dy \tag{2.4}$$

For example, if the applied pressure is uniformly distributed over the plate surface, i.e. $p = p_0$ (p_0 being constant), $p_{mn} = \frac{16p_0}{mn\pi^2}$. If the applied load is represented by a concentrated force $P = P(x = \xi, y = \eta)$, $p_{mn} = \frac{4P}{ab}\sin\alpha_m\xi \sin\beta_n\eta$.

Obviously, the number of terms in series (2.3) cannot be infinite, i.e. in reality, these series should be truncated. The convergence of series (2.3) and accordingly the number of terms retained in these series depend on the applied load function p(x,y). If this is a monotonous function of coordinates, several terms are sufficient to accurately represent it in double Fourier series. On the other hand, in the case of a concentrated force or if pressure is applied over a limited area of the plate, the number of terms necessary to retain in (2.3) may be quite large.

The choice of the deflection function is guided by two requirements, i.e. it must satisfy both the equation of equilibrium as well as the boundary conditions. In the case of a simply supported plate boundary conditions (2.2) are identically satisfied if the deflection is assumed in the form

$$w = \sum_{m=1}^{M}\sum_{n=1}^{N} W_{mn}\sin\alpha_m x \sin\beta_n y \tag{2.5}$$

where M and N correspond to the terms retained in truncated series (2.3).

The substitution of (2.5) and (2.3) into the equation of equilibrium (2.1) yields

$$\sum_{m=1}^{M}\sum_{n=1}^{N}\left[D\left(\alpha_m^2 + \beta_n^2\right)^2 W_{mn} - p_{mn}\right]\sin\alpha_m x \sin\beta_n y = 0 \tag{2.6}$$

Equation 2.6 must be satisfied at all points within the domain $0 \leq x \leq a,\ 0 \leq y \leq b$. Obviously, this can be achieved only if the coefficients at every product of trigonometric functions are equal to zero. This yields the values of amplitudes of harmonics in (2.5):

$$W_{mn} = \frac{p_{mn}}{D\left(\alpha_m^2 + \beta_n^2\right)^2} \tag{2.7}$$

Accordingly,

$$w = \frac{1}{D}\sum_{m=1}^{M}\sum_{n=1}^{N}\frac{p_{mn}}{\left(\alpha_m^2 + \beta_n^2\right)^2}\sin\alpha_m x\sin\beta_n y \tag{2.8}$$

In numerous applications the analysis of deformations is important only to ensure that the geometrically linear theory adopted in the analysis is applicable (the analysis is usually concerned with the stresses and the check of strength). At large deflections, typically exceeding one-half the plate thickness, a discrepancy between the linear and nonlinear solutions cannot be ignored and the analysis should be carried numerically using a nonlinear software package.

The constitutive relations for an isotropic plate (1.19) can be simplified utilizing the assumption of plane stress, so that $\sigma_3 = \tau_{13} = \tau_{23} = 0$. Then, using the Lame constants according to (1.20) and replacing the subscripts $1 \to x,\ 2 \to y,\ 3 \to z$ yields the bending and in-plane shear stresses:

$$\begin{aligned}
\sigma_x &= \frac{E}{1-\nu^2}\left(\varepsilon_x + \nu\varepsilon_y\right) \\
\sigma_y &= \frac{E}{1-\nu^2}\left(\varepsilon_y + \nu\varepsilon_x\right) \\
\tau_{xy} &= G\gamma_{xy}
\end{aligned} \tag{2.9a}$$

In the case where the linear solution is acceptable, deflections found by (2.8) can be substituted into the changes of curvature and twist (1.29), while in-plane strains given by (1.28) are equal to zero. The substitution of (1.29) into constitutive relations (2.9a) yields the bending and in-plane shear stresses that are linear functions of the thickness coordinate:

$$\sigma_x = \frac{Ez}{1-\nu^2}\sum_{m=1}^{M}\sum_{n=1}^{N}\left[\alpha_m^2 + \nu\beta_n^2\right]W_{mn}\sin\alpha_m x\sin\beta_n y$$

$$\sigma_y = \frac{Ez}{1-\nu^2}\sum_{m=1}^{M}\sum_{n=1}^{N}\left[\beta_n^2 + \nu\alpha_m^2\right]W_{mn}\sin\alpha_m x\sin\beta_n y$$

$$\tau_{xy} = -\frac{Ez}{1+\nu}\sum_{m=1}^{M}\sum_{n=1}^{N}\alpha_m\beta_n W_{mn}\cos\alpha_m x\cos\beta_n y \tag{2.9b}$$

Alternatively, using (2.8)

$$\sigma_x = \frac{12z}{h^3}\sum_{m=1}^{M}\sum_{n=1}^{N}\frac{\alpha_m^2+\nu\beta_n^2}{\left(\alpha_m^2+\beta_n^2\right)^2}p_{mn}\sin\alpha_m x\sin\beta_n y$$

$$\sigma_y = \frac{12z}{h^3}\sum_{m=1}^{M}\sum_{n=1}^{N}\frac{\nu\alpha_m^2+\beta_n^2}{\left(\alpha_m^2+\beta_n^2\right)^2}p_{mn}\sin\alpha_m x\sin\beta_n y$$

$$\tau_{xy} = -\frac{12(1-\nu)z}{h^3}\sum_{m=1}^{M}\sum_{n=1}^{N}\frac{\alpha_m\beta_n}{\left(\alpha_m^2+\beta_n^2\right)^2}p_{mn}\cos\alpha_m x\cos\beta_n y \tag{2.9c}$$

The stresses given by (2.9b) and (2.9c) reach extreme (algebraically maximum and minimum) values on the plate surfaces $z = \pm\frac{h}{2}$ and are equal to zero at the middle plane. The convergence of series (2.9b) and (2.9c) is generally worse than that of series (2.5).

Although the stresses can directly be calculated by (2.9), numerous references present stress couples, instead of the stresses. The stress couples are obtained by the substitution of (2.8) into (1.58):

$$M_x = \sum_{m=1}^{M}\sum_{n=1}^{N}\frac{\alpha_m^2+\nu\beta_n^2}{\left(\alpha_m^2+\beta_n^2\right)^2}p_{mn}\sin\alpha_m x\sin\beta_n y$$

$$M_y = \sum_{m=1}^{M}\sum_{n=1}^{N}\frac{\nu\alpha_m^2+\beta_n^2}{\left(\alpha_m^2+\beta_n^2\right)^2}p_{mn}\sin\alpha_m x\sin\beta_n y$$

$$M_{xy} = -(1-\nu)\sum_{m=1}^{M}\sum_{n=1}^{N}\frac{\alpha_m\beta_n}{\left(\alpha_m^2+\beta_n^2\right)^2}p_{mn}\cos\alpha_m x\cos\beta_n y \tag{2.10}$$

The extreme stresses given by (2.9c) with $z = \pm\frac{h}{2}$ are related to the stress couples by

$$\sigma_x^{extreme}\left(z=\pm\frac{h}{2}\right) = \pm\frac{6M_x}{h^2}$$

$$\sigma_y^{extreme}\left(z=\pm\frac{h}{2}\right) = \pm\frac{6M_y}{h^2}$$

$$\tau_{xy}^{extreme}\left(z=\pm\frac{h}{2}\right) = \pm\frac{6M_{xy}}{h^2} \tag{2.11}$$

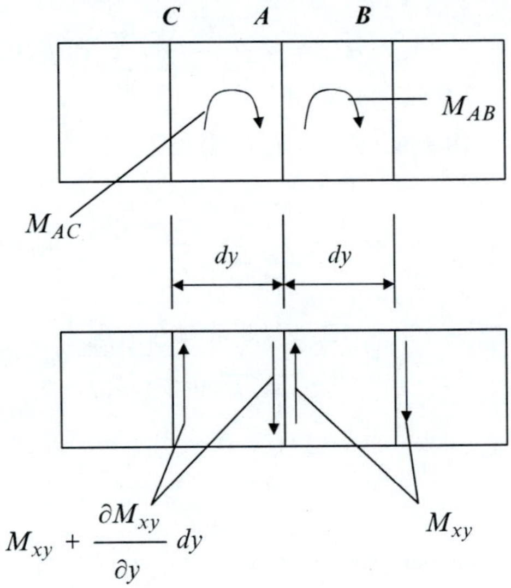

Fig. 2.2 Contribution of twisting stress couples to the force applied by the plate to the edge support structure

The previous part of this paragraph presents the comprehensive solution for a plate subject to an arbitrary pressure. In addition, it is sometimes necessary to estimate the load applied by the plate to the supporting structures. One of the reasons is the proof of the assumption that these structures provide simple support to the plate, i.e. the deflections of the plate at the edges really approach zero. If the rib supporting the edge is relatively flexible, such assumption becomes unacceptable and it is necessary to analyze the plate with elastically supported edges.

Consider two adjacent infinitesimal elements of the edge $x = const$, each of them of length dy. As is reflected in Fig. 2.2, the twisting moment varies along the edge, so that it is equal to $M_{AB} = M_{xy}dy$ for element AB and $M_{AC} = (M_{xy} + \frac{\partial M_{xy}}{\partial y}dy)dy$ for element CA. The moment acting on each of two elements can be replaced with a couple of forces shown in Fig. 2.2. Accordingly, the force per unit length of the edge at point A that includes the contributions of both the transverse shear stress resultant as well as the twisting moment becomes

$$V_x = Q_x + \frac{\partial M_{xy}}{\partial y} \tag{2.12a}$$

Similarly, along the edges $y = const$, the edge force per unit length is given by

$$V_y = Q_y + \frac{\partial M_{xy}}{\partial x} \tag{2.12b}$$

Transverse shear stress resultants can be expressed in terms of stress couples from the static versions of the last two equations (1.81). Subsequently, substituting the stress couples in terms of deflections according to (1.58), we obtain

$$Q_x = -D\left(\frac{\partial^3 w}{\partial x^3} + \frac{\partial^3 w}{\partial x \partial y^2}\right)$$

$$Q_y = -D\left(\frac{\partial^3 w}{\partial y^3} + \frac{\partial^3 w}{\partial y \partial x^2}\right) \tag{2.13}$$

The combination of Eqs. 2.10, 2.12 and 2.13 yields the expressions for the forces per unit length of each edge of the plate shown in Fig. 2.1:

$$x = 0,\ x = a:\ V_x = \sum_{m=1}^{M}\sum_{n=1}^{N} \frac{p_{mn}\alpha_m\left[\alpha_m^2 - (2-\nu)\,\beta_n^2\right]}{\left(\alpha_m^2 + \beta_n^2\right)^2} \cos\alpha_m\hat{x}\sin\beta_n y$$

$$y = 0,\ y = b:\ V_y = \sum_{m=1}^{M}\sum_{n=1}^{N} \frac{p_{mn}\beta_n\left[\beta_n^2 - (2-\nu)\,\alpha_m^2\right]}{\left(\alpha_m^2 + \beta_n^2\right)^2} \sin\alpha_m x\cos\beta_n \hat{y} \tag{2.14}$$

where $\hat{x} = 0\ or\ a,\ \hat{y} = 0\ or\ b$. Note that Eq. 2.14 determine the load applied by the edge of the simply supported plate to the supporting structure and vice versa, the force applied by the structure to the plate.

The reactions at the corners may be of particular interest to designers of rectangular hatch covers. As was shown in numerous references on theory of plates (e.g., Timoshenko and Woinowsky-Krieger 1959, and Ventsel and Krauthammer 2001), at the corner of two supported edges such reactions are obtained as

$$R = 2M_{xy} = -\left[2D\,(1-\nu)\,\frac{\partial^2 w}{\partial x \partial y}\right]_{\breve{x},\breve{y}} \tag{2.15}$$

where $\left(\breve{x},\ \breve{y}\right)$ are the coordinates of the corner. The substitution of (2.8) yields

$$R = -\left[2\,(1-\nu)\sum_{m=1}^{M}\sum_{n=1}^{N}\frac{p_{mn}\alpha_m\beta_n}{\left(\alpha_m^2+\beta_n^2\right)^2}\cos\alpha_m\breve{x}\cos\beta_n\breve{y}\right] \tag{2.16}$$

There are numerous tabulated results available to engineers designing simply supported rectangular plates under a variety of loads (e.g., Timoshenko and Woinowsky-Krieger 1959 and Roark 1965). We illustrate here results shown in the former reference. For a simply supported rectangular plate with the Poisson ratio equal to 0.3 that is subject to a uniform pressure, the maximum deflection can be represented by

$$w_{\max} = w\left(\frac{a}{2},\ \frac{b}{2}\right) = k_1\frac{p_0 a^4}{D} \tag{2.17}$$

Table 2.1 Coefficients in the expressions for maximum deflections and bending stress couples for a simply supported plate subject to a uniform pressure (Poisson ratio $\nu = 0.3$)

$\frac{b}{a}$	1.0	1.5	2.0	3.0	4.0	∞
k_1	0.00406	0.00772	0.01013	0.01223	0.01282	0.01302
k_2	0.0479	0.0812	0.1017	0.1189	0.1235	0.1250
k_3	0.0479	0.0498	0.0464	0.0406	0.0384	0.0375

The maximum bending stress couples occur at the center of the plate and are given by ($b \geq a$):

$$M_{x,\max} = k_2 p_0 a^2, \quad M_{y,\max} = k_3 p_0 a^2 \tag{2.18}$$

The coefficients in (2.17) and (2.18) are shown for several representative plate aspect ratios in Table 2.1.

As is obvious from Table 2.1, increasing the length of a rectangular plate subject to a uniform pressure results in larger deflections and a larger maximum principal stress σ_x. However, as the plate becomes longer, the detrimental effect of further separation of short edges on maximum deflections and principal stresses at the center weakens since they are mostly limited by the support provided by the long edges of the plate (x = 0 and x = a). At large values b/a the plate bends into a cylindrical surface, so that the support provided by short edges to the central section of the plate becomes negligible.

It is also useful to refer here to tabulated results for the maximum transverse shear stress resultants, edge reactions, and corner forces provided by Timoshenko and Woinowsky-Krieger (1959). Maximum edge reactions vary in a narrow range $V_{x,\max} = \delta p_0 a$, $V_{y,\max} = \delta_1 p_0 a$ where a is the shorter edge of the plate, while $0.42 \leq (\delta, \delta_1) \leq 0.50$, the lower limit corresponding to a square plate and the higher limit referring to the plate with an infinite aspect ratio. The corner reaction varies dependent on the aspect ratio in the range $0.065pb^2 \leq |R| \leq 0.095pb^2$.

Example 2.1: Estimate of the Validity of the Assumption of Simple Support for a Plate Subject to Uniform Pressure. The estimate of the accuracy of the solution obtained by assumption that support structures prevent deflections of the edges of the plate can be obtained by considering a plate that is simply supported along two opposite edges and elastically supported by flexible beams along the other couple of edges. Such results were tabulated for square plates by Timoshenko and Woinowsky-Krieger (1959, p. 218). The results for deflections and stress couples presented in this reference depend on the coefficient

$$\lambda = \frac{EI}{Da}$$

where I is the moment of inertia of the flexible beams.

Consider the case where the side of a 10 mm thick square plate (and the length of the supporting beams) is equal to 1.0 m. A light rolled steel S shape profile

S75 × 8.5 has the height d $= 76mm$, moment of inertia I $= 1.05 \times 10^6 mm^4$ and weighs 8.5 kg/m (Beer and Johnston 1991). The modulus of elasticity of the plate and supporting beams is E $= 200GPa$ and the Poisson ratio is equal to $\nu = 0.3$. Accordingly,

$$\lambda = \frac{EI}{Da} = \frac{12\left(1-\nu^2\right)I}{h^3 a} = \frac{12 \times 0.91 \times 1.05 \times 10^{-6}}{10^{-6} \times 1.0} = 11.466$$

Comparing the maximum deflection and bending couples corresponding to such structure (see the table on page 218 of Timoshenko and Woinowsky-Krieger 1959) to those of the plate that is simply supported along all edges ($\lambda = \infty$)we observe that the error is less than 5%, even though the beams are very flexible. Accordingly, the assumption of simply supported edges is justified, with the exception of designs that involve very long and slender beams supporting the edges of the plate.

Example 2.2: Convergence of Series Representing Deflections of a Simply Supported Square Plate Subject to Uniform Pressure. Consider a square plate ($a = b$) subject to uniform pressure $p = p_0$. The solution (2.8) yields the maximum deflection at the center of the plate:

$$w\left(\frac{a}{2}, \frac{a}{2}\right) = \frac{1}{\pi^4 D} \sum_{m=1}^{M} \sum_{n=1}^{N} \frac{p_{mn} a^4}{(m^2+n^2)^2} \sin\frac{m\pi}{2} \sin\frac{n\pi}{2}$$

$$= \frac{16 p_0 a^4}{\pi^6 D} \sum_{m=1}^{M} \sum_{n=1}^{N} \frac{1}{mn(m^2+n^2)^2} \sin\frac{m\pi}{2} \sin\frac{n\pi}{2}$$

where m and n are odd numbers. It is easy to observe that even values of these integers correspond to asymmetric about the center of the plate pressure and deflections. Such asymmetry is physically meaningless and the corresponding terms are equal to zero as also follows from (2.4).

The corresponding stresses and stress couples can be obtained from (2.9c) and (2.10). It is evident that the convergence of the series for the stresses and stress couples is identical; accordingly, we will present here the results for the latter. At the center of the plate the shear stress couple is equal to zero, while the bending stress couples are

$$M_x\left(\frac{a}{2}, \frac{a}{2}\right) = \frac{1}{\pi^2} \sum_{m=1}^{M} \sum_{n=1}^{N} \frac{m^2+\nu n^2}{(m^2+n^2)^2} p_{mn} a^2 sin\frac{m\pi}{2} sin\frac{n\pi}{2}$$

$$= \frac{16 p_0 a^2}{\pi^4} \sum_{m=1}^{M} \sum_{n=1}^{N} \frac{m^2+\nu n^2}{mn(m^2+n^2)^2} sin\frac{m\pi}{2} sin\frac{n\pi}{2}$$

$$M_y\left(\frac{a}{2}, \frac{a}{2}\right) = \frac{16 p_0 a^2}{\pi^4} \sum_{m=1}^{M} \sum_{n=1}^{N} \frac{n^2+\nu m^2}{mn(m^2+n^2)^2} sin\frac{m\pi}{2} sin\frac{n\pi}{2}$$

Naturally, it is appropriate to use $M = N$, since in a square plate $M_x\left(\frac{a}{2}, \frac{a}{2}\right) = M_y\left(\frac{a}{2}, \frac{a}{2}\right)$.

Even a casual inspection of the expressions for deflections and stress couples at the center of the plate leads to the conclusion that the convergence of deflections should be better than that of stress couples or stresses. The reason is that the terms in the series for stress couples contain squares of integrals m and n in the numerator, while the series for deflections does not include such terms. Limiting the analysis to the first four terms, i.e. the combinations $(m, n) = (1, 1),\ (3, 1),\ (1, 3),\ (3, 3)$, we obtain

$$\frac{\pi^6 D}{16 p_0 a^4} w\left(\frac{a}{2}, \frac{a}{2}\right) = \sum_{m=1}^{3}\sum_{n=1}^{3} \frac{1}{mn[m^2+n^2]^2} \sin\frac{m\pi}{2} \sin\frac{n\pi}{2}$$
$$= 0.25 - 2 \times 0.00333 + 0.0003429 = 0.2437$$

$$\frac{\pi^4}{16 p_0 a^2} M_x\left(\frac{a}{2}, \frac{a}{2}\right) = \frac{\pi^4}{16 p_0 a^2} M_y\left(\frac{a}{2}, \frac{a}{2}\right) = \sum_{m=1}^{3}\sum_{n=1}^{3} \frac{m^2+0.3n^2}{mn[m^2+n^2]^2} \sin\frac{m\pi}{2}$$
$$\times \sin\frac{n\pi}{2} = 0.325 - 0.03097 - 0.01232 + 0.00401 = 0.28572$$

As was indicated by Timoshenko and Woinowsky-Krieger (1959) and as follows from the present computations, the result obtained retaining the first term in the expansion for deflections is only 2.5% larger that the exact solution. However, we can also observe that the convergence of the series for bending couples is not as good as that for deflections, i.e. it is necessary to retain more terms in the corresponding series. Thus, it may be necessary to retain a large number of terms in the solution of the Navier problem to accurately evaluate stresses in the plate.

Example 2.3: Navier's Solution in Cases of Concentrated Loads and Loads Applied Over a Limited Area (Timoshenko and Woinowsky-Krieger 1959). In numerous applications the load is distributed over a limited area of the plate. An example is a wheel of a vehicle supported by the plate of a deck. Such plates are designed for the most unfavorable loading case assuming that the pressure from the wheel is either uniformly distributed over its footprint or simply replacing it with a concentrated force.

Consider a simply supported plate loaded by a force F that is uniformly distributed over the rectangular region $\xi - \frac{u}{2} \le x \le \xi + \frac{u}{2}$, $\eta - \frac{v}{2} \le y \le \eta + \frac{v}{2}$ (Fig. 2.3). The coefficients in series (2.4) representing the load become

$$p_{mn} = \frac{4F}{abuv} \int_{\xi-\frac{u}{2}}^{\xi+\frac{u}{2}} \int_{\eta-\frac{v}{2}}^{\eta+\frac{v}{2}} \sin\alpha_m x \sin\beta_n y dy dx$$
$$= \frac{16F}{\pi^2 mnuv} \sin\alpha_m\xi \sin\beta_n\eta \sin\frac{\alpha_m u}{2} \sin\frac{\beta_n v}{2}$$

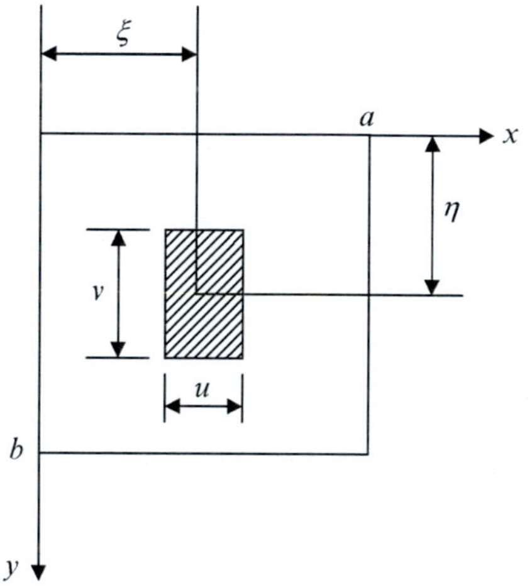

Fig. 2.3 Plate loaded over a rectangular region $\xi - \frac{u}{2} \le x \le \xi + \frac{u}{2}$, $\eta - \frac{v}{2} \le y \le \eta + \frac{v}{2}$

Subsequently, the analysis is identical to that for the case where pressure is uniformly distributed over the entire surface of the plate.

In case where the force can be assumed concentrated at the point (ξ, η), so that $u \to 0,\ v \to 0$, the coefficients in series (2.4) are

$$p_{mn} = \frac{4F}{ab} \sin \alpha_m \xi \sin \beta_n \eta$$

The deflections of the plate subject to a concentrated force F are given by

$$w = \frac{4F}{abD} \sum_{m=1}^{M} \sum_{n=1}^{N} \frac{\sin \alpha_m \xi \sin \beta_n \eta}{\left(\alpha_m^2 + \beta_n^2\right)^2} \sin \alpha_m x \sin \beta_n y$$

The convergence of series (2.4) approximating a concentrated force or a force distributed over a small area is inferior to that in the case where the series represent a distributed pressure. Similarly, the series for deflections, and particularly series for bending stresses or couples, do not exhibit good convergence observed in Example 2.2. For example, the deflections of a square plate subject to the force applied at the center are given by

$$w\left(\frac{a}{2}, \frac{a}{2}\right) = \frac{4Fa^2}{\pi^4 D} \sum_{m=1}^{M} \sum_{n=1}^{N} \frac{1}{(m^2 + n^2)^2} = k' \frac{Fa^2}{D}$$

where m and n are odd integrals and k' is a numerical coefficient.

Retaining four terms $(m, n = 1, 3)$ Timoshenko and Krieger (1959) obtained $k' = 0.01121$ that is 3.5% smaller than the "exact" solution. In case where nine first terms are retained $(m, n = 1, 3, 5)$ the coefficient $k' = 0.01142$ that is 1.56%

smaller than the accurate solution (Ugural 1999). Predictably, the convergence of series for the stresses is even worse.

It should be noted that a "concentrated force" applied to the plate is physically impossible. Such force would create an infinite pressure at the point of application. Even if the force was applied over a very small area, it would simply penetrate the plate. It is obvious that the state of stress in the vicinity to the concentrated force is three-dimensional, i.e. the plane-stress assumption is not applicable due to high transverse shear and normal stresses acting in the thickness direction. Approximate analytical solutions of the corresponding problems have been suggested by a number of researchers (the reference to these solutions can be found in the books by Timoshenko and Woinowsky-Krieger 1959; Ugural 1999; Ventsel and Krauthammer 2001). An alternative approach is based on using a three-dimensional finite element analysis in the vicinity to the point of application of the concentrated force accounting for the local stresses.

2.2 Boundary Conditions in Realistic Structures

The solution shown in the previous paragraph is both simple and explicit. However, it is applicable only if the plate is simply supported. Such boundary conditions are quite typical in applications.

Simple support corresponds to the situation where the edge is prevented from transverse deflections, while its rotations are unconstrained. The former requirement implies that the support structure has an infinite stiffness in the plane of the edge. For example, a bulkhead shown in Fig. 2.4 almost completely prevents deflections in the z-direction (the bulkhead is shown with stringers since an unsupported bulkhead would be too vulnerable to lateral loads and to in-plane compression). At the same time, the torsional resistance of the bulkhead to rotations of the plate around the edge (around the y-axis) is relatively small. While this resistance could be evaluated

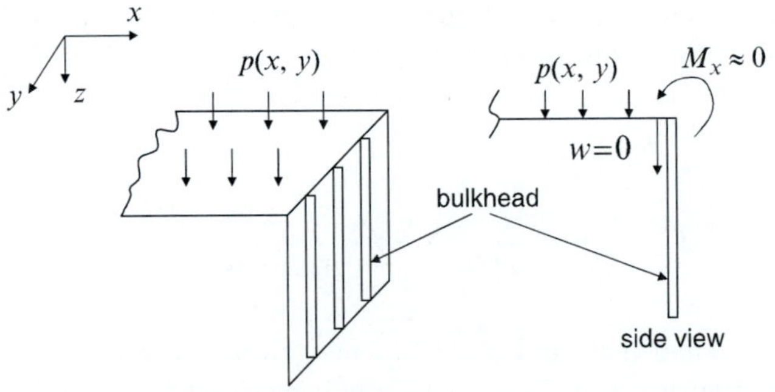

Fig. 2.4 Support provided to the plate by a bulkhead

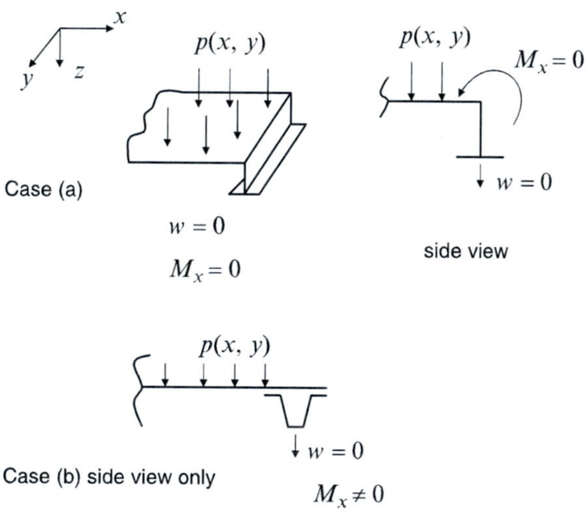

Fig. 2.5 Support provided to the plate by a stringer. Case (**a**): open profile stringer provides simple support, case (**b**): closed profile stringer provides elastic clamping

numerically (or even analytically), it is customary to assume that the conditions at the edge of the plate are $w = 0,\ M_x = 0$ corresponding to simple support. Note that neglecting rotational stiffness of the bulkhead implies that elastic clamping of the plate supported by the bulkhead and associated with such stiffness is disregarded. This is a conservative assumption that results in an overestimate of deflections and stresses in the plate.

The other case where the edge of the plate can be assumed simply supported may occur if the support structure represents a stringer or frame as shown in Fig. 2.5. While the stringer is almost always sufficiently stiff to prevent deflections of the plate ($w = 0$), its torsional stiffness that defines a degree of resistance to rotation of the plate about the support structure depends on geometry. The torsional stiffness of open-profile beams is much smaller than that of closed-profile counterparts. Accordingly, while the edge supported by an open profile stringer or frame can usually be modeled by assumption of simple support (Fig. 2.5, case (a)), this assumption may be too conservative if the stringer or frame has a closed profile (Fig. 2.5, case (b)). On the other hand, the assumption that closed-profile stringers and frames provide complete clamping may lead to an underestimation of deflections and stresses. Accordingly, it is preferable to solve such problem numerically or develop an analytical solution accurately modeling torsional stiffness of the support.

Finally, it is necessary to discuss the case of continuous plates supported by identical stringers (Fig. 2.6). An interesting situation occurs if the plate is subject to uniform pressure and the spacing between the stringers is equal. As is reflected in Fig. 2.6, in such case each section of the plate between stringers deforms as if it was clamped ($w = w_{,x} = 0$). Such clamping is not due to special structural arrangements (in fact, stringers can have open profile and negligible torsional

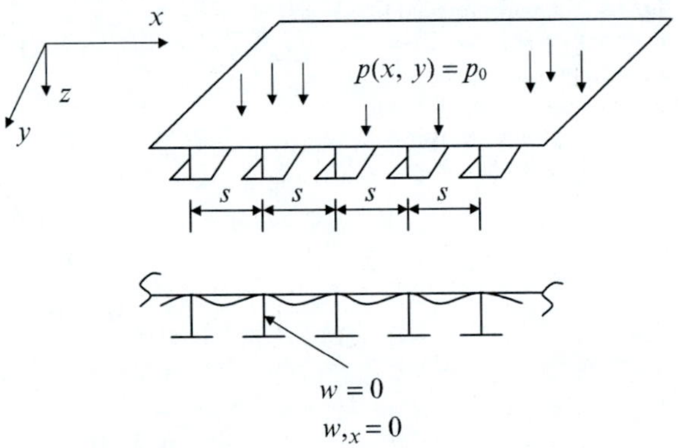

Fig. 2.6 Continuous plate supported by equally-spaced stringers and subject to uniform pressure. As a result of symmetry of load and geometry, the plate is effectively clamped at each stringer

stiffness). Instead, it is a result of symmetry of both the load and geometry about each stringer. However, the designer has to be cautious adopting the assumption that the sections of the plate are clamped and accordingly, reducing its thickness based on the analysis of stresses in clamped plates that are smaller than those in otherwise identical plates with simply supported edges. This is because if the load becomes nonuniform (for example, only one of adjacent sections of the plate is subject to pressure), symmetry is lost and the corresponding section of the plate can rotate about the stringer. It is necessary to remember that a simply supported edge results in a conservative design that is often justified in engineering applications, while the assumption that the edge is clamped may lead to underestimating the stresses in the plate.

A further discussion on the analysis of continuous plates is presented below in Example 2.6. As follows from this example, if symmetry of loading and/or structural symmetry about the supports is violated, the exact solution is still available, although it involves a time consuming procedure.

2.3 Representative Analytical Solution: Levy's Method

As is clear from the previous discussion, there are numerous cases where the assumption of simply supported edges is not acceptable. For example, if a deck supported by identical parallel stringers (Figs. 2.6 and 2.7) is subject to uniform pressure, the symmetry of the structure and the load about each stringer enables us to model the edges of each section supported by these stringers as clamped. If the other edges are supported by frames with a negligible torsional stiffness, they can be assumed simply supported. Therefore, in this case the plates of the deck are clamped at two opposite edges and simply supported at the other two edges.

The Levy method deals with plates that have a couple of simply supported parallel edges, while the other couple of edges can be arbitrary supported. In

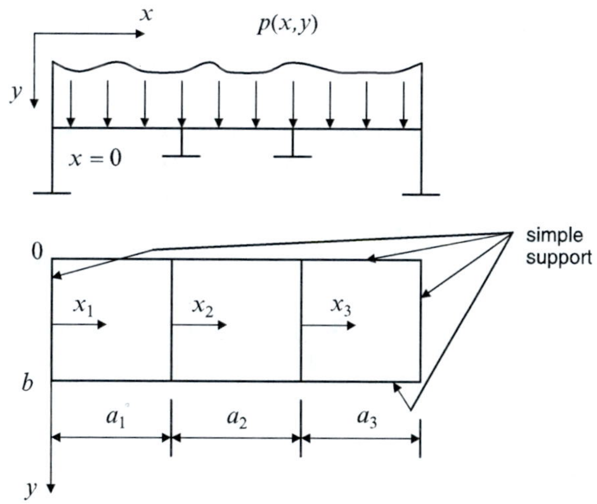

Fig. 2.7 Continuous plate supported by parallel stringers

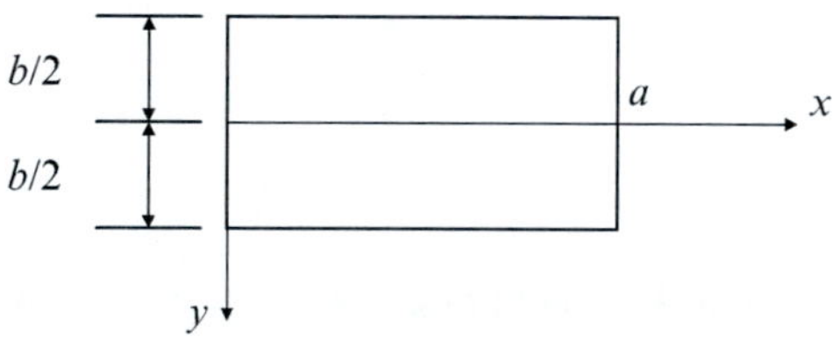

Fig. 2.8 Plate is simply supported at the edges $x = 0$ and $x = a$. The boundary conditions along $y = \pm\frac{b}{2}$are arbitrary

particular, if the edges x = 0 and x = a are simply supported, while the boundary conditions along the edges $y = -\frac{b}{2}$ and $y = \frac{b}{2}$ are arbitrary (Fig. 2.8), the solution of (2.1) is sought in the form

$$w = w_h + w_p = \sum_{m=1}^{M} (f_m(y) + g_m(y)) \sin \alpha_m x \tag{2.19}$$

where w_h and w_p denote the solution of the homogeneous equation obtained from (2.1) and a particular integral of (2.1), respectively, while $f_m(y)$ and $g_m(y)$ are functions that have to satisfy the boundary conditions along the y-edges. The coordinate system in Fig. 2.8 is different from those in Fig. 2.1. This choice is in compliance with standard coordinate systems and notations employed in the solution of the present problem and in the analyses of continuous plates in literature. If the edges $y = \pm\frac{b}{2}$ have identical boundary conditions and the load is symmetric about the plate centerline, this choice of the coordinate system enables us to simplify the solution as shown below. Obviously, series (2.19) satisfy the boundary conditions along the simply supported edges x = 0 and x = a. The substitution of the first series in (2.19) into the homogeneous version of (2.1) yields

$$\sum_{m=1}^{M}\left[\frac{d^4 f_m}{dy^4} - 2\alpha_m^2\frac{d^2 f_m}{dy^2} + \alpha_m^4 f_m\right]\sin\alpha_m x = 0 \tag{2.20}$$

This equation has to be satisfied at any point within the interval $0 < x < a$. Accordingly, the terms in square brackets must be equal to zero. The solution of the resulting ordinary fourth-order differential equation yields

$$f_m = C_{1m}\cosh\alpha_m y + C_{2m} y\sinh\alpha_m y + C_{3m}\sinh\alpha_m y + C_{4m} y\cosh\alpha_m y \tag{2.21}$$

where C_{im} $(i = 1, 2, 3, 4)$ are constants that should be specified from the boundary conditions. Note that the first two terms are symmetric about $y = 0$ and the last two terms are antisymmetric about this axis.

The particular integral can be specified if the load is represented in series

$$p(x, y) = \sum_{m=1}^{M} p_m(y)\sin\alpha_m x$$

$$p_m(y) = \frac{2}{a}\int_0^a p(x, y)\sin\alpha_m x dx \tag{2.22}$$

Now the functions $g_m(y)$ have to be found from (2.1) which, upon the substitution of (2.22), and the corresponding terms in (2.19) yields

$$\sum_{m=1}^{M}\left[\frac{d^4 g_m}{dy^4} - 2\alpha_m^2\frac{d^2 g_m}{dy^2} + \alpha_m^4 g_m - \frac{p_m(y)}{D}\right]\sin\alpha_m x = 0 \tag{2.23}$$

Equation 2.23 is satisfied at every point within $0 < x < a$ if

$$\frac{d^4 g_m}{dy^4} - 2\alpha_m^2\frac{d^2 g_m}{dy^2} + \alpha_m^4 g_m - \frac{p_m(y)}{D} = 0 \tag{2.24}$$

The solution of (2.24) depends on the form of the functions $p_m(y)$. Once this solution is available, constants of integration C_{im} $(i = 1, 2, 3, 4)$ are obtained by the substitution of (2.19) where $f_m(y)$ is given by (2.21) into the boundary conditions along the edges $y = -\frac{b}{2}$ and $y = \frac{b}{2}$.

A simplification is possible if the load is symmetric about the x-axis and the boundary conditions along $y = \frac{b}{2}$ and $y = -\frac{b}{2}$ are identical. Then the deflections of the plate are symmetric about the x-axis, so that $w(y) = w(-y)$. Obviously, this can only be achieved if $C_{3m} = C_{4m} = 0$ in (2.21). Accordingly, for such symmetric problems

$$f_m = C_{1m}\cosh\alpha_m y + C_{2m} y\sinh\alpha_m y \tag{2.25}$$

The following examples concentrate on plates with various representative boundary conditions. In all examples, the plate is subject to uniform pressure as is rather typical in applications.

Example 2.4: Plate with Two Simply Supported and Two Clamped Edges Subject to Uniform Pressure p_0. Consider the case shown in Fig. 2.8 where the edges $y = \pm\frac{b}{2}$ are clamped. Equation 2.22 yields $p_m = \frac{4p_0}{m\pi}$, m being an odd number. A particular integral of (2.24) is

$$g_m = \frac{4p_0a^4}{m^5\pi^5D}$$

Now using the condition of symmetry about y = 0, the substitution of $f_m + g_m$, the former function being given by (2.25), into the boundary conditions $y\left(\frac{b}{2}\right) = \frac{\partial w\left(\frac{b}{2}\right)}{\partial y} = 0$ yields the system of two equations with respect to constants of integration:

$$\begin{bmatrix} \cosh m' & \frac{b}{2}\sinh m' \\ \alpha_m \sinh m' & m'\cosh m' + \sinh m' \end{bmatrix} \begin{Bmatrix} C_{1m} \\ C_{2m} \end{Bmatrix} = -\begin{Bmatrix} g_m \\ 0 \end{Bmatrix} \quad \text{(a)}$$

where $m' = \frac{m\pi b}{2a}$.

Once the constants of integration are found from (a), the stresses can be determined from equations (1.31) where the middle plane strains are equal to zero, (1.29) and (1.19). Subsequently, a strength criterion should be applied to predict whether the plate fails or remains capable of carrying the prescribed applied pressure. Alternatively, it is possible to solve the problem in terms of unknown applied pressure and specify its allowable value from the strength criterion.

Example 2.5: Plate with Two Simply Supported and Two Elastically Supported Edges Subject to Uniform Pressure (Timoshenko and Woinowsky-Krieger 1959). Consider now the problem shown in Fig. 2.8 assuming that the edges $x = 0, a$ are simply supported while the edges $y = \pm\frac{b}{2}$ are supported by ribs of a finite flexural stiffness EI that have negligible torsional stiffness (open-profile flexible stringers).

The boundary conditions along $y = \pm\frac{b}{2}$ are identical and since the load is symmetric about the centerline y = 0 it is justified to represent the homogeneous solution by (2.25). Then two conditions that have to be satisfied are

$$M_y\left(\frac{b}{2}\right) = 0$$

$$V_y\left(\frac{b}{2}\right) = EI\frac{\partial^4 w\left(\frac{b}{2}\right)}{\partial x^4} \quad \text{(b)}$$

The substitution of the expressions for the stress couple and transverse shear stress resultant given by (1.58), (2.12b) and (2.13) in terms of the deflection into boundary conditions (b) yields

$$\frac{\partial^2 w\left(\frac{b}{2}\right)}{\partial y^2} + \nu\frac{\partial^2 w\left(\frac{b}{2}\right)}{\partial x^2} = 0$$

$$[6pt]\frac{\partial^3 w\left(\frac{b}{2}\right)}{\partial y^3} + (2-\nu)\frac{\partial^3 w\left(\frac{b}{2}\right)}{\partial x^2 \partial y} = \frac{EI}{D}\frac{\partial^4 w\left(\frac{b}{2}\right)}{\partial x^4} \tag{c}$$

The substitution of (2.19) with f_m given by (2.25) and g_m by the formula in Example 2.4 into boundary conditions (c) yield the system of two equations with respect to constants C_{1m} and C_{2m}. The solution of this system is

$$C_{1m} = \frac{4p_0 a^4}{m^5\pi^5 D} \times \frac{\nu(1+\nu)\sinh m' - \nu(1-\nu)m'\cosh m' - \lambda m\pi\,(2\cosh m' + m'\sinh m')}{(3+\nu)\,(1-\nu)\sinh m'\cosh m' - (1-\nu)^2 m' + 2\lambda m\pi\cosh^2 m'}$$

$$C_{2m} = \frac{4p_0 a^3}{m^4\pi^5 D} \times \frac{\nu\,(1-\nu)\sinh m' + \lambda m\pi\,\cosh m'}{(3+\nu)\,(1-\nu)\sinh m'\cosh m' - (1-\nu)^2 m' + 2\lambda m\pi\cosh^2 m'}$$

where the coefficient λ is defined as in Example 2.1. Tabulated results for a square plate can be found in the book of Timoshenko and Woinowsky-Krieger (1959).

Example 2.6: Approach to the Analysis of Continuous Plates. Continuous plates shown in Figs. 2.6 and 2.7 may be subject to different loads within each section. Furthermore, the spacing of parallel stringers may be unequal. In such cases, the assumption that the stringers provide clamping to the adjacent sections of the plate discussed in Sect. 2.2 becomes invalid and the analysis has to rely on the approach similar to the three-moment equation method for continuous beams.

Consider the three-span plate shown in Fig. 2.7 where the edges are simply supported. The plate is reinforced by two stringers that are sufficiently stiff to prevent deflections but whose torsional stiffness is insignificant (open-profile beams). Then, if the lengths of three sections of the plate are unequal or if the load applied to the plate differs from one section to the other, the plate will experience rotations at the cross sections $x_1 = a_1,\ x_2 = 0$ and $x_2 = a_2,\ x_3 = 0$.

The kinematic continuity of the sections of the plate over the stringers dictates the following conditions:

$$x_1 = a_1,\ x_2 = 0: \quad \frac{\partial w_1}{\partial x_1} = \frac{\partial w_2}{\partial x_2}, \qquad w_1 = w_2 = 0$$
$$x_2 = a_2,\ x_3 = 0: \quad \frac{\partial w_2}{\partial x_2} = \frac{\partial w_3}{\partial x_3}, \qquad w_2 = w_3 = 0 \tag{d}$$

where deflections w_i refer to the i-th section of the plate.

Additional conditions can be formulated by considering the stress couples that are applied to each adjacent section over the stringer. The torsional stiffness of the stringers being assumed negligible, the equilibrium of these stress couples yields

$$x_1 = a_1,\ x_2 = 0: \quad M_x^{(1)} = M_x^{(2)}$$
$$x_2 = a_2,\ x_3 = 0: \quad M_x^{(2)} = M_x^{(3)} \tag{e1}$$

where the superscripts identify the corresponding section. Note that the expressions for the stress couples along the stringers can be simplified. This is because $\frac{\partial^2 w}{\partial y^2} = 0$ along the stringers where deflections remain constant and equal to zero. Accordingly, conditions (e1) are replaced with

$$x_1 = a_1,\ x_2 = 0: \quad \frac{\partial^2 w_1}{\partial x_1^2} = \frac{\partial^2 w_2}{\partial x_2^2}$$
$$x_2 = a_2,\ x_3 = 0: \quad \frac{\partial^2 w_2}{\partial x_2^2} = \frac{\partial^2 w_3}{\partial x_3^2} \tag{e2}$$

The outer edges parallel to the stringers are simply supported, so that

$$x_1 = 0: \quad w_1 = M_x^{(1)} = 0$$
$$x_3 = a_3: \quad w_3 = M_x^{(3)} = 0 \tag{f1}$$

These conditions can be replaced with

$$x_1 = 0: \quad w_1 = \frac{\partial^2 w_1}{\partial x_1^2} = 0$$
$$x_3 = a_3: \quad w_3 = \frac{\partial^2 w_3}{\partial x_3^2} = 0 \tag{f2}$$

The solution is now obtained by applying the Levy method to each plate section. For simply supported edges $y = 0$ and $y = b$ the solution for the i-th section can be represented by

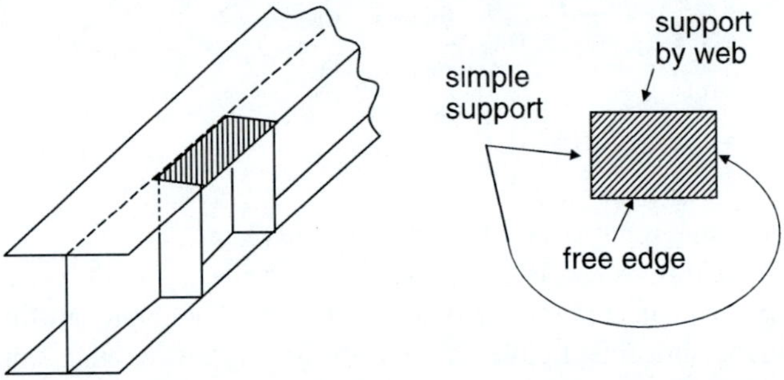

Fig. 2.9 Section of the flange of an I-beam between the web and two brackets (*shaded*). The flange may be clamped or simply supported at the web. Brackets usually provide simple support

$$w_i = \sum_{m=1}^{M} \left(C_{1m}^{(i)} \cosh \alpha_i x_i + C_{2m}^{(i)} x_i \sinh \alpha_i x_i + C_{3m}^{(i)} \sinh \alpha_i x_i + C_{4m}^{(i)} x_i \cosh \alpha_i x_i \right.$$
$$\left. + g_m^{(i)} (x_i) \right) \sin \beta_m y \tag{g}$$

where $\alpha_{mi} = \frac{m\pi}{a_i}$, $g_m^{(i)}(x)$ is a particular integral for the corresponding section that depends on the applied load and $C_{rm}^{(i)}$ $(r = 1, 2, 3, 4)$ are constants of integration.

There are 12 constants of integration in Eq. g for each index m since the plate consists of three sections. These constants can be determined from six equations (d), two equations (e2) and four equations (f2). Other continuous plates can be treated using the same approach. The advantage of the present solution is that it is exact. However, in the case of a larger number of sections or different boundary conditions along the edges $y = 0$ and $y = b$, the solution may become too cumbersome so that it may be preferable to apply numerical methods. A finite torsional stiffness of stringers could also be incorporated in the solution.

Once the deflections of each span are determined, the stresses can be found using the strain-displacement and constitutive relations. The deflections being affected by the constraint on rotations that adjacent spans superimpose on each other, the corresponding solution for the stresses also reflects this constraint.

There are numerous problems where the Levy method is useful, in spite of the necessity to conduct straightforward but lengthy transformations. An example of such problems is demonstrated in Fig. 2.9 where the flange of an I-beam supported by the web and two brackets can be modeled as a plate clamped or simply supported along one edge (web), simply supported by the brackets and free along the fourth edge. Brackets similar to those in Fig. 2.9 are often employed in welded I-beams to increase the stability of the web and of the compressed flange. The assumption that the brackets provide simple support to the flange is usually justified. If the transverse load is symmetric about the web, the corresponding edge of the flange can be assumed clamped. If the symmetry of the transverse load about the web is in question, the corresponding edge should be treated as simple support. The analyses of both cases can be conducted by the Levy method.

2.4 Plates on Elastic Foundation

There are numerous applications where the plate is continuously supported within the span. In the case where the support is linear elastic, its reaction is proportional to the local deflection of the structure (so-called Winkler's elastic foundation). Accordingly, if the plate supported by an elastic foundation experiences a local deflection w, the reaction (counter-pressure) applied by the foundation to the plate is kw where k is a proportionality coefficient called the modulus of the foundation. It is evident that the units of the modulus of foundation are force per unit area, per unit deflection, i.e. N/m^3 or lbf/in^3. It should be noted that the linear elastic model of the foundation is often only a computationally convenient simplification.

Soil provides an example of the elastic foundation in civil engineering structures. For example, if a storage tank is resting on the ground, the plates of the bottom will be subject to the reaction of the ground that can be modeled as an elastic foundation. Another example of a linear elastic foundation is found in concrete pavements and foundation slabs of buildings (Ventsel and Krauthammer 2001).

Predictably, the moduli of elastic foundations found in civil engineering applications vary within a broad range. For example, the measurements on the stiffness of subgrades taken in 63 tests conducted by the New York Department of Transportation varied from 27 to 680 kPa/mm (Bryden et al. 1971).

Consider the problem where a simply supported plate subject to an arbitrary load $p(x, y)$ as shown in Fig. 2.1 is resting on a continuous nonuniform linear elastic foundation with the modulus $k(x, y)$. Then the equation of equilibrium becomes

$$D\nabla^4 w = p(x, y) - k(x, y)\,w \tag{2.26}$$

Naturally, boundary conditions are not affected by the foundation. Therefore, the solution can be sought in series (2.5) that satisfy these conditions.

The solution is particularly simple if the foundation is uniform, i.e. $k(x, y) = k = const$. Then the substitution of (2.5) into (2.26) yields

$$W_{mn} = \frac{p_{mn}}{D\left(\alpha_m^2 + \beta_n^2\right)^2 + k} \tag{2.27}$$

As could be anticipated, the presence of an elastic foundation reduces the deflections. Bending stresses in the plate that are proportional to deflections are also reduced by the foundation. Although the foundation introduces transverse normal stresses at the interface with the plate, these stresses cannot overcome the beneficial effect of reduced bending stresses, i.e. foundations improve the strength and stiffness of plates.

In the case where the stiffness of the foundation is variable, the solution can be sought using the Galerkin procedure (Sect. 1.5). For a simply supported plate this procedure results in the system of linear algebraic equations where the mn-th equation is:

$$\int_0^b \int_0^a \left[D\nabla^4 w - p(x,y) + k(x,y)\,w\right] \sin\alpha_m x \sin\beta_n y\,dx\,dy = 0 \tag{2.28}$$

where the integers m and n are within the range $1 \le m \le M,\ 1 \le n \le N$ used in truncated series (2.5).

Substituting series (2.3) and (2.5) into (2.28) and recalling that sine functions are orthogonal, one obtains

$$D\left(\alpha_m^2 + \beta_n^2\right)^2 W_{mn} - p_{mn} + \frac{4}{ab}\sum_{i=1}^{M}\sum_{j=1}^{N}\int_0^b\int_0^a k(x,y)\,W_{ij}\sin\alpha_i x \sin\beta_j y \sin\alpha_m x \sin\beta_n y\,dx\,dy = 0 \tag{2.29}$$

The integrals in the right side of (2.29) can be obtained analytically or numerically. Subsequently, the amplitudes of series (2.5) can be determined from the system of linear algebraic equations (2.29).

An alternative approach to the analysis of the plate supported by a nonuniform elastic foundation is based on the Rayleigh-Ritz approach (Sect. 1.5). The advantage of this approach is that static boundary conditions can be violated, though the kinematic boundary conditions must be satisfied. The presence of the elastic foundation is accounted for by adding its strain energy to the total energy of the plate:

$$U_k = \frac{1}{2}\int_0^b\int_0^a k(x,y)\,w^2\,dx\,dy \tag{2.30}$$

Accordingly, if the deflection is represented by

$$w = \sum_i W_i \Phi_i(x,y) \tag{2.31}$$

where functions $\Phi_i(x,y)$ satisfy the kinematic boundary conditions, the Rayleigh-Ritz method implies

$$\frac{\partial}{\partial W_p}\left(U + U_k + V\right) = 0 \tag{2.32}$$

where V is the energy of the applied pressure and external in-plane stress resultants given by (1.68) and (1.89). The strain energy of the plate U is obtained from (1.70) that is simplified to account for the Kirchhoff-Love hypothesis $\left(\tau_{xz} = \tau_{yz} = 0\right)$ and for the fact that middle-plane strains $\varepsilon_x^0 = \varepsilon_y^0 = \gamma_{xy}^0 = 0$. Then substituting the expressions for bending and twisting stress couples and for the changes of curvatures and twist in terms of deflections according to (1.58) and (1.29), respectively, we obtain

$$U = \frac{1}{2}\int_0^b\int_0^a D\left[\left(\frac{\partial^2 w}{\partial x^2}\right)^2 + \left(\frac{\partial^2 w}{\partial y^2}\right)^2 + 2\nu\frac{\partial^2 w}{\partial x^2}\frac{\partial^2 w}{\partial y^2} + 2(1-\nu)\left(\frac{\partial^2 w}{\partial x \partial y}\right)^2\right]dxdy \tag{2.33}$$

Using the expressions for the components of energy referred to above and the expression for the deflections (2.31) in the Rayleigh-Ritz formulation (2.32) is straightforward. Besides the variable over the plate surface elastic foundation, this method can also be applied to the case of a variable-stiffness plate where $D = D\,(x,\, y)$.

The linear elastic foundation model discussed above may not be accurate in a number of practical situations. Accordingly, several alternative foundation models have been suggested. While the discussion of these models is outside the scope of this book, mentioned here is the review provided in the paper by Kerr (1984) and research of Reissner (1967). In particular, models that represent the reaction of the foundation in terms of power series can be introduced by a differential operator including only even order derivatives, so that the foundation response is given by

$$\bar{p} = k_0 w + k_1 \nabla^2 w + k_2 \nabla^4 w + \ldots \tag{2.34}$$

where $k_i \ \ (i = 0,\ 1,\ 2, \ldots)$ are coefficients.

The Pasternak foundation model that represents an extension of the Winkler foundation accounting for the effect of in-plane shear may be accurate in certain applications (Pasternak 1954). According to this model, the response of the foundation is given by

$$\bar{p} = kw - G_f \nabla^2 w \tag{2.35}$$

where both k and G_f are foundation constants. Of course, the Pasternak model can be characterized as a particular case of the more general formulation (2.34).

It is also worth mentioning that besides models based on a representation of the foundation reaction via differential operators with coefficients available from experimental data, an analytical approach based on the theory of elasticity may be applied as long as the foundation remains elastic. If the foundation is "sufficiently deep," the analysis can be conducted modeling it by an elastic semi-infinite half-sphere supporting the plate. Then the Boussinesq solution relates the local deflection at a point (x_i, y_j) to the reaction of the foundation (Vlasov and Leont'ev 1966):

$$w\left(x_i, y_j\right) = \int_0^b \int_0^a K\left[(x - x_i),\, (y - y_j)\right]\bar{p}\,(x, y)\,dxdy \tag{2.36}$$

where $K\left[(x - x_i),\, (y - y_j)\right]$ is the Green function.Detailed analyses based on the foundation models reflected in Eqs. 2.34, 2.35 and 2.36 are outside the scope of this book.

2.5 Combined Lateral and In-Plane Loading

There are a number of applications where the plate is subject to a combination of lateral loads and in-plane tension, compression or shear. For example, a bottom plate of the central section of a cargo ship in rough seas experiences bending from cargo combined with in-plane tension or compression. In-plane loads are generated by bending of the hull due to a nonuniform distribution of buoyancy forces and weight along the hull. In-plane loads reach maximum values when the length of the waves is equal to that of the vessel. In particular, the mid-section bottom structure is subject to tension when the ship is sagging due to the waves that peak at the bow and stern. However, when a wave peaks at the midsection (hogging), the bottom experiences compressive stresses (see Fig. 2.10). Dynamic effects of in-plane loads could be accounted for through a magnification factor.

A rectangular plate subject to a combination of lateral and in-plane loads is shown in Fig. 2.11. The equations of equilibrium correspond to the static version of (1.92), i.e.

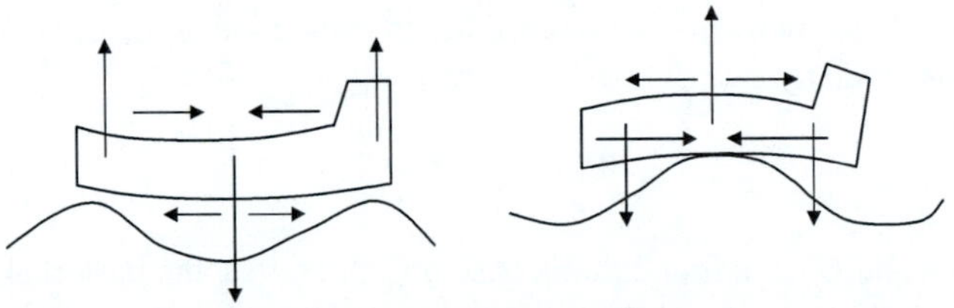

Fig. 2.10 Schematic illustration of bending of a ship in rough seas. *Left*: upper deck is compressed, while the bottom is in tension. *Right*: upper deck works in tension, while the bottom is compressed

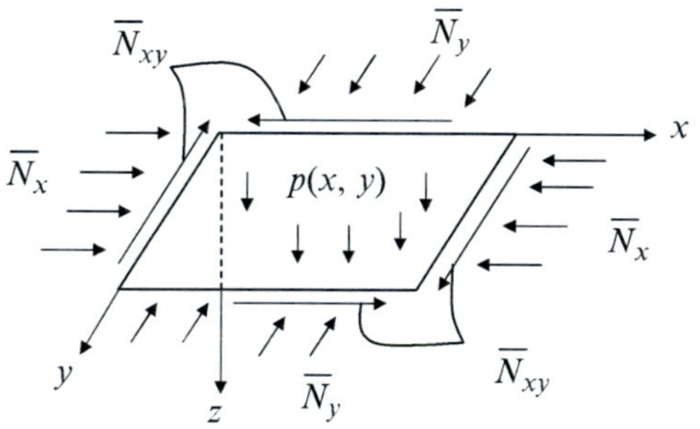

Fig. 2.11 A plate subject to a combination of transverse pressure and in-plane loads

$$\frac{\partial N_x}{\partial x} + \frac{\partial N_{xy}}{\partial y} = 0$$
$$\frac{\partial N_{xy}}{\partial x} + \frac{\partial N_y}{\partial y} = 0$$
$$\frac{\partial^2 M_x}{\partial x^2} + 2\frac{\partial^2 M}{\partial x \partial y} + \frac{\partial^2 M_y}{\partial y^2} + p + N_x \frac{\partial^2 w}{\partial x^2} + 2N_{xy} \frac{\partial^2 w}{\partial x \partial y} + N_y \frac{\partial^2 w}{\partial y^2} = 0 \quad (2.37)$$

It is important to discuss here the principal difference between geometrically linear and nonlinear problems. In the case of large deflections, i.e. if the problem is geometrically nonlinear, in-plane stresses are affected by deflections w, (in other words, stretching and bending of the plate are coupled). Accordingly, in-plane stress resultants are not constant but location-dependent, so that $N_i = N_i\,(x, y)$, $i = x, y, xy$. Furthermore, in the nonlinear problem $N_i = N_i\,(u, v, w)$, $i = x, y, xy$. This implies that three equations (2.37) are coupled, so that each of them written in terms of displacements includes all three displacements u, v and w. In the geometrically nonlinear problem, the last three terms in the third equation (2.37) are nonlinear functions of displacements.

The situation is much simpler in the linear case where deflections remain small, so that their effect on in-plane stress resultants is negligible. Then the in-plane stress resultants are constants and equal to the applied external loads, $N_i = \bar{N}_i$. Then substituting the expression for bending stress couples in terms of deflections (1.58) into the last equation (2.37) yields the following linear equation:

$$D\nabla^4 w = p\,(x, y) + \bar{N}_x \frac{\partial^2 w}{\partial x^2} + 2\bar{N}_{xy} \frac{\partial^2 w}{\partial x \partial y} + \bar{N}_y \frac{\partial^2 w}{\partial y^2} \quad (2.38)$$

The solution of equation (2.38) can be sought by the same methods as those applied to the case where in-plane loads are absent. In the simple but rather typical case where a simply supported plate is undergoing a combination of lateral pressure and in-plane loads $N_x = \bar{N}_x,\ N_y = \bar{N}_y$, a modified Navier solution is applicable as follows.

Recognizing that in-plane loads do not affect boundary conditions (2.2) corresponding to simply supported edges, the solution is sought in the form (2.5) that identically satisfies these conditions. Furthermore, the pressure is represented in series (2.3). Then the substitution of (2.3) and (2.5) into (2.38) yields the amplitudes of terms in the series for deflections:

$$W_{mn} = \frac{p_{mn}}{D\left(\alpha_m^2 + \beta_n^2\right)^2 + \bar{N}_x \alpha_m^2 + \bar{N}_y \beta_n^2} \tag{2.39}$$

Compressive in-plane stress resultants ($\bar{N}_x < 0,\ \bar{N}_y < 0$) increase the bending deflection, while tension results in smaller deformations.

Bending stresses throughout the plate can now be evaluated from (2.9b). These stresses are superimposed on constant in-plane membrane stresses produced by applied loads, so that the total stress is

$$\sigma_x = \frac{\bar{N}_x}{h} + \frac{Ez}{1-\nu^2} \sum_{m=1}^{M} \sum_{n=1}^{N} \left[\alpha_m^2 + \nu \beta_n^2\right] W_{mn} \sin \alpha_m x \sin \beta_n y$$

$$\sigma_y = \frac{\bar{N}_y}{h} + \frac{Ez}{1-\nu^2} \sum_{m=1}^{M} \sum_{n=1}^{N} \left[\beta_n^2 + \nu \alpha_m^2\right] W_{mn} \sin \alpha_m x \sin \beta_n y$$

$$\tau_{xy} = -\frac{Ez}{1+\nu} \sum_{m=1}^{M} \sum_{n=1}^{N} \alpha_m \beta_n W_{mn} \cos \alpha_m x \cos \beta_n y \tag{2.40}$$

It is worth emphasizing that in-plane loads have a dual effect on the stresses in the plate. First of all, they produce uniform throughout the plate stresses as reflected in the first terms in the expressions for σ_x and σ_y in (2.40). In addition, these stresses affect bending deflections of the plate since the amplitudes W_{mn} are influenced by in-plane stress resultants according to (2.39). Therefore, bending stresses given by the terms dependent on these amplitudes in (2.40) are affected by in-plane loading.

Finally, it is noted that in the case where applied in-plane loading is compressive, the denominator in (2.39) may become equal to zero, signifying infinite deflections of the plate. This may occur even if the lateral pressure is very small. As we will see in the next paragraph, such situation occurs when the loads reach a buckling combination. However, the validity of the present linear solution in the case of combined bending and nearly buckling in-plane loads is limited since if the plate is flexible and experiences large deformations, it is necessary to use a geometrically nonlinear formulation. On the other hand, maximum stresses in a more rigid plate reach the yield limit even at small deformations, necessitating a physically nonlinear analysis. Accordingly, the present solution is applicable only if the deformations remain relatively small and the stresses are within the elastic range.

2.6 Buckling of Rectangular Isotropic Plates

Consider the case where a rectangular plate is subject to a combination of in-plane compression along both x- and y-axes as well as in-plane shear. In the absence of lateral loads the equation of equilibrium (2.38) is reduced to

$$D\nabla^4 w = N_x \frac{\partial^2 w}{\partial x^2} + 2N_{xy}\frac{\partial^2 w}{\partial x \partial y} + N_y \frac{\partial^2 w}{\partial y^2} \tag{2.41}$$

where in-plane stress resultants differ from the applied loads in a nonlinear formulation. The other two equilibrium equations coincide with the first two equations (2.37).

Note that (2.41) is a homogeneous equation, so that the trivial solution $w = 0$ is always possible. This solution corresponds to the prebuckling state where the plate remains flat but experiences in-plane displacements (u and v). These displacements are very small due to a large in-plane stiffness of flat plates. Prebuckling displacements can be determined from the linear version of the strain-displacement relationship (1.28) where $w = 0$ and the strains are expressed in terms of applied stress resultants by:

$$\begin{aligned}\tilde{\varepsilon}_x &= \frac{1}{Eh}\left(\bar{N}_x - \nu \bar{N}_y\right)\\ \tilde{\varepsilon}_y &= \frac{1}{Eh}\left(\bar{N}_y - \nu \bar{N}_x\right)\\ \tilde{\gamma}_{xy} &= \frac{\bar{N}_{xy}}{Gh}\end{aligned} \tag{2.42}$$

In the above equations, the wave over the strain identifies the prebuckling state.

The integration of the prebuckling strain-displacement equations is straightforward if the plate is subject to in-plane compression only. Then the solution of the first two linearized equations (1.28) combined with the symmetry requirement to prebuckling displacements $\tilde{u}_0\left(x = \frac{a}{2}\right) = 0,\ \tilde{v}_0\left(y = \frac{b}{2}\right) = 0$ yields

$$\begin{aligned}\tilde{u}_0 &= \frac{1}{Eh}\left(\bar{N}_x - \nu \bar{N}_y\right)\left(x - \frac{a}{2}\right)\\ \tilde{v}_0 &= \frac{1}{Eh}\left(\bar{N}_y - \nu \bar{N}_x\right)\left(y - \frac{b}{2}\right)\end{aligned} \tag{2.43}$$

The first two equations (2.37) representing the conditions of in-plane equilibrium are identically satisfied since in the linear prebuckling state $N_x = \bar{N}_x,\ N_y = \bar{N}_y$ and $N_{xy} = 0$.

As the load applied to the plate increases, there are the following possibilities:

1. A thin and flexible plate remains within the elastic range. However, at a certain load the state is reached where the energy balance permits more than one

equilibrium position, i.e. there is an alternative deflected configuration, besides $w = 0$. The plate invariably assumes such deflected configuration, exhibiting the phenomenon that we call buckling.

2. A thicker plates may reach plastic state prior to buckling. Note that prebuckling in-plane stresses are uniformly distributed throughout the plate, i.e. plasticity will develop simultaneously at all points of the plate. Subsequently, as the load continues to increase, plastic stresses at every point of the plate will increase at the same rate. Such elasto-plastic plates can either collapse due to the loss of strength or become unstable in the plastic range.

In this paragraph we concentrate on elastic buckling.

If the plate buckles, the corresponding displacements u, v, w are superimposed on the prebuckling state, so that the total deformations become $\tilde{U} = \tilde{u}_0 + u, \ \tilde{V} = \tilde{v}_0 + v, \ \tilde{W} = w$. Equations of equilibrium of a plate experiencing buckling and boundary conditions for such plate can be written as

$$\begin{aligned} F\left(\tilde{U}, \tilde{V}, \tilde{W}\right) &= 0 \\ B\left(\tilde{U}, \tilde{V}, \tilde{W}\right) &= 0 \end{aligned} \tag{2.44}$$

where F and B are differential operators (the former representing the system of equations of equilibrium and the latter being the set of boundary conditions). Applied in-plane loads are incorporated in these equations. In general, the operators in (2.44) can include both linear and nonlinear terms, so that $F = FL + FN, \ B = BL + BN$ where linear and nonlinear contributions are $FL, \ BL$ and $FN, \ BN$, respectively.

The substitution of total displacements, including prebuckling and buckling deformations, into (2.44) yields

$$\begin{aligned} \underline{FL\left(\tilde{u}_0, \tilde{v}_0\right)} + FL\left(u, v, w\right) + FN\left(\tilde{U}, \tilde{V}, \tilde{W}\right) &= 0 \\ \underline{BL\left(\tilde{u}_0, \tilde{v}_0\right)} + BL\left(u, v, w\right) + BN\left(\tilde{U}, \tilde{V}, \tilde{W}\right) &= 0 \end{aligned} \tag{2.45}$$

In the case where we are concerned with linear buckling, i.e. limit the analysis to specifying the buckling load, rather than studying the postbuckling response of the plate, nonlinear operators in (2.45) can be omitted. Furthermore, the terms underlined in (2.45) are identically satisfied by the solution of the problem of prebuckling deformations. Accordingly, the linear formulation of the buckling problem is reduced to

$$\begin{aligned} FL\left(u, v, w\right) &= 0 \\ BL\left(u, v, w\right) &= 0 \end{aligned} \tag{2.46}$$

This implies that linear buckling of a mechanically loaded plate can be studied without a reference to prebuckling deformations. Of course, as is obvious from

(2.45), prebuckling deformations affect nonlinear postbuckling behavior, i.e. the corresponding problems (prebuckling and postbuckling responses) are coupled.

The formulation of linear buckling problems can be reduced to the simple statement: "Find the combination of applied loads that makes a nontrivial equilibrium configuration possible." Obviously, prebuckling deformations affect the configuration of the plate, but it remains flat. Thus, a "nontrivial equilibrium" referred to above implies deflections from the flat configuration.

The approach leading to a closed-form solution of equations of equilibrium usually utilizes the assumption that the mode shape of the buckled plate can be represented in the form of products of functions of coordinates. These functions should identically satisfy all boundary conditions. In addition, they should be cancelled out in the equations of equilibrium leading to an algebraic equation for the buckling load (see Example 2.7 below). Unfortunately, such approach is applicable only to a limited number of boundary conditions, such as simply supported plates. An analytical solution of the buckling problem for plates with two opposite simply supported edges utilizing equations of equilibrium is also possible along the lines of the Levy's method. Other analytical procedures can utilize energy methods, such as the Rayleigh-Ritz method or the Galerkin procedure. In numerous cases geometry and/or boundary conditions are such that the solution has to be obtained by one of numerical methods, such as the finite element method.

In view of the previous discussion, it is obvious that linear buckling of a rectangular plate can be studied using Eq. 2.41 where in-plane stress resultants represent applied loads identified below with an overbar and w is a deflection from the prebuckling flat shape:

$$D\nabla^4 w - \bar{N}_x \frac{\partial^2 w}{\partial x^2} - 2\bar{N}_{xy} \frac{\partial^2 w}{\partial x \partial y} - \bar{N}_y \frac{\partial^2 w}{\partial y^2} = 0 \tag{2.47}$$

The other two equilibrium conditions given by first two equations (2.37) depend only on in-plane stress resultants. As follows from (1.58), as long as the problem is linear, these stress resultants do not include transverse deflections w. Accordingly, they do not affect the analysis of buckling.

Example 2.7: Buckling of a Simply Supported Plate Subject to Uniaxial Compressive Load. This classical example appears in every book on the theory of plates or on stability of structures. This example is worth considering since simply supported plates are encountered in numerous applications. For example, the plates of the upper deck of the middle section of a ship shown in Fig. 2.10 experience compression every time the bow and stern are supported by the waves, while the middle section is sagging.

The formulation of the problem is reduced to finding the stress resultant $\bar{N}_x$ that allows a nonzero solution of the equation of equilibrium (2.47) where $\bar{N}_y = \bar{N}_{xy} = 0$, subject to boundary conditions (2.2). The buckling deflection is sought in the form

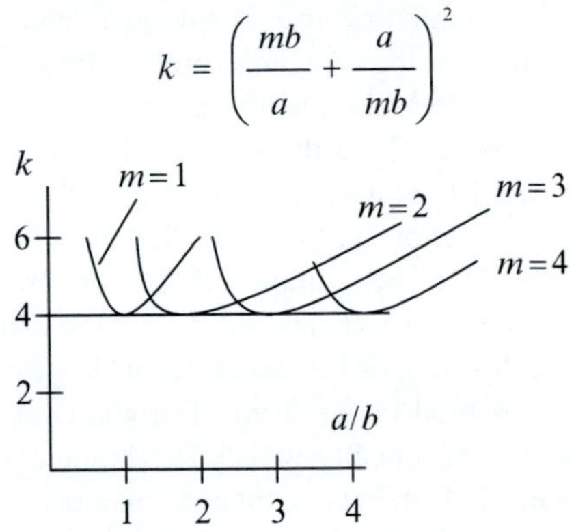

Fig. 2.12 Buckling load coefficient for a simply supported plate compressed along the x-axis as a function of the plate aspect ratio and the number of half-waves in the buckling mode shape (m)

$$w = W_{mn} \sin \alpha_m x \sin \beta_n y \tag{h}$$

where W_{mn} is unknown amplitude of the buckling deflection.

It is evident that integers m and n represent the number of half-waves in the mode shape of buckling. The buckling mode (h) identically satisfies boundary conditions (2.2). The substitution of (h) into the equilibrium equation (2.47) yields

$$\left[D\left(\alpha_m^2 + \beta_n^2\right)^2 + \bar{N}_x \alpha_m^2\right] W_{mn} \sin \alpha_m x \sin \beta_n y = 0 \tag{i}$$

This equation must be satisfied at every point within the plate. Accordingly, trigonometric functions cannot be assumed equal to zero. Furthermore, instability implies that $W_{mn} \neq 0$. Therefore, the buckling equation (i) is satisfied by prescribing the values of the applied stress resultant, such that the term in the square brackets is equal to zero:

$$\bar{N}_{x,cr} = -\frac{\pi^2 D}{b^2}\left(\frac{mb}{a} + \frac{n^2 a}{mb}\right)^2 = -k\frac{\pi^2 D}{b^2}, \quad k = \left(\frac{mb}{a} + \frac{n^2 a}{mb}\right)^2 \tag{j}$$

The actual buckling load corresponds to the smallest absolute value of those given by (j). It can be observed that the plate buckles forming only one half-wave in the direction perpendicular to the direction of the applied load, i.e. $n = 1$. The value of m corresponding to buckling depends on the plate aspect ratio as shown schematically in Fig. 2.12. In numerous engineering applications, the buckling stress resultant for plates with the aspect ratio $\frac{a}{b} \geq 1$ is taken using $k = 4$ resulting in a very simple formula $\bar{N}_{x,cr} = -\frac{4\pi^2 D}{b^2}$.

Note that the condition of buckling determined by (j) does not provide information about the amplitude of deflection, i.e. W_{mn}. This is a common feature of eigenvalue problems, such as buckling and free vibrations. The linear solution can only establish the eigenvalues and eigenvectors (mode shapes), without specifying the amplitude of the eigenvectors. The magnitude of postbuckling deformations

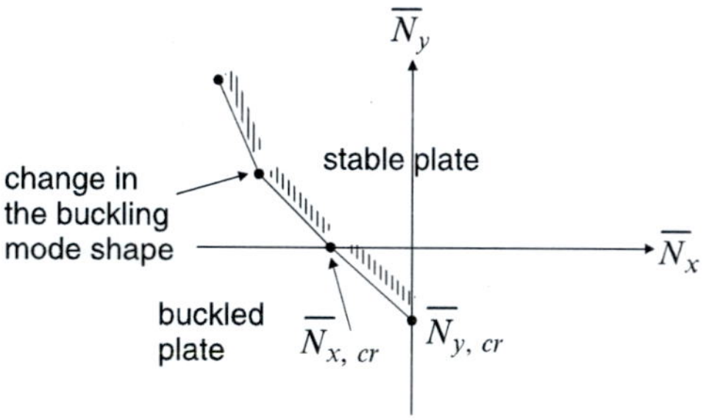

Fig. 2.13 Schematic illustration of buckling loads combinations for biaxially compressed rectangular plates

can be determined solving a geometrically nonlinear problem. Nevertheless, in numerous applications where buckling is disallowed the linear solution is sufficient.

Example 2.8: Buckling of a Simply Supported Plate Subject to Biaxial Load. The previous solution can easily be extended to the case where the plate is subject to a combination of in-plane stress resultants oriented along both x- and y-axes. Retaining the corresponding terms in (2.47) and assuming the mode shape of buckling in the form (k) we obtain the following equation for critical combinations of applied loads:

$$D\left(\alpha_m^2 + \beta_n^2\right)^2 + \bar{N}_x\alpha_m^2 + \bar{N}_y\beta_n^2 = 0 \tag{k}$$

Equation j can be obtained as a particular case of (k) where $\bar{N}_y = 0$. As is easily observed from (m), instability is possible even if the load along one of the axes is tensile. The results available from (k) are schematically shown in Fig. 2.13 that illustrates the combinations of applied stress resultants corresponding to instability. The combinations of the applied stress resultants to the right and above the boundary correspond to a stable plate. The combinations to the left and below the boundary cause instability. The buckling combinations can be specified by assuming the value of one of the stress resultants, say $\bar{N}_y$. Then the couples of indices $(m,\ n)$ are varied to find the smallest absolute value $\bar{N}_x$. The change in the buckling mode shape referred to in Fig. 2.13 implies that the combinations of applied loads corresponding to buckling occur at a different couple $(m,\ n)$ than that associated with the adjacent branch of the boundary. The values $\bar{N}_{x,cr}$ and $\bar{N}_{y,cr}$ represent buckling loads for the plate uniaxially compressed along the x- and y-axes, respectively.

2.7 Application of the Rayleigh-Ritz Method and Galerkin Procedure to Bending and Buckling Problems

In numerous situations it is impossible to exactly satisfy both the equation of equilibrium as well as the boundary conditions. An example is found in plates without a couple of parallel simply supported edges that cannot be analyzed by the Levy method. Structures that deny a closed form or exact analytical solution are often analyzed using approximate methods, such as the Rayleigh-Ritz and Galerkin methods introduced in Chap. 1. These methods are illustrated below for the cases where the plate is subject to an arbitrary pressure and in-plane loads and accounting for a possible elastic foundation.

The strain energy of a rectangular plate experiencing elastic, geometrically linear deformations is given by (2.33). The potential energy of pressure $p(x, y)$ and the energy of applied in-plane stress resultants are given by (1.68) and (1.89), respectively. These terms are employed in solutions utilizing the Rayleigh-Ritz method.

The Galerkin procedure was briefly described in Sect. 1.5. The linear equations of equilibrium being uncoupled, only the last equation (1.66) reflecting equilibrium of forces in the transverse to the plate direction has to be considered. For example, if a rectangular plate ($0 \le x \le a,\ 0 \le y \le b$) resting on a linear elastic foundation is subject to a combination of transverse pressure and in-plane stress resultants, this equation results in a system of equations

$$\int_0^b \int_0^a \left(D\nabla^4 w - p - \bar{N}_x \frac{\partial^2 w}{\partial x^2} - 2\bar{N}_{xy} \frac{\partial^2 w}{\partial x \partial y} - \bar{N}_y \frac{\partial^2 w}{\partial y^2} + kw \right) f_{mn}(x, y) dx dy = 0 \tag{2.48}$$

where

$$w = \sum_{m=1}^{M} \sum_{n=1}^{N} W_{mn} f_{mn}(x, y) \tag{2.49}$$

In (2.49), $f_{mn}(x, y)$ are functions that satisfy all boundary conditions.

The following two examples illustrate the application of the Rayleigh-Ritz method to the analysis of buckling of a plate subject to in-plane axial and shear loads and the Galerkin procedure for a clamped plate subject to an arbitrary distributed pressure.

Example 2.9: Buckling of a Plate Subject to In-Plane Axial and Shear Loads. Consider a simply supported plate subject to a combination of in-plane axial stress resultant $\bar{N}_x$ and in-plane shear stress resultant $\bar{N}_{xy}$. An example of such situation is encountered in panels of the web of an I-beam subject to bending load and supported by flanges and brackets as is shown in Fig. 2.14 (the brackets are used to enhance stability of the web and the compressed flange). In such case, each web panel is subject to a combination of variable through the height axial stresses and in-plane shear stresses. In numerous applications the bending moment varies with the axial x-coordinate. However, the analysis can often be conducted adopting a conservative approach and using the maximum value of the bending moment acting on the particular panel. In such case, the corresponding stress resultants $\bar{N}_x$ are

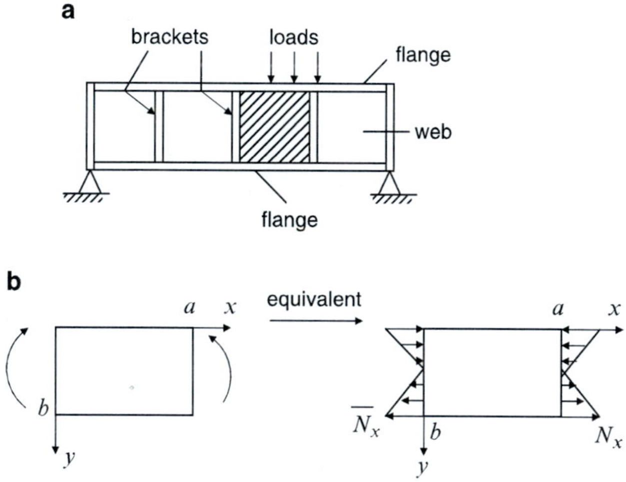

Fig. 2.14 Simply supported section of the web of an I-beam between flanges and brackets (case (**a**): *shaded*) that can buckle as a result of in-plane bending loads. The computational scheme is shown in case (**b**). Shear stresses acting in the plane of the web as a result of transverse shear force are not shown

independent of the axial coordinate. The shear stress reaches the maximum at the middle axis of the web $\left(y = \frac{b}{2}\right)$ but the variations of this stress through the depth of the web are relatively small. Accordingly, it is possible to conservatively use the maximum value of the shear stress resultant $\bar{N}_{xy}\left(\frac{b}{2}\right)$ and assume it constant through the depth of the web. In the following analysis we consider the case where the stress resultant $\bar{N}_x(y)$ is a linear function of the y-coordinate, while the shear stress resultant $\bar{N}_{xy}$ is constant.

The approach to the solution of the stability problem can combine the results for particular cases where the plate is subject to in-plane bending and compression and where the plate is subject to in-plane shear (Timoshenko and Gere 1981). The applied in-plane bending stress resultant can be represented by

$$\bar{N}_x = \bar{N}\left(1 - \alpha\frac{y}{b}\right) \tag{1}$$

where the coefficient $\alpha = 2$. If $\alpha \neq 2$ we obtain various cases where in-plane tension or compression is applied simultaneously with bending. In particular, $\alpha = 0$ corresponds to a uniformly distributed compressive load without bending, $0 < \alpha < 2$ represent various combinations of compression and in-plane bending, while $\alpha > 2$ describes cases of tension combined with in-plane bending.

The plate being simply supported, all boundary conditions are satisfied representing the buckled surface in series (2.5). The presence of in-plane shear makes it impossible to satisfy the equation of equilibrium, even if in-plane bending and tension/compression are absent. This is easily observed if we substitute (2.5) into the equilibrium equation $D\nabla^4 w = 2\bar{N}_{xy}\frac{\partial^2 w}{\partial x \partial y}$. The result is

$$D\sum_{m=1}^{M}\sum_{n=1}^{N}\left[\alpha_m^2+\beta_n^2\right]^2 W_{mn}\sin\alpha_m x\sin\beta_n y = 2\bar{N}_{xy}\sum_{m=1}^{M}\sum_{n=1}^{N} W_{mn}\cos\alpha_m x\cos\beta_n y$$

Obviously, it is impossible to eliminate trigonometric functions in the above equation, i.e. the exact solution is impossible. A similar conclusion could be obtained in case of in-plane bending, without shear. Therefore, the solution is sought using the Rayleigh-Ritz method.

The substitution of (2.5) into the expression for the strain energy (2.33) and integration yield

$$U = \frac{Dab}{8}\sum_{m=1}^{M}\sum_{n=1}^{N}\left(\alpha_m^2+\beta_n^2\right)^2 W_{mn}^2 \tag{m1}$$

The potential energy of the in-plane bending load shown here for the general case, i.e. for an arbitrary value of $\alpha \geq 0$ can be obtained from (1.89) by the substitution of (l) and (2.5):

$$\begin{aligned} V_1 = &-\frac{\bar{N}ab}{8}\sum_{m=1}^{M}\sum_{n=1}^{N}\alpha_m^2 W_{mn}^2 \\ &+\frac{\bar{N}\alpha ab}{4}\sum_{m=1}^{M}\alpha_m^2\left[\frac{1}{4}\sum_{n=1}^{N}W_{mn}^2-\frac{8}{\pi^2}\sum_{n=1}^{N}\sum_{i=1}^{N}\frac{ni}{\left(n^2-i^2\right)^2}W_{mn}W_{mi}\right] \end{aligned} \tag{m2}$$

where $n \pm i$ are odd numbers.

The potential energy of the applied shear load is available from (1.89) and (2.5) as

$$V_2 = -4\bar{N}_{xy}\sum_{m=1}^{M}\sum_{n=1}^{N}\sum_{p=1}^{M}\sum_{q=1}^{N}\frac{mnpq}{\left(m^2-p^2\right)\left(q^2-n^2\right)}W_{mn}W_{pq} \tag{m3}$$

where $m \pm p$ and $n \pm q$are odd numbers.

The buckling load combinations are available using the Rayleigh-Ritz method (Sect. 1.5) that in the present problem implies that $\frac{\partial(U+V_1+V_2)}{\partial W_{mn}} = 0$. The result is the system of linear algebraic equations (Timoshenko and Gere 1981 and Ugural 1999):

$$\begin{aligned} \frac{Dab}{4}\left(\alpha_m^2+\beta_n^2\right)^2 W_{mn} = &\frac{\bar{N}ab}{4}\alpha_m^2\left\{W_{mn}-\frac{\alpha}{2}\left[W_{mn}-\frac{16}{\pi^2}\sum_{i=1}^{N}\frac{ni}{\left(n^2-i^2\right)^2}W_{mi}\right]\right\} \\ &+8\bar{N}_{xy}\sum_{p=1}^{M}\sum_{q=1}^{N}\frac{mnpq}{\left(m^2-p^2\right)\left(q^2-n^2\right)}W_{pq} \end{aligned} \tag{n}$$

This is a system of linear homogeneous algebraic equations with respect to amplitudes W_{mn}. The combinations of buckling loads can be determined from the requirement of a nonzero solution of the system of equations (n). This requirement implies that the determinant of the system of equations (n) must be equal to zero. It is evident that applied stress resultants, i.e. $\bar{N}$ and $\bar{N}_{xy}$, are included in the coefficients of the system of equations (n), so that their buckling combinations are available from the equation equating the determinant to zero. As a result of the presence of in-plane shear loads and a dependence of the in-plane load on the y-coordinate, series (2.5) can approximate the mode shape of buckling only if they include a significant number of terms. The solutions for particular cases where the plate is subject to a single type of loads, i.e. in-plane bending or shear are available in the book of Timoshenko and Gere (1981).

Example 2.10: Bending of a Clamped Rectangular Plate on a Linear Elastic Foundation Subject to an Arbitrary Distributed Pressure. Consider a clamped rectangular plate of in-plane dimensions $2a \times 2b$ subject to an arbitrary distributed pressure $p(x, y)$. This problem was solved by the Galerkin procedure for a clamped plate subject to a uniform pressure by Timoshenko and Woinowsky-Krieger (1959) and by Ventsel and Krauthammer (2001) whose solution is reproduced here.

Considering a plate with the origin of coordinates at the center, it can easily be verified that a deflection represented by the following expression identically satisfies the boundary conditions along the edges $x = \pm a,\ \ y = \pm b$:

$$w = \sum_{m=0}^{M}\sum_{n=0}^{N} W_{mn} f_{mn}(x, y) = \sum_{m=0}^{M}\sum_{n=0}^{N} W_{mn}\left(x^2 - a^2\right)^{2+m}\left(y^2 - b^2\right)^{2+n} \tag{o}$$

Limiting the analysis to the first term in series (o) and substituting it into (2.48) we obtain

$$\begin{aligned} f_1 W_{00} &= f_2 \\ f_1 &= 20.805 D a^5 b^5 \left(a^4 + \frac{4}{7}a^2 b^2 + b^4\right) \\ f_2 &= \int_0^b \int_0^a p(x, y)\left(x^2 - a^2\right)^2\left(y^2 - b^2\right)^2 dx dy \end{aligned} \tag{p}$$

In particular, if a square clamped plate without a foundation is subject to a uniform pressure, the maximum deflection at the center of the plate is $w_{\max} = 0.0213\frac{p_0 a^4}{D}$ exceeding the exact solution by about 5%.

The stress couples corresponding to the deflection modeled by the first term in series (o) can be determined from (1.58):

$$M_x = -4DW_{00}\left[\left(3x^2 - a^2\right)\left(y^2 - b^2\right)^2 + \nu\left(x^2 - a^2\right)^2\left(3y^2 - b^2\right)\right]$$

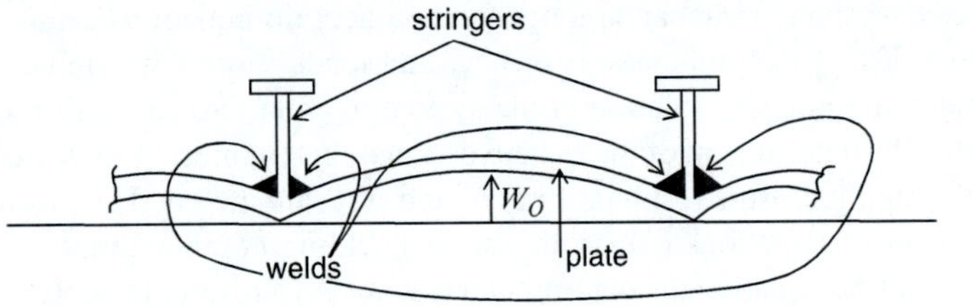

Fig. 2.15 Initial imperfections in fillet welded plates (plate welded to stringers)

$$M_y = -4DW_{00}\left[\left(x^2-a^2\right)^2\left(3y^2-b^2\right)+\nu\left(3x^2-a^2\right)\left(y^2-b^2\right)^2\right]$$

$$M_{xy} = -16D\left(1-\nu\right)W_{00}xy\left(x^2-a^2\right)\left(y^2-b^2\right) \tag{q}$$

The maximum stresses are given by (2.11). The presence of an elastic foundation does not explicitly affect the stresses in (2.11). However, the foundation is incorporated in the stress analysis through its effect on the amplitude value of the deflection, i.e. W_{00}.

2.8 Effect of Initial Imperfections on Bending and Buckling of Rectangular Plates

Initial imperfections in metallic plates often result from postwelding deformations. A detailed discussion of imperfections in welded plates was presented in the monograph of Masubuchi (1980). It should be noted that fillet welds are typical joints in plated structures. Postwelding deformations in the plate that is fillet welded to unidirectional stringers are schematically shown in Fig. 2.15.

The analysis of a plate with initial imperfections $w_0(x, y)$ that is subject to transverse pressure and/or in-plane compression requires adjustments to the governing equations. Bending stresses in an unloaded imperfect plate are equal to zero. Accordingly, the terms contributed by bending stress couples in the equilibrium equation are not affected by the presence of imperfections if deflections w are counted from the imperfect shape. However, the situation is different when we account for the effect of the applied in-plane stress resultants. The difference can easily be deduced from Fig. 1.10 replacing the deviation from the flat surface w for the perfect plate with $w_0 + w$ for the plate with the initial imperfection. In the latter case, stress resultants produce the projections on the z-axis that account for the total deviation of the plate from the flat configuration. In particular, the projection of the stress resultant $\bar{N}_x$ becomes

$$-\bar{N}_x\frac{\partial\left(w_0+w\right)}{\partial x}dy+\bar{N}_x\left[\frac{\partial\left(w_0+w\right)}{\partial x}+\frac{\partial^2\left(w_0+w\right)}{\partial x^2}dx\right]dy$$

$$= \bar{N}_x \frac{\partial^2 (w_0 + w)}{\partial x^2} dxdy \tag{2.50}$$

where we assume that in-plane forces remain independent of the in-plane coordinate and equal to the applied loads (this assumption is valid in a geometrically linear problem). The contributions of other applied stress resultants can be obtained in a similar manner.

The substitution of the projections of the applied stress resultants into the equation of equilibrium of an imperfect plate results in

$$D\nabla^4 w - \bar{N}_x \frac{\partial^2 (w_0 + w)}{\partial x^2} - 2\bar{N}_{xy} \frac{\partial^2 (w_0 + w)}{\partial x \partial y} - \bar{N}_y \frac{\partial^2 (w_0 + w)}{\partial y^2} = p(x, y) \tag{2.51}$$

It is impractical to search for the analytical solution in case where the plate is subject to in-plane shear. However, we can obtain a closed-form result if the plate is subject to axial loads acting in the x- and y-directions. Accordingly, the following solution refers to the situation where $\bar{N}_{xy} = 0$.

Boundary conditions are not affected by the presence of initial imperfections. Therefore, if the plate is simply supported the solution can be sought in the form (2.5). The imperfect shape of the plate can be represented in double Fourier series similar to (2.5):

$$w_0 = \sum_{m=1}^{M} \sum_{n=1}^{N} \overset{0}{W}_{mn} \sin \alpha_m x \sin \beta_n y \tag{2.52}$$

where amplitudes $\overset{0}{W}_{mn}$ are assumed known.

The substitution of series (2.5), (2.52) and (2.3) into (2.51) yields uncoupled equations for the amplitudes of harmonics in the series for deflections:

$$W_{mn} = \frac{p_{mn} - \left(\bar{N}_x \alpha_m^2 + \bar{N}_y \beta_n^2\right) \overset{0}{W}_{mn}}{D\left(\alpha_m^2 + \beta_n^2\right)^2 + \bar{N}_x \alpha_m^2 + \bar{N}_y \beta_n^2} \tag{2.53}$$

Several conclusions are available from (2.53):

1. If the transverse pressure is absent and the perfect plate is subject to compression, the buckling combinations of in-plane stress resultants are available from the requirement that the denominator of the right side of this equation must be equal to zero. This implies that the linear buckling problem represents a particular case of bending of an imperfect plate subject to compression (and/or in-plane shear) and transverse loads.

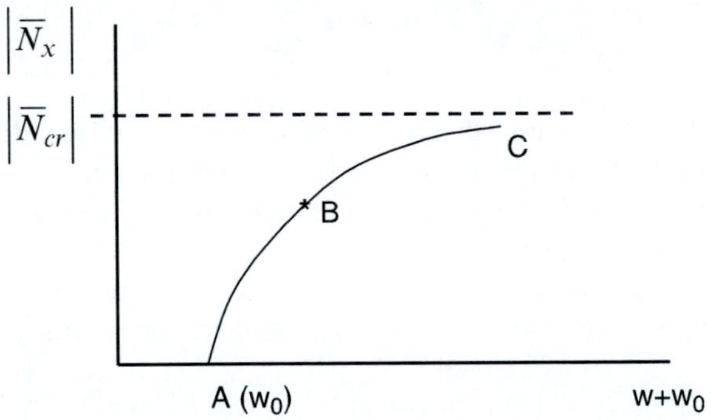

Fig. 2.16 A qualitative response of an imperfect plate to in-plane compression. $A =$ imperfect unloaded plate, $C =$ as the load approaches the buckling value, deflections asymptotically approach infinite values, $B =$ in reality, the plate collapses due to a load that is smaller than the buckling value as a result of the loss of strength

2. If the plate without imperfections is subject to transverse pressure and in-plane compression equation (2.53) reduces to (2.39).
3. Even if transverse pressure is absent, an imperfect plate subject to in-plane compression ($\bar{N}_i < 0$) experiences bending, rather than buckling. However, if the applied in-plane loads are tensile ($\bar{N}_i > 0$), the plate actually becomes "flatter". While the physical reason for such "flattening" of the plate is obvious, mathematically, the phenomenon is due to opposite signs of $\overset{0}{W}_{mn}$ and W_{mn} in case of tensile in-plane loads.
4. If in-plane loads are absent, deflections from the imperfect surface of the plate subject to transverse pressure are found from the correspondingly simplified equation (2.53) that yields the result identical with (2.8). Thus, initial imperfections do not affect bending as long as in-plane loads are absent and the problem is linear.

The stresses in the plate with imperfections subject to a combination of transverse pressure and in-plane axial compression represent the sum of stresses due to applied compressive loads that are uniformly distributed through the thickness and bending stresses. The corresponding expressions for the stresses are identical to (2.40) where the effect of initial imperfections is incorporated through the amplitudes W_{mn} determined by (2.53).

Extreme values of stresses occur on the surfaces of the plate, i.e. at $z = \pm\frac{h}{2}$, as was the case in perfectly flat plates. However, in-plane coordinates of the location of the extreme stresses can be influenced by the shape of the initial imperfection. For example, if a flat plate fails at $x = \frac{a}{2}$, $y = \frac{b}{2}$, this does not mean that an imperfect plate of the same geometry and subject to the same loading fails at the same location.

A qualitative illustration of the response of an imperfect plate subject to in-plane compression is depicted in Fig. 2.16. As the load increases, the imperfect plate experienced increasing deflections from the initial shape following the path ABC. If the entire load-deflection response was linear, the deflections would approach infinity as the load asymptotically approaches the buckling value. However, in reality, the deflection cannot increase indefinitely due to physical and geometric nonlinearities. As schematically shown in the figure, the plate collapses at point B as a result of yielding of the material. Such collapse may occur at geometrically linear deformations (i.e., if the deflections are small enough to neglect geometric nonlinearity), if the plate is stiff. In a thin and flexible plate, the analysis may require using nonlinear strain-displacement relationships as the plate approaches the collapse state.

If the magnitude of the imperfection can be predicted from experience or statistical data, while the exact shape is too difficult or too expensive to measure, it is safe to assume that this shape corresponds to the shape of deformation of the otherwise identical perfect plate subject to the same load as the imperfect plate. In the case of in-plane compression or shear the mode shape of imperfections should coincide with the buckling mode of the perfect plate. Usually, this represents the "worst case scenario," predicting failure at the smallest combination of applied loads and resulting in the safest design.

2.9 Effect of Stringers on Bending and Buckling of Plates

Plates employed in aerospace, mechanical, marine and civil engineering structures are often reinforced by ribs. Accordingly, a typical plated structure consist of a plate supported by "rigid" frames and/or bulkheads and numerous ribs (stringers) providing local support (Fig. 2.17). These stringers are usually flexible, so that they bend (or buckle) together with the plate. Accordingly, the total deformation of the plate includes:

1. Deformation of the reinforced plate jointly with the stringers;
2. Local deformation of sections of the plate between parallel stringers.

These deformation components are not independent. "Local" deformations of the sections of the plate between parallel stringers are superimposed on the "global" deformation of the plate together with the stringers. Note that the failure of stringers

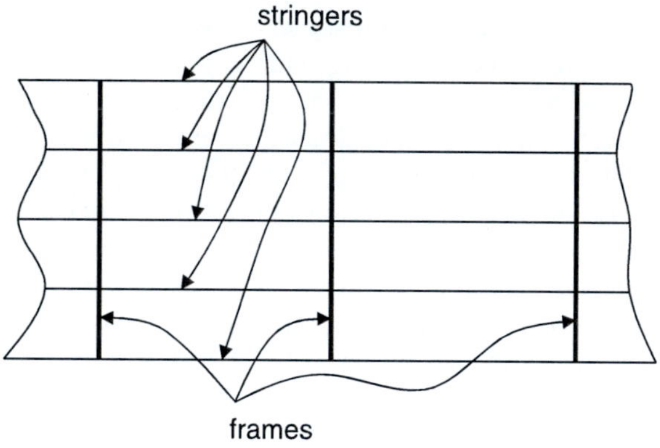

Fig. 2.17 Plate supported by transverse frames and longitudinal stringers

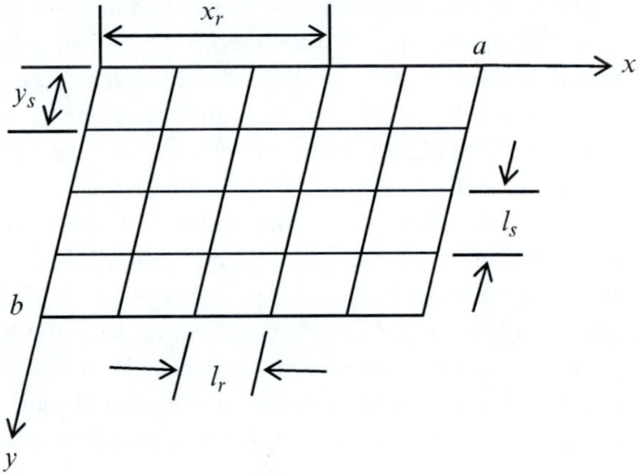

Fig. 2.18 Rectangular plate supported by stringers parallel to the edges. The edges of the plate are simply supported by frames or bulkheads

is usually unlikely in reinforced structures, but it can be analyzed by evaluating maximum stresses in the stringers that are available from either numerical or analytical solution.

In the buckling problem, it is necessary to compare buckling loads corresponding to global and local buckling and select the smaller result as the actual buckling load. The subsequent postbuckling response has to be analyzed using a nonlinear theory and accounting for the interaction between global and local deformations. Such postbuckling analysis is usually carried by a numerical (finite element or finite difference) method.

Note that the solution of the problem of buckling of a simply supported rectangular plate reinforced by one stringer oriented along the direction of applied compressive stresses incorporating both global and local deformations was reported by Timoshenko in 1915. This solution developed by Timoshenko is illustrated in Sect. 2.10. The formulation presented in the present paragraph is suitable to the solution of the global problem, although local deformations of the plate between the stringers may be incorporated through an appropriate choice of the expression for the displacements of the plate.

Consider a rectangular plate reinforced by stringers oriented in two mutually perpendicular directions parallel to the edges (Fig. 2.18). The following formulation can address both bending and buckling problems, i.e. the load can be either transverse pressure or axial compression in one or both x- and y-directions. Stringers usually have an open profile and negligible stiffness in the direction perpendicular to their axis. Accordingly, neglecting torsional stiffness of such stringers, it is justified to assume that they contribute only to the stiffness in the plane of their axes. The approach demonstrated below can be applied to the case of a small number of stringers ("discrete" stringers approach) as well as if stringers are numerous and closely spaced. In the latter situation, the so-called "smearing technique" is applicable smearing the stiffness of stringers over the plate surface. The corresponding solutions have been developed for both plates and shells; mentioned here are some of the first contributions made by Baruch and Singer (1963), Hedgepeth and Hall

(1965), Block et al. (1965) and Singer et al. (1965). While these studies refer to cylindrical shells, the solution for flat plates is available as a particular case (e.g., Birman 1993).

If the stringers are not symmetric about the middle plane of the plate, as is typical in applications, it is necessary to account for coupling between bending and in-plane displacements. Linear equations of equilibrium of a reinforced plate subject to a combination of static pressure and in-plane tensile or compressive stress resultants are available from (2.37):

$$\begin{aligned}
&\frac{\partial N_x}{\partial x}+\frac{\partial N_{xy}}{\partial y}=0\\
&\frac{\partial N_{xy}}{\partial x}+\frac{\partial N_y}{\partial y}=0\\
&\frac{\partial^2 M_x}{\partial x^2}+2\frac{\partial^2 M}{\partial x\partial y}+\frac{\partial^2 M_y}{\partial y^2}+\bar{N}_x\frac{\partial^2 w}{\partial x^2}+\bar{N}_y\frac{\partial^2 w}{\partial y^2}=-p
\end{aligned}\tag{2.54}$$

The stress resultants and stress couples are written in the form that is usually employed in the analysis of anisotropic or composite structures (for more details, see Chap. 5):

$$\begin{Bmatrix} N \\ M \end{Bmatrix}=\begin{bmatrix} A & B \\ B & D \end{bmatrix}\begin{Bmatrix} \varepsilon^0 \\ \kappa \end{Bmatrix}\tag{2.55}$$

where the vectors of stress resultants, stress couples, middle-plane strains and changes of curvature and twist are introduced in Chap. 1. The stiffness terms are composed of the contributions of the skin and stringers (in the following, single and double primes refer to the skin and stringers, respectively):

$$\{A_{ij},B_{ij},D_{ij}\}=\{A'_{ij},B'_{ij},D'_{ij}\}+\{A''_{ij},B''_{ij},D''_{ij}\}\tag{2.56}$$

In (2.56), the stiffness terms contributed by the isotropic skin are defined as

$$A'_{ij}=\begin{bmatrix} A'_{11} & A'_{12} & 0 \\ A'_{12} & A'_{22} & 0 \\ 0 & 0 & A'_{66} \end{bmatrix}=\int_{-\frac{h}{2}}^{\frac{h}{2}}\begin{bmatrix} \frac{E}{1-\nu^2} & \frac{\nu E}{1-\nu^2} & 0 \\ \frac{\nu E}{1-\nu^2} & \frac{E}{1-\nu^2} & 0 \\ 0 & 0 & G \end{bmatrix}dz=h\begin{bmatrix} \frac{E}{1-\nu^2} & \frac{\nu E}{1-\nu^2} & 0 \\ \frac{\nu E}{1-\nu^2} & \frac{E}{1-\nu^2} & 0 \\ 0 & 0 & G \end{bmatrix}$$

$$B'_{ij}=\begin{bmatrix} B'_{11} & B'_{12} & 0 \\ B'_{12} & B'_{22} & 0 \\ 0 & 0 & B'_{66} \end{bmatrix}=\int_{-\frac{h}{2}}^{\frac{h}{2}}\begin{bmatrix} \frac{E}{1-\nu^2} & \frac{\nu E}{1-\nu^2} & 0 \\ \frac{\nu E}{1-\nu^2} & \frac{E}{1-\nu^2} & 0 \\ 0 & 0 & G \end{bmatrix}zdz=0$$

$$D'_{ij}=\begin{bmatrix} D'_{11} & D'_{12} & 0 \\ D'_{12} & D'_{22} & 0 \\ 0 & 0 & D'_{66} \end{bmatrix}=\int_{-\frac{h}{2}}^{\frac{h}{2}}\begin{bmatrix} \frac{E}{1-\nu^2} & \frac{\nu E}{1-\nu^2} & 0 \\ \frac{\nu E}{1-\nu^2} & \frac{E}{1-\nu^2} & 0 \\ 0 & 0 & G \end{bmatrix}z^2dz=\frac{h^3}{12}\begin{bmatrix} \frac{E}{1-\nu^2} & \frac{\nu E}{1-\nu^2} & 0 \\ \frac{\nu E}{1-\nu^2} & \frac{E}{1-\nu^2} & 0 \\ 0 & 0 & G \end{bmatrix}\tag{2.57}$$

where the integrals are taken with respect to the z-coordinate counted from the middle plane of the plate.

Physically, terms A'_{ij} represent extensional stiffness of the plate (without accounting for the contribution of the stringers), terms B'_{ij} reflect so-called coupling stiffness between bending and in-plane stretching and shear contributions, and D'_{ij} are bending stiffness terms. The reason coupling terms are absent in the isotropic plate is related to their symmetry about the middle plane. It is evident that the contribution of the skin to the stress resultants and couples is identical to that in the unreinforced plate being given by (1.58) where the bending stiffness corresponds to (1.59).

It is now possible to evaluate the contribution of the stringers to the bending response of the plate. Based on the previous discussion, the stringers contribute to the stiffness exclusively in their axial direction. Thus, the contribution of the stringers is

$$
\begin{aligned}
N''_x &= \sum_s \delta\left(y - y_s\right) \int_z \sigma''_x dz \\
N''_y &= \sum_r \delta\left(x - x_r\right) \int_z \sigma''_y dz \\
M''_x &= \sum_s \delta\left(y - y_s\right) \int_z \sigma''_x z dz \\
M''_y &= \sum_r \delta\left(x - x_r\right) \int_z \sigma''_y z dz
\end{aligned}
\tag{2.58}
$$

where σ''_x and σ''_y are the stresses in the stringers oriented in the x- and y-directions, respectively, $\delta\left(y - y_s\right)$ and $\delta\left(x - x_r\right)$ are Dirac delta functions, and y_s and x_r the coordinates of the corresponding stringers. Note that torsional stiffness of the stringers being neglected, they do not affect in-plane shear stress resultant and the twist stress couple.

The stringers deform together with the plate; accordingly, the stresses in (2.58) are

$$
\begin{aligned}
\sigma''_x &= E\varepsilon''_x = E\left(\frac{\partial u_0}{\partial x} - z\frac{\partial^2 w}{\partial x^2}\right) \\
\sigma''_y &= E\varepsilon''_y = E\left(\frac{\partial v_0}{\partial y} - z\frac{\partial^2 w}{\partial y^2}\right)
\end{aligned}
\tag{2.59}
$$

The substitution of (2.59) into (2.58) yields

$$N''_x = \sum_s \delta(y - y_s) E \left(A_s \frac{\partial u_0}{\partial x} - F_s \frac{\partial^2 w}{\partial x^2} \right)$$

$$N''_y = \sum_r \delta(x - x_r) E \left(A_r \frac{\partial v_0}{\partial y} - F_r \frac{\partial^2 w}{\partial y^2} \right)$$

$$M''_x = \sum_s \delta(y - y_s) E \left(F_s \frac{\partial u_0}{\partial x} - I_s \frac{\partial^2 w}{\partial x^2} \right)$$

$$M''_y = \sum_r \delta(x - x_r) E \left(F_r \frac{\partial v_0}{\partial y} - I_r \frac{\partial^2 w}{\partial y^2} \right) \tag{2.60}$$

In (2.60) $A_s,\ F_s,\ I_s$ are the cross sectional area, and the first and second moments of the stringer oriented along $y = y_s$ about the middle plane of the skin, respectively, and $A_r,\ F_r,\ I_r$ are the area and moments of the stringer oriented along $x = x_r$. Accordingly, the expressions for the total stress resultants and stress couples, accounting for the contributions of both the skin and the stringers, are

$$\begin{aligned} N_x &= N'_x + N''_x \\ N_y &= N'_y + N''_y \\ M_x &= M'_x + M''_x \\ M_y &= M'_y + M''_y \end{aligned} \tag{2.61}$$

where the contribution of the skin (terms with a single prime) is given by equations (2.56) with extensional, coupling and stiffness terms as per (2.57).

Now we can evaluate the stiffness terms contributed by the stringers, i.e. $\{A''_{ij},\ B''_{ij},\ D''_{ij}\}$. As follows from (2.56) and (2.60), the contribution of the stringers oriented in the x-direction to the stiffness of the plate is

$$\{A''_{11},\ B''_{11},\ D''_{11}\} = \sum_s \delta(y - y_s) E \{A_s,\ F_s,\ I_s\} \tag{2.62}$$

Similarly, a system of stringers oriented in the y-direction along the lines $x = x_r$ contributes the stiffness

$$\{A''_{22},\ B''_{22},\ D''_{22}\} = \sum_r \delta(x - x_r) E \{A_r,\ F_r,\ I_r\} \tag{2.63}$$

In a particular case of a large number of closely spaced identical stringers in one or both directions, the so-called smeared stiffeners technique can be used. According to this technique

$$\delta\left(y-y_s\right)=\frac{1}{l_s}$$

$$\delta\left(x-x_r\right)=\frac{1}{l_r} \tag{2.64}$$

l_s and l_r being the spacings of the corresponding stringers (see Fig. 2.18). This simplification affects terms in Eqs. 2.58, 2.60, 2.62 and 2.63. Note that using the smeared stiffeners techniques automatically results in studying only the global response of the plate since local deformations between the stringers cannot be modeled.

Now, combining the stress resultants and couples contributed by the skin and by the stringers, we obtain the equations of equilibrium that coincide with those for an orthotropic plate with a different stiffness in the directions of the x- and y-axes:

$$A_{11}u_{,xx}+A_{66}u_{,yy}+\left(A_{12}+A_{66}\right)v_{,xy}-B_{11}w_{,xxx}-\left(B_{12}+2B_{66}\right)w_{,xyy}=0$$

$$\left(A_{12}+A_{66}\right)u_{,xy}+A_{66}v_{,xx}+A_{22}v_{,yy}-\left(B_{12}+2B_{66}\right)w_{,xxy}-B_{22}w_{,yyy}=0$$

$$D_{11}w_{,xxxx}+2\left(D_{12}+2D_{66}\right)w_{,xxyy}+D_{22}w_{,yyyy}-B_{11}u_{,xxx}-\left(B_{12}+2B_{66}\right)$$

$$\times\left(u_{,xyy}+v_{,xxy}\right)-B_{22}v_{,yyy}-\bar{N}_xw_{,xx}-\bar{N}_yw_{,yy}=p \tag{2.65}$$

In (2.65), $(\ldots)_{,i}\equiv\frac{\partial(\ldots)}{\partial i}$, $i=x,y$. This notation for differential operators is often employed in the theory of plates. Accordingly, we present it here in addition to the notation used throughout the book.

Boundary conditions written in terms of displacements, stress resultants and stress couples are not explicitly affected by the presence of stringers. However, if the stringers are extended to the edges, as they usually are, the stress resultants and stress couples in these conditions are given by (2.61) accounting for the contribution of the stringers.

In rare situations where stringers are joined to the plate symmetrically, i.e. identical stringers are placed against each other on the opposite surfaces of the plate, coupling between in-plane and transverse displacements disappears. In such case, only the third equation of equilibrium (2.65) should be considered. Upon the corresponding simplifications (eliminating the coupling terms), this equation becomes

$$\left[\bar{D}_{11}\frac{\partial^4}{\partial x^4}+2D\frac{\partial^4}{\partial x^2\partial y^2}+\bar{D}_{22}\frac{\partial^4}{\partial y^4}\right]w-\bar{N}_x\frac{\partial^2 w}{\partial x^2}-\bar{N}_y\frac{\partial^2 w}{\partial y^2}=p \tag{2.66}$$

where D is the bending stiffness of the plate and

$$\bar{D}_{11} = D + \sum_{s} \delta\,(y - y_s) EI_s$$

$$\bar{D}_{22} = D + \sum_{r} \delta\,(x - x_r) EI_r \tag{2.67}$$

In the case where the plate is simply supported, the solution of the equation of equilibrium (2.66) is sought in the form (2.5). If the pressure is represented by series (2.3), the solution is

$$W_{mn} = \frac{p_{mn}}{\bar{D}_{11}\alpha_m^4 + 2D\alpha_m^2\beta_n^2 + \bar{D}_{22}\beta_n^4 + \bar{N}_x\alpha_m^2 + \bar{N}_y\beta_n^2} \tag{2.68}$$

Consider the case where stringers are located on one surface of the plate as is typical in applications. Then we have to integrate the system of equations (2.65) subject to boundary conditions. The analytical solution is available that satisfies the conditions of simple support along all boundaries:

$$\begin{aligned} x = 0,\ x = a \quad & w = M_x = N_x = v_0 = 0 \\ y = 0,\ y = b \quad & w = M_y = N_y = u_0 = 0 \end{aligned} \tag{2.69}$$

These conditions are satisfied if the mode shape of deformation is represented by the following double Fourier series:

$$u_0 = \sum_{m=1}^{M}\sum_{n=1}^{N} U_{mn} \cos\alpha_m x \sin\beta_n y$$

$$v_0 = \sum_{m=1}^{M}\sum_{n=1}^{N} V_{mn} \sin\alpha_m x \cos\beta_n y$$

$$w = \sum_{m=1}^{M}\sum_{n=1}^{N} W_{mn} \sin\alpha_m x \sin\beta_n y \tag{2.70}$$

Pressure applied to the plate can be represented in series (2.3).

The substitution of (2.70) and (2.3) into (2.65) yields

$$\begin{bmatrix} S_{11mn} & S_{12mn} & S_{13mn} \\ S_{12mn} & S_{22mn} & S_{23mn} \\ S_{13mn} & S_{23mn} & S_{33mn} \end{bmatrix} \begin{Bmatrix} U_{mn} \\ V_{mn} \\ W_{mn} \end{Bmatrix} = \begin{Bmatrix} 0 \\ 0 \\ p_{mn} \end{Bmatrix} \tag{2.71}$$

where

$$S_{11mn} = \alpha_m^2 A_{11} + \beta_n^2 A_{66} \qquad S_{12mn} = \alpha_m \beta_n (A_{12} + A_{66})$$
$$S_{13mn} = -\alpha_m \left[\alpha_m^2 B_{11} + \beta_n^2 (B_{12} + 2B_{66})\right]$$
$$S_{22mn} = \alpha_m^2 A_{66} + \beta_n^2 A_{22}$$
$$S_{23mn} = -\beta_n \left[\alpha_m^2 (B_{12} + 2B_{66}) + \beta_n^2 B_{22}\right]$$
$$S_{33mn} = \alpha_m^4 D_{11} + 2\alpha_m^2 \beta_n^2 (D_{12} + 2D_{66}) + \beta_n^4 D_{22} + \alpha_m^2 \bar{N}_x + \beta_n^2 \bar{N}_y \tag{2.72}$$

Given the applied loads, the amplitudes of deflections in series (2.70) can be evaluated from (2.71). Subsequently, the strains and stresses can be determined throughout the plate. In the case of in-plane loading, the buckling combinations of the applied stress resultants $\bar{N}_x$ and $\bar{N}_y$ corresponding to the mode shape of instability characterized by m and n deformation half-waves in the x- and y-directions, respectively, are available from the nonzero requirements to the solution of the homogeneous version of (2.71). This requirement is satisfied if the determinant of the system of equations (2.71) is equal to zero yielding the buckling equation:

$$\begin{vmatrix} S_{11mn} & S_{12mn} & S_{13mn} \\ S_{12mn} & S_{22mn} & S_{23mn} \\ S_{13mn} & S_{23mn} & S_{33mn} \end{vmatrix} = 0 \tag{2.73}$$

While buckling combinations of loads corresponding to the global overall instability of the plate involving both the skin and stringers deforming together can be determined from (2.73), local buckling of sections of the plate between stringers and buckling of stringers is not addressed by this solution.

2.10 Stability of a Simply Supported Plate Reinforced with a Single Longitudinal Stringer

Consider a plate subject to axial compression $\bar{N}_x$ and supported by a single centrally located stringer along $y = 0$ (the coordinate system is shown in Fig. 2.8). The stringer is assumed to have negligible torsional and out-of-axis stiffness, so that we account only for its flexural stiffness EI. As indicated above, this problem was first analyzed by Timoshenko (1915). The present solution is based on the monograph of Vol'mir (1967).

The solution of the equation of equilibrium for the section of the plate limited by $x = 0, x = a$ and $y = 0, y = \frac{b}{2}$ is obtained assuming simply supported edges $x = 0,\ x = a$ and $y = \frac{b}{2}$. Deformations of each section of the plate buckling together with the stringer are identical, so that the slope $\frac{\partial w}{\partial y} = 0$ along $y = 0$.

The solution is obtained by the Levy method, so that the shape of the buckled plate is represented by

$$w = Y(y)\sin\alpha_m x \tag{2.74}$$

where the function $Y(y)$ must satisfy both the equation of equilibrium as well as the boundary conditions along the edges $y = \frac{b}{2}$ and $y = 0$.

The expression (2.74) is substituted into the equation of equilibrium for the section of the plate, i.e.

$$D\nabla^4 w + \bar{N}_x \frac{\partial^2 w}{\partial x^2} = 0 \tag{2.75}$$

where the sign of the second term reflects the rule of signs adopted by Vol'mir (1967), i.e. the compressive stress resultant is positive. This yields an ordinary differential equation

$$\frac{d^4Y}{dy^4} - 2\alpha_m^2 \frac{d^2Y}{dy^2} + \alpha_m^2\left[\alpha_m^2 - \frac{\bar{N}_x}{D}\right]y = 0 \tag{2.76}$$

The solution of (2.76) can be sought as $Y(y) = e^{ky}$. The roots of the characteristic equation obtained by the substitution of $Y(y)$ into (2.76) are

$$k_{1,2} = \sqrt{\alpha_m\left(\alpha_m + \sqrt{\frac{\bar{N}_x}{D}}\right)}, \quad k_{3,4} = \sqrt{\alpha_m\left(\alpha_m - \sqrt{\frac{\bar{N}_x}{D}}\right)} \tag{2.77}$$

The analysis of the roots in (2.77) yields the conclusion that $k_{3,4}$ are imaginary numbers. Hence the solution of (2.76) can be written as

$$Y(y) = C_{1m}\cosh ky + C_{2m}\sinh ky + C_{3m}\cos\bar{k}y + C_{4m}\sin\bar{k}y \tag{2.78}$$

where C_{im} are constants of integration, $k = k_{1,2}$ and $\bar{k} = \sqrt{\alpha_m\left(\sqrt{\frac{\bar{N}_x}{D}} - \alpha_m\right)}$

The critical value of the applied compressive load can be determined by substituting (2.78) into the boundary conditions along $y = 0$ and $y = \frac{b}{2}$. Three of these conditions have already been specified:

$$Y\left(\frac{b}{2}\right) = 0, \quad \frac{d^2Y\left(\frac{b}{2}\right)}{dy^2} = 0, \quad \frac{dy(0)}{dy} = 0 \tag{2.79}$$

The forth boundary condition is derived by considering deformations of the stringer. The stringer is subject to the reaction from two adjacent sections of the plate that can be evaluated from

$$R = -D\left[\frac{\partial^3 w}{\partial y^3} + (2-\nu)\frac{\partial^3 w}{\partial x^2 \partial y}\right] \tag{2.80}$$

where the second term in the brackets in the right side is equal to zero by virtue of the other boundary condition along the same edge being $\frac{\partial w}{\partial y} = 0$. Accordingly, the equation for the stringer that deforms as a column subject to compression is

$$EI_s \frac{\partial^4 w}{\partial x^4} + \sigma_s A_s \frac{\partial^2 w}{\partial x^2} + 2D\frac{\partial^3 w}{\partial y^3} = 0 \tag{2.81}$$

where E, A_s and I_s are the modulus of elasticity, cross sectional area and the moment of inertia of the stringer, respectively, and σ_s is the compressive stress applied to the stringer. The latter stress is related to the stress resultant acting on the plate, i.e. $\bar{N}_x$, so that the corresponding term in (2.81) can be written as a function of the applied stress resultant. The factor 2 in the last term in the left side of (2.81) reflects the presence of two adjacent sections of the plate supported by the stringer.

Substituting (2.74) into (2.81) yields the fourth boundary condition

$$EI_s \alpha_m^4 Y - \sigma_s A_s \alpha_m^2 Y + 2D\frac{\partial^3 Y}{\partial y^3} = 0 \tag{2.82}$$

The buckling equation corresponding to the mode shape of deformation with m halfwaves along the x-axis is now derived by substituting (2.78) into (2.79) and (2.82) and requiring nonzero solutions of the resulting system of four homogeneous equations with respect to constants of integration C_{im}.

An alternative solution of this problem is obtained by the Rayleigh-Ritz method (e.g., Timoshenko and Gere 1981). This reference also contains tabulated results that may be useful in design of plates reinforced by one or two unidirectional stringers and subject to compression or shear loads. Whichever method is applied to the analysis, the advantage of the analytical approach considering both the deformations of the plate sections between the stringers as well as deformations of the stringers is evident since the solution automatically accounts for both global and local deformations.

2.11 Postbuckling Response of Plates

While plates subject to compression and in-plane shear loads are often considered unsafe if they buckle, in reality, this phenomenon is less dangerous to flat plates than to either columns or shells. The reason is that plates possess stable postbuckling behavior, i.e. they do not snap through to an alternative equilibrium position, but rather gradually develop postbuckling deflections as a result of an increasing load. This behavior represents an important potential safety margin for plates that is absent in columns or shells, as discussed below.

The analysis of postbuckling behavior of plates has been conducted since the pioneering work of von Karman and his associates (Karman et al. 1932). This analysis requires an analytical solution of nonlinear partial differential equations subject to boundary conditions. The accuracy of the solution sought in double Fourier series is limited by the number of terms retained in these series. Besides, it appears difficult to satisfy boundary conditions corresponding to realistic engineering problems. Accordingly, while early solutions retain their importance due to qualitative results and conclusions that are still valid today, the nonlinear analysis of plates, including their postbuckling behavior, is usually conducted using numerical methods.

Equations of equilibrium and compatibility considered in the case where the plate is subject to axial compression are available from (1.96) and (1.99), respectively. These equations are reproduced here, for convenience:

$$\frac{D}{h}\nabla^4 w = \frac{\partial^2 \varphi}{\partial x^2}\frac{\partial^2 w}{\partial y^2} + \frac{\partial^2 \varphi}{\partial y^2}\frac{\partial^2 w}{\partial x^2} - 2\frac{\partial^2 \varphi}{\partial x \partial y}\frac{\partial^2 w}{\partial x \partial y}$$

$$\frac{1}{E}\nabla^4 \varphi = \frac{\partial^2 w}{\partial x^2}\frac{\partial^2 w}{\partial y^2} - \left(\frac{\partial^2 w}{\partial x \partial y}\right)^2 \tag{2.83}$$

Kinematic boundary conditions are not affected by nonlinearity. However, static boundary conditions are affected as linearized constitutive expressions for the stress resultants and stress couples have to be replaced with their nonlinear counterparts (1.58).

If the plate is simply supported, its postbuckling deflections can be represented in double Fourier series. The complication is that while the terms in these series are uncoupled upon the substitution in the linear equation of equilibrium (e.g., Sect. 2.1), they remain coupled in the nonlinear problem. This results in a complicated and time consuming solution procedure that requires a researcher to retain a limited number of terms in the series, reducing the accuracy of the analytical solution. Furthermore, it appears that some of the boundary conditions can be satisfied only in the integral sense as is demonstrated below.

Consider the case of a simply supported plate subject to a uniform compression $\bar{N}_x$ along the x-axis. The deflections are sought in double Fourier series (2.5) satisfying the conditions of zero deflection and bending stress couple along the plate boundary. The substitution of (2.5) into the second (compatibility) equation (2.83) yields

$$\varphi = \sum_{r=0}\sum_{s=0} F_{rs}\left(W_{mn}\right)\cos\alpha_r x \cos\beta_s y + \frac{\bar{N}_x y^2}{2h} \tag{2.84}$$

where $F_{rs}\left(W_{mn}\right)$ are quadratic algebraic functions and the number of terms in the series in the right side of (2.84) is dependent on the number of terms in series (2.5).

The problem with the in-plane boundary conditions becomes evident upon the analysis of (2.84). At the loaded edges of the plate, i.e. $x = 0$ and $x = a$, the

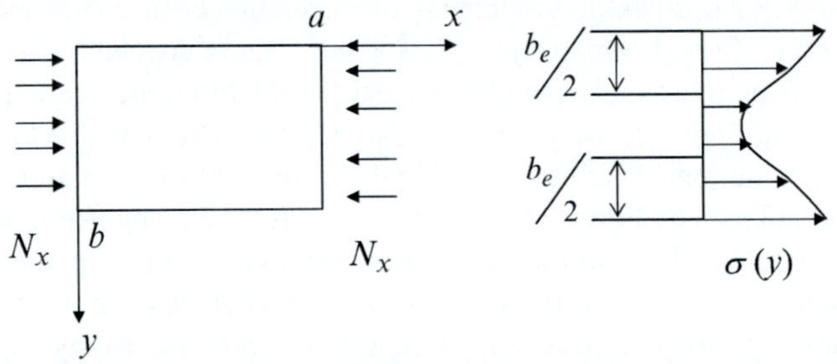

Fig. 2.19 Axially compressed plate. The stresses do not remain uniform in the postbuckling regime: they are higher in the vicinity to the edge supports and lower in the central region of the plate. The effective width introduced according to von Karman et al. (1932) is denoted by b_e

condition $N_x = \bar{N}_x$ can be satisfied only in the integral sense, i.e. according to (1.97),

$$N_x = \frac{h}{b}\int_0^b \frac{\partial^2 \varphi}{\partial y^2} dy = -\bar{N}_x \tag{2.85}$$

The second boundary condition superimposed by this solution is apparently, $N_{xy} = -h\frac{\partial^2 \varphi}{\partial x \partial y} = 0$ along all edges. However, this condition makes little sense in real structures where the edges of a simply supported plate are welded, bolted or bonded to rigid ribs or frames that constrain displacements along the edge and apply in-plane shear stresses to the plate. Thus, it is preferable to rely on numerical procedures, such as finite elements or finite difference methods that can provide a more accurate solution for nonlinear postbuckling deformations and stresses. The same comment remains valid in cases where the plate experiences nonlinear bending.

Both the analysis of postbuckling behavior of rectangular plates and experimental evidence have demonstrated that these plates do not collapse if compressive or in-plane shear loads exceed the buckling values. This indication of the capacity of plates to operate in the postbuckling load range generated significant interest, particularly, in view of the fact that columns and shells cannot operate beyond the buckling load. The reason was explained in the work of Karman et al. (1932) who pointed to a nonuniform postbuckling in-plane stress distribution across the width of a compressed plate (Fig. 2.19). Accordingly, an approach to the estimate of the ultimate load-carrying capacity of rectangular plates subject to compression is based on the replacement of the actual plate with two identical strips adjacent to its edges $y = 0$ and $y = b$, while the central underloaded strip of the plate is discounted. According to this approach, the strips of the plate adjacent to the edges are uniformly stressed and failure occurs when this stress reaches the ultimate or allowable value for the material (σ_{ult}), i.e.,

Fig. 2.20 Difference in postbuckling response of plates (*1*), columns (*2*) and cylindrical shells (*3*) subject to axial compression

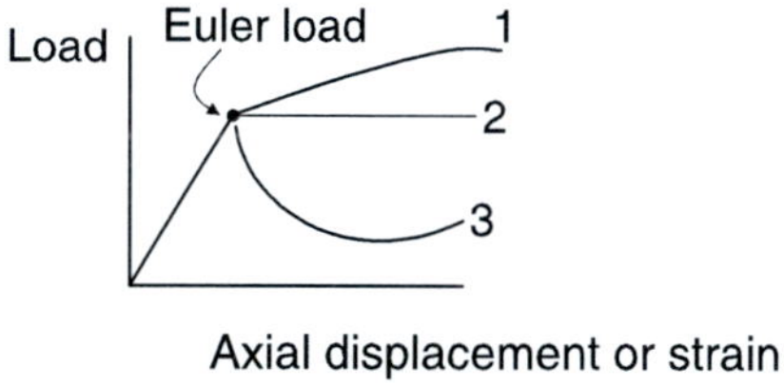

$$\bar{N}_{ult} = \sigma_{ult} b_e h \tag{2.86}$$

where b_e is the width of two strips adjacent to the edges.

Equation 2.86 can be used to evaluate the ultimate (postbuckling) load-carrying capacity of the plate if the effective width is known. There are several methods of estimating this width. For example, von Karman et al (1932) suggested the following formula:

$$b_e = b\sqrt{\frac{\sigma_{cr}}{\sigma_{ult}}} \tag{2.87}$$

where $\sigma_{cr} = \frac{\bar{N}_{x,cr}}{h}$ is the buckling value of the applied stress.

The empirical formula obtained by Winter (e.g., Bedair 2009) is

$$b_e = b\sqrt{\frac{\sigma_{cr}}{\sigma_{ult}}}\left(1 - 0.22\sqrt{\frac{\sigma_{cr}}{\sigma_{ult}}}\right) \tag{2.88}$$

Other formulas for the effective width have also been proposed.

As emphasized above, the qualitative response of flat plates in the postbuckling regime is different from those of columns and shells. This is reflected in Fig. 2.20. Upon reaching the buckling load, shells snap through to a buckled equilibrium position. The descending branch of curve 3 that schematically represents the response of a shell is unstable. Instead of following this branch the shell abruptly snaps through to the ascending branch of curve 3. This phenomenon is often accompanied by large stresses, far exceeding those allowable in the shell structure. Accordingly, the buckling load is often associated with the ultimate load of the shell. The presence of almost unavoidable initial imperfections makes shells even more vulnerable as their instability (snap through) occurs at an even lower load than the theoretical buckling value (e.g., see for details Jones 2006).

Straight columns subject to compression may support a small additional load upon buckling, but the corresponding postbuckling deformations are very large even at very small additional loads in excess of the buckling value. This implies that curve 2 characterizing the response of a column is nearly horizontal, i.e. the buckling load serves as the ultimate load for columns.

The situation is different in the case of plates that exhibit gradually increasing deformations as the load exceeds the buckling value (curve 1). Eventually, the stresses reach the yield limit and the plate collapses. However, the additional postbuckling load-carrying capacity may be quite significant, providing a designer with an implicit margin of safety.

2.12 Design Philosophy and Recommendations

Rectangular plates are encountered in more engineering applications than plates of other shapes. Exact analytical solutions for rectangular plates are available in numerous design problems. In this paragraph we comment on several aspects of design and analysis of rectangular plates the engineer should be aware of.

Boundary conditions of rectangular plates can often be modeled as simple support. In such model, the torsional stiffness of stringers, frames or bulkheads supporting the edges is neglected. This simplification is usually justified if support structures have an open profile (closed profile beams have a higher torsional stiffness, providing an elastic clamping to the plate). The assumption that the deflection of a simply supported edge is negligible is usually acceptable, except for the case where the support structure is very flexible and experiences a noticeable bending. While the analysis of plates by assumption that the edges provide simple support is often justified, a notable exception is the case where the corresponding edge supports two adjacent identical plates subject to identical loading. Then the support can be modeled as clamping, by virtue of symmetry of both geometry and load.

Plates found in applications often have initial imperfections. Such imperfections are acquired as a result of rough handling and/or postwelding deformations. The presence of initial imperfections has a relatively little effect on bending of the plate as long as deflections remain small. This is because the plate is not stressed in the imperfect position, so that only additional deformations from this position cause stresses in the plate. However, if deflections from the imperfect state are large, justifying the nonlinear analysis, the interaction between the imperfections and deflections becomes essential and it has to be accounted for.

While initial imperfections affect the bending response only in the nonlinear range, their effect in case of in-plane loading (either compression or shear) is significant, even if the imperfections are small. The classical buckling phenomenon is replaced in imperfect plates with progressive bending from the imperfect position, increasing with a larger load. The geometrically linear load-deflection curve asymptotically approaches the buckling value of the load as deflections increase to infinity. In reality, infinite deflections cannot occur since as deflections (and stresses) become large, the geometrically linear solution becomes invalid. Moreover, at large stresses, the material becomes plastic and the plate collapses at lower loads than the buckling values.

It should be emphasized that the shape and magnitude of initial imperfections are difficult and expensive to measure. Sometimes, the relevant information is available based on statistics. If the magnitude of the imperfection is known or estimated, while its shape is not determined, it is preferable to assume the most unfavorable shape. Such shape usually corresponds to the bending shape of the plate without imperfections or, in buckling problems, the buckling shape of the plate.

Stringers significantly increase the overall stiffness of the plate and are often employed to enhance its response to loading. Additionally, the size of unreinforced sections of the plate between the stringers is reduced compared to that of the same plate without stringers. Both these effects result in the capacity to sustain larger bending loads as well as a better buckling response (both a higher buckling load as well as a higher post-buckling load-carrying capacity). The weight added by stringers is usually much lower than the additional weight if the thickness of the plate is increased to achieve an equivalent improvement in the load-carrying capacity of the plate.

In the case of a plate reinforced by flexible stringers, the sections between the stringers cannot be analyzed as simply supported. Instead, the deformation of the plate includes the overall bending of the reinforced structure and local bending between the stringers. The former problem can usually be approached assuming that the boundaries of the reinforced plate are simply supported. Bending of sections between the stringers is superimposed on the global deformation. With the exception of situations where it is possible to analytically account for the combination of the global and local response, the superposition of these response patterns often requires the application of numerical models that yield the overall solution (see Sect. 2.10 for the representative case where the analytical overall solution is available). In the problem where the response is nonlinear, it is necessary to use a numerical approach since the combination of global and local nonlinear analytical solutions is not feasible.

The effect of large deformations, i.e. geometric nonlinearity, on the bending response is usually "beneficial." This is because deflections and stresses in the plate analyzed accounting for nonlinearity are smaller than those in the counterpart evaluated using the linear theory. The adaption of the linear approach introduces an additional margin of safety since the response of real plates is more accurately modeled using a geometrically nonlinear theory. In buckling problems geometric nonlinearity has to be accounted for in the postbuckling analysis. Contrary to beams and shells, the postbuckling response of plates is stable, i.e. the buckling phenomenon does not automatically imply the collapse of the structure. While designers usually avoid buckling, such stable postbuckling response implies an additional load-carrying capacity that may be important in the ultimate load analysis. As postbuckling deflections accumulate in the plate as a result of additional load, the stresses also increase, eventually reaching the yield limit. Accordingly, the ultimate load-carrying capacity should be specified accounting for a combination of a geometrically and physically nonlinear behavior of the plate (nonlinear elasto-plastic response). In less flexible plates subject to compression or in-plane shear yielding may occur prior to reaching the elastic buckling load. Then the analysis should be concerned with elasto-plastic buckling.

References

Baruch, M., & Singer, J. (1963). The effect of eccentricity on the general instability of stiffened cylindrical shells under hydrostatic pressure. *Journal of Mechanical Engineering Science, 5*, 22–27.

Bedair, O. (2009). Analytical effective width equations for limit state design of thin plates under non-homogeneous in-plane loading. *Archives of Applied Mechanics, 79*, 1173–1189.

Beer, F. A., & Johnston, E. R. (1981). *Mechanics of materials*. New York: McGraw-Hill.

Birman, V. (1993). Active control of composite plates using piezoelectric stringers. *International Journal of Mechanical Science, 35*, 387–396.

Block, D. L., Card, M. F., & Mikulas, M. M. (1965). *Buckling of eccentrically stiffened orthotropic cylinders* (NASA TN D-2960).

Bryden, J. E. et al. (1971). *The catskill-cairo experimental rigid pavement construction and material testing* (Research Report 2). New York Department of Transportation. In T. J. Pasco, Jr. Concrete pavements – Past, present, and future, Public Roads, July/August 1998, Vol. 62, No. 1.

Hendepeth, J. M., & Hall, D. B. (1965). Stability of stiffened cylinders. *AIAA Journal, 3*, 2275–2286.

Jones, R. M. (2006). *Buckling of bars, plates and shells*. Blacksburg, VA: Bull Ridge Publishing.

Kerr, A. D. (1984). On the formal development of elastic foundation models. *Ingenieur Archiv, 54*, 455–464.

Masubuchi, K. (1980). *Analysis of welded structures*. Oxford, UK: Pergamon Press.

Pasternak, P. L. (1954). *On a new method of analysis of an elastic foundation by means of two foundation constants*. Moscow: State Publishers of Literature on Building and Architecture (Gosudarstvenoe Izdatelstvo Literaturi po Stroitelstvu i Architecture) (In Russian).

Reissner, E. (1967). Note on the formulation of the plate on an elastic foundation. *Acta Mechanica, 4*, 88–91.

Roark, R. J. (1965). *Formulas for stress and strain* (4th ed.). New York: McGraw-Hill.

Singer, J., Baruch, M., & Harari, O. (1965). On the stability of eccentrically stiffened cylindrical shells under axial compression. *International Journal of Solids and Structures, 3*, 445–470.

Timoshenko, S. R. (1915). Proceedings of the Railroad Institute, Petersburg, Russia (In Russian). This work is cited in the monograph by Vol'mir, A.S. (1967). *Stability of deformable systems*, Moscow: Nauka Publishers, (In Russian).

Timoshenko, S. P., & Gere, J. M. (1981). *Stability of elastic structures* (2nd ed.). Auckland, New Zealand: McGraw-Hill.

Timoshenko, S., & Woinowsky-Krieger, S. (1959). *Theory of plates and shells* (2nd ed.). New York: McGraw-Hill.

Ugural, A. G. (1999). *Stresses in plates and shells* (2nd ed.). Boston: McGraw-Hill.

Ventsel, E., & Krauthammer, T. (2001). *Thin plates and shells*. New York: Marcel Dekker.

Vlasov, V. Z., & Leont'ev, N. N. (1966). *Beams, plates, and shells on elastic foundation* (NASA TTF-357). Washington, DC: NASA.

Vol'mir, A. S. (1967). *Stability of deformable systems*. Moscow: Nauka Publishing House (in Russian).

Von Karman, T., Sechler, E. E., & Donnel, L. H. (1932). The strength of thin plates in compression. *Transactions of ASME, 54*, 53–57.

Chapter 3
Static Problems in Isotropic Circular Plates and in Plates of Other Shapes

Although rectangular plates considered in Chap. 2 are often found in applications, other shapes of plates are also encountered in various branches of engineering. Examples of circular plates include bulkheads in submersible vehicles and in aerospace applications and end closures of pressure vessels. If circular bulkheads are reinforced by ring and radial stringers, the sections of these bulkheads between stringers represent sector plates. Triangular plates are found in isogrid structures. Elliptical plates may serve as covers for hatches in decks.

In this chapter, we concentrate on physically linear (elastic) thin plates of various shapes. Accordingly, the Kirchhoff-Love assumption of the classical (technical) plate theory remains valid.

Governing equations of plates of various nonrectangular shapes may differ from those for rectangular plates reflecting the actual geometry of the structure and the convenient coordinate system. However, the list of equations necessary for the analysis remains without a change, including:

- Kinematic relations;
- Strain-displacement relations;
- Constitutive law, i.e. elastic material Hookean relations, elasto-plastic equations, etc.;
- Expressions for stress couples and stress resultants;
- Equations of motion or equilibrium;
- Boundary conditions.

In the energy formulation, equations of motion or equilibrium are replaced with the expressions for the strain energy, the energy of the applied load, and the kinetic energy (in dynamic problems). The other alternative formulation utilizing the stress function requires us to specify the compatibility equation in the appropriate coordinate system. An outline of the governing equations in the case of a circular plate is demonstrated in the next paragraph.

V. Birman, *Plate Structures*, Solid Mechanics and Its Applications 178,
DOI 10.1007/978-94-007-1715-2_3, © Springer Science+Business Media B.V. 2011

3.1 Governing Equations of Circular Plates

Consider a thin circular plate shown in Fig. 1.6. Kinematic equations representing displacements as functions of polar coordinates are given by (1.32). The strain-displacement relations are (1.39) for geometrically linear asymmetric deformations and (1.40) for geometrically nonlinear deformations. If the strains are decomposed into the components in the middle plane of the plate and the changes of curvature and twist, the corresponding components are given by (1.41) and (1.42), respectively. The simplified strain-displacement relationships for the case of axisymmetric deformations, such that the displacements, strains and stresses are independent of the circumferential θ coordinate, are (1.43) and (1.44).

All fundamental equations necessary for the analysis of plates in the polar coordinate system can be directly obtained from the corresponding equations in the Cartesian coordinate system introduced in Chap. 1 replacing the derivatives of the functions in Cartesian coordinates with their counterparts in the polar coordinate system. The relationships between the first derivatives of functions in two coordinate systems have already been given by Eq. 1.36. Additionally, we illustrate here without a derivation the laws governing the transformation from the Cartesian to the polar coordinate system for second derivatives as well as for the Laplacian operator $\nabla^2(\ldots)$:

$$
\begin{aligned}
\frac{\partial^2(\ldots)}{\partial x^2} &= \frac{\partial^2(\ldots)}{\partial r^2}\cos^2\theta - \frac{\partial^2(\ldots)}{\partial r\partial\theta}\frac{\sin 2\theta}{r} + \frac{\partial(\ldots)}{\partial r}\frac{\sin^2\theta}{r} + \frac{\partial(\ldots)}{\partial\theta}\frac{\sin 2\theta}{r^2} \\
&\quad + \frac{\partial^2(\ldots)}{\partial\theta^2}\frac{\sin^2\theta}{r^2} \\
\frac{\partial^2(\ldots)}{\partial y^2} &= \frac{\partial^2(\ldots)}{\partial r^2}\sin^2\theta + \frac{\partial^2(\ldots)}{\partial r\partial\theta}\frac{\sin 2\theta}{r} + \frac{\partial(\ldots)}{\partial r}\frac{\cos^2\theta}{r} - \frac{\partial(\ldots)}{\partial\theta}\frac{\sin 2\theta}{r^2} \\
&\quad + \frac{\partial^2(\ldots)}{\partial\theta^2}\frac{\cos^2\theta}{r^2} \\
\frac{\partial^2(\ldots)}{\partial x\partial y} &= \frac{\partial^2(\ldots)}{\partial r^2}\sin\theta\cos\theta + \frac{\partial^2(\ldots)}{\partial r\partial\theta}\frac{\cos 2\theta}{r} - \frac{\partial(\ldots)}{\partial\theta}\frac{\cos 2\theta}{r^2} - \frac{\partial(\ldots)}{\partial r}\frac{\sin 2\theta}{2r} \\
&\quad - \frac{\partial^2(\ldots)}{\partial\theta^2}\frac{\sin 2\theta}{2r^2} \\
\nabla^2(\ldots) &= \frac{\partial^2(\ldots)}{\partial x^2} + \frac{\partial^2(\ldots)}{\partial y^2} = \frac{\partial^2(\ldots)}{\partial r^2} + \frac{1}{r}\frac{\partial(\ldots)}{\partial r} + \frac{1}{r^2}\frac{\partial^2(\ldots)}{\partial\theta^2}
\end{aligned}
\tag{3.1}
$$

The Hookean relationships for circular isotropic plates are available from (1.19). For thin plates in the state of plane stress, these relations are reduced to

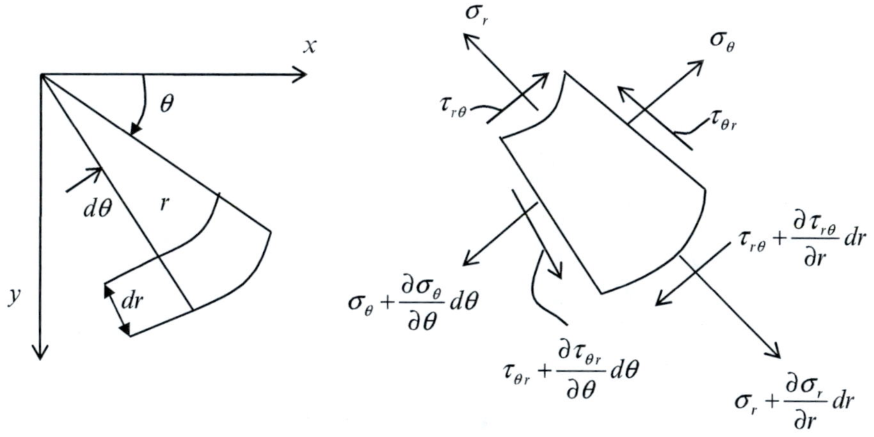

Fig. 3.1 Stresses applied to the faces of an infinitesimal element of a circular plate in the state of plane stress

$$\begin{Bmatrix} \sigma_r \\ \sigma_\theta \\ \tau_{r\theta} \end{Bmatrix} = \begin{bmatrix} \frac{E}{1-\nu^2} & \frac{\nu E}{1-\nu^2} & 0 \\ \frac{\nu E}{1-\nu^2} & \frac{E}{1-\nu^2} & 0 \\ 0 & 0 & G \end{bmatrix} \begin{Bmatrix} \varepsilon_r \\ \varepsilon_\theta \\ \gamma_{r\theta} \end{Bmatrix} \tag{3.2}$$

Radial σ_r, tangential σ_θ and in-plane shear stresses $\tau_{r\theta}$ that appear in (3.2) are applied to an infinitesimal element of the plate formed by two infinitely close circumferences and two radial lines forming an infinitesimal angle as shown in Fig. 3.1. The substitution of strain-displacement relations enables us to express the stresses in terms of displacements. Obviously, dependent on the strain-displacement relationships, the expressions for the stresses can reflect either geometrically linear or nonlinear formulations.

Stress couples and stress resultants in a polar coordinate system are defined by (1.53). The substitution of the expressions for the stresses (3.2) and strains given by (1.41), (1.42) into (1.53) yields a counterpart to Eq. 1.58 that presented stress couples and stress resultants in the Cartesian coordinate system. In particular, employing a nonlinear formulation according to (1.41) and (1.42), the stress couples and stress resultants in the polar coordinate system are:

$$N_r = \frac{Eh}{1-\nu^2}\left\{\frac{\partial u_0}{\partial r} + \frac{1}{2}\left(\frac{\partial w}{\partial r}\right)^2 + \frac{\nu}{r}\left[u_0 + \frac{\partial v_0}{\partial \theta} + \frac{1}{2r}\left(\frac{\partial w}{\partial \theta}\right)^2\right]\right\}$$

$$N_\theta = \frac{Eh}{1-\nu^2}\left\{\frac{1}{r}\left[u_0 + \frac{\partial v_0}{\partial \theta} + \frac{1}{2r}\left(\frac{\partial w}{\partial \theta}\right)^2\right] + \nu\left[\frac{\partial u_0}{\partial r} + \frac{1}{2}\left(\frac{\partial w}{\partial r}\right)^2\right]\right\}$$

$$N_{r\theta} = G\left(\frac{1}{r}\frac{\partial u_0}{\partial \theta} + \frac{\partial v_0}{\partial r} - \frac{v_0}{r} + \frac{1}{r}\frac{\partial w}{\partial r}\frac{\partial w}{\partial \theta}\right)$$

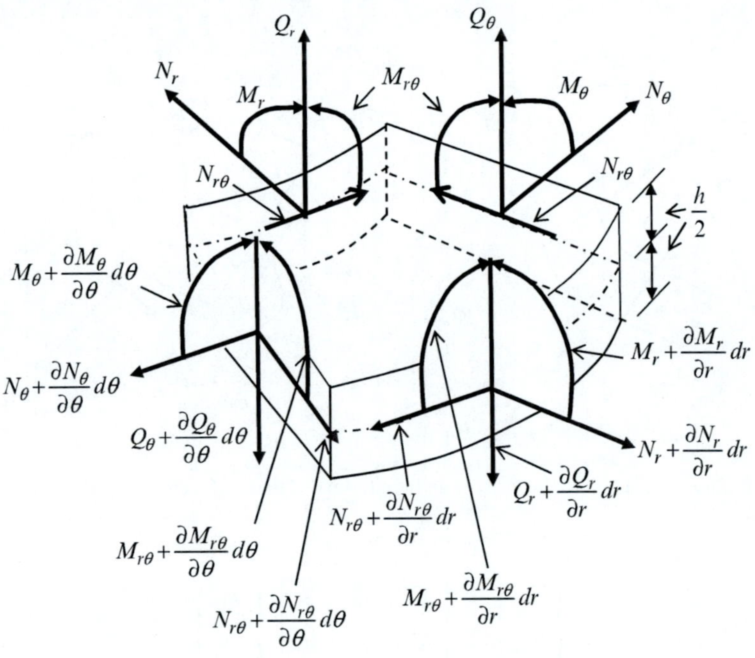

Fig. 3.2 Stress resultants and stress couples acting on an infinitesimal element of a circular plate

$$
\begin{aligned}
M_r &= -D\left[\frac{\partial^2 w}{\partial r^2} + \frac{\nu}{r}\left(\frac{\partial w}{\partial r} + \frac{1}{r}\frac{\partial^2 w}{\partial \theta^2}\right)\right] \\
M_\theta &= -D\left[\frac{1}{r}\left(\frac{\partial w}{\partial r} + \frac{1}{r}\frac{\partial^2 w}{\partial \theta^2}\right) + \nu\frac{\partial^2 w}{\partial r^2}\right] \\
M_{r\theta} &= -(1-\nu)\frac{D}{r}\left(\frac{\partial^2 w}{\partial r \partial \theta} - \frac{1}{r}\frac{\partial w}{\partial \theta}\right) \\
Q_r &= -D\frac{\partial\left(\nabla^2 w\right)}{\partial r} \\
Q_\theta &= -D\frac{1}{r}\frac{\partial\left(\nabla^2 w\right)}{\partial \theta}
\end{aligned}
\tag{3.3}
$$

The stress resultants and stress couples defined above are shown in Fig. 3.2. Transverse shear stress resultants Q_r and Q_θ that are also depicted in Fig. 3.2 were determined from the analysis of equilibrium of moments acting on the infinitesimal element in the planes rz and θz, respectively. Alternatively, these stress resultants could be obtained in terms of displacements from their definition (1.53) by integrating transverse shear stresses through the thickness of the plate.

The equations of motion or equilibrium can be derived by three methods:

1. Derivation of both the equations of equilibrium as well as the boundary conditions using the Hamilton principle, similarly to the procedure for rectangular plates (e.g., Reddy 1999, 2007).
2. Transformation of equations of equilibrium in Cartesian coordinates into equations in polar coordinates using the transformation of coordinates relationships (e.g., Timoshenko and Woinowsky-Krieger 1959).
3. Analysis of equilibrium of forces and moments acting on an infinitesimal element of the plate shown in Fig. 3.2.

In particular, the following equations of motion of thin plates were derived using the Hamilton principle by Reddy (2007):

$$\begin{aligned}
&\frac{\partial (rN_r)}{\partial r} + \frac{\partial N_{r\theta}}{\partial \theta} - N_\theta = rI_0\frac{\partial^2 u_0}{\partial t^2}\\
&\frac{\partial (rN_{r\theta})}{\partial r} + \frac{\partial N_\theta}{\partial \theta} + N_{r\theta} = rI_0\frac{\partial^2 v_0}{\partial t^2}\\
&\frac{\partial^2 (rM_r)}{\partial r^2} - \frac{\partial M_\theta}{\partial r} + \frac{1}{r}\frac{\partial^2 M_\theta}{\partial \theta^2} + 2\frac{\partial^2 M_{r\theta}}{\partial r\partial \theta} + \frac{2}{r}\frac{\partial M_{r\theta}}{\partial \theta} + \frac{\partial \left(rN_r\frac{\partial w}{\partial r}\right)}{\partial r}\\
&\quad + \frac{1}{r}\frac{\partial \left(N_\theta\frac{\partial w}{\partial \theta}\right)}{\partial \theta} + \frac{\partial}{\partial r}\left(N_{r\theta}\frac{\partial w}{\partial \theta}\right) + \frac{1}{r}\frac{\partial}{\partial \theta}\left(rN_{r\theta}\frac{\partial w}{\partial r}\right) = -pr + rI_0\frac{\partial^2 w}{\partial t^2}\\
&\quad - rI_2\frac{\partial^2}{\partial t^2}\left(\frac{1}{r}\frac{\partial \left(r\frac{\partial w}{\partial r}\right)}{\partial r} + \frac{1}{r^2}\frac{\partial^2 w}{\partial \theta^2}\right)
\end{aligned}\tag{3.4}$$

where $I_0 = \hat{m} = \rho h,\ \ I_2 = \frac{\rho h^3}{12}$.

In addition, transverse shear stress resultants can be determined from the analysis of the equilibrium of moments:

$$\begin{aligned}
Q_r &= \frac{1}{r}\left[\frac{\partial (rM_r)}{\partial r} + \frac{\partial M_{r\theta}}{\partial \theta} - M_\theta\right]\\
Q_\theta &= \frac{1}{r}\left[\frac{\partial (rM_{r\theta})}{\partial r} + \frac{\partial M_\theta}{\partial \theta} + M_{r\theta}\right]
\end{aligned}\tag{3.5}$$

Boundary conditions obtained from the Hamilton principle resemble those in the Cartesian coordinate system specified in Chap. 1:

$$\begin{array}{llll}
r = const: & N_r = \hat{N}_r & or & u_0 = \hat{u}_0\\
& N_{r\theta} = \hat{N}_{r\theta} & or & v_0 = \hat{v}_0\\
& V_r = \hat{V}_r & or & w = \hat{w}
\end{array}$$

$$M_r = \hat{M}_r \quad or \quad \frac{\partial w}{\partial r} = \frac{\partial \hat{w}}{\partial r}$$

$$\theta = const: \quad N_\theta = \hat{N}_\theta \quad or \quad v_0 = \hat{v}_0$$

$$N_{r\theta} = \hat{N}_{r\theta} \quad or \quad u_0 = \hat{u}_0$$

$$V_\theta = \hat{V}_\theta \quad or \quad w = \hat{w}$$

$$M_\theta = \hat{M}_\theta \quad or \quad \frac{\partial w}{\partial \theta} = \frac{\partial \hat{w}}{\partial \theta} \tag{3.6}$$

The conditions for straight boundaries ($\theta = const$) are valid for a sector plate, such as the section of an annular plate between two radial and two ring stiffeners. The Kirchhoff shear forces in (3.6) are:

$$V_r = Q_r + \frac{1}{r}\frac{\partial M_{r\theta}}{\partial \theta} + I_2\frac{\partial \ddot{w}}{\partial r} + N_r\frac{\partial w}{\partial r} + \frac{N_{r\theta}}{r}\frac{\partial w}{\partial \theta}$$

$$V_\theta = Q_\theta + \frac{\partial M_{r\theta}}{\partial r} + I_2\frac{\partial \ddot{w}}{r\partial \theta} + \frac{N_\theta}{r^2}\frac{\partial w}{\partial \theta} + \frac{N_{r\theta}}{r}\frac{\partial w}{\partial r} \tag{3.7}$$

If the problem is geometrically linear, the first two equations (3.4) characterizing in-plane motion or equilibrium are uncoupled from the last equation. It is also easy to show that the boundary conditions are uncoupled in linear problems. Accordingly, linear bending problems require us to specify transverse deflections only as was also the case in rectangular plates (Chap. 2).

The alternative to solving three equations of motion (3.4) or using an energy method is based on the introduction of the stress function that can satisfy the first two equations (3.4) leaving us with the single equation of motion and the compatibility equation. The latter equation can be written in the polar coordinate system by substituting the strains given by (1.40) into (1.97) and applying the coordinate transformation relationships (3.1). The application of the stress function in dynamic problems of circular plates requires us to adopt the assumption that in-plane inertias in the first two equations (3.4) are negligible. Since this simplification is acceptable in most problems, the stress function that identically satisfies the corresponding equations is introduced by

$$\sigma_r = \frac{N_r}{h} = \frac{1}{r}\frac{\partial \varphi}{\partial r} + \frac{1}{r^2}\frac{\partial^2 \varphi}{\partial \theta^2}, \quad \sigma_\theta = \frac{N_\theta}{h} = \frac{\partial^2 \varphi}{\partial r^2}, \quad \tau_{r\theta} = -\frac{\partial}{\partial r}\left(\frac{1}{r}\frac{\partial \varphi}{\partial \theta}\right) \tag{3.8}$$

The remaining third equation (3.4) and the compatibility equation are shown below for the case where the plate is subject to transverse pressure (static in-plane loads can also be incorporated in this formulation):

$$D\nabla^4 w = p + hL(w, \varphi)$$

$$\frac{1}{E}\nabla^4 \varphi = -\frac{1}{2}L(w, w) \tag{3.9}$$

The differential operators in (3.9) are derived using (3.1):

$$\nabla^4(...) = \nabla^2(...)\,\nabla^2(...) = \left(\frac{\partial^2}{\partial r^2} + \frac{1}{r}\frac{\partial}{\partial r} + \frac{1}{r^2}\frac{\partial^2}{\partial \theta^2}\right)\left(\frac{\partial^2(...)}{\partial r^2} + \frac{1}{r}\frac{\partial(...)}{\partial r} + \frac{1}{r^2}\frac{\partial^2(...)}{\partial \theta^2}\right) \tag{3.10}$$

and

$$L(w,\varphi) = \left(\frac{1}{r}\frac{\partial \varphi}{\partial r} + \frac{1}{r^2}\frac{\partial^2 \varphi}{\partial \theta^2}\right)\frac{\partial^2 w}{\partial r^2} + \left(\frac{1}{r}\frac{\partial w}{\partial r} + \frac{1}{r^2}\frac{\partial^2 w}{\partial \theta^2}\right)\frac{\partial^2 \varphi}{\partial r^2} - 2\frac{\partial}{\partial r}\left(\frac{1}{r}\frac{\partial \varphi}{\partial \theta}\right)\frac{\partial}{\partial r}\left(\frac{1}{r}\frac{\partial w}{\partial \theta}\right) \tag{3.11}$$

3.2 Axisymmetric Bending Problem

If a circular or annular plate is subject to an axisymmetric load that depends only on the radial coordinate, deformations, strains and stresses are also axisymmetric. Accordingly, they are independent of the circumferential coordinate and all derivatives with respect to this coordinate are equal to zero. Furthermore, tangential displacements $v_0 = 0$, though strains and stresses in the circumferential direction are present. Several representative axisymmetric bending problems are considered in this section.

First of all, it is useful to reproduce axisymmetric versions of some of the governing equations presented or referred to in the previous section. In particular, the axisymmetric version of the strain-displacement relationships is given by (1.43) and (1.44). The stress resultants and stress couples, including nonlinear terms, are obtained from (3.3):

$$N_r = \frac{Eh}{1-\nu^2}\left\{\frac{du_0}{dr} + \frac{1}{2}\left(\frac{dw}{dr}\right)^2 + \frac{\nu}{r}u_0\right\}$$

$$N_\theta = \frac{Eh}{1-\nu^2}\left\{\frac{u_0}{r} + \nu\left[\frac{du_0}{dr} + \frac{1}{2}\left(\frac{dw}{dr}\right)^2\right]\right\}$$

$$N_{r\theta} = 0 \qquad M_{r\theta} = 0$$

$$M_r = -D\left[\frac{d^2w}{dr^2} + \frac{\nu}{r}\frac{dw}{dr}\right]$$

$$M_\theta = -D\left[\frac{1}{r}\frac{dw}{dr} + \nu\frac{d^2w}{dr^2}\right] \tag{3.12}$$

$$Q_r = \frac{1}{r}\left[\frac{\partial (rM_r)}{\partial r} - M_\theta\right] = -D\frac{d}{dr}\left(\frac{d^2w}{dr^2} + \frac{1}{r}\frac{dw}{dr}\right)$$
$$= -D\frac{d}{dr}\left[\frac{1}{r}\frac{d}{dr}\left(r\frac{dw}{dr}\right)\right] \quad Q_\theta = 0$$

Equations of motion in terms of stress resultants and stress couples (3.4) are now reduced to two equations:

$$\frac{\partial (rN_r)}{\partial r} - N_\theta = rI_0\frac{\partial^2 u_0}{\partial t^2}$$
$$\frac{\partial^2 (rM_r)}{\partial r^2} - \frac{\partial M_\theta}{\partial r} + \frac{\partial \left(rN_r \frac{\partial w}{\partial r}\right)}{\partial r} = -pr + rI_0\frac{\partial^2 w}{\partial t^2} - rI_2\frac{\partial^2}{\partial t^2}\left[\frac{1}{r}\frac{\partial \left(r\frac{\partial w}{\partial r}\right)}{\partial r}\right] \tag{3.13}$$

If the problem is static, equations of equilibrium obtained from (3.13) are ordinary differential equations.

The expression for the stress function (3.8) is also simplified in an axisymmetric problem:

$$\sigma_r = \frac{N_r}{h} = \frac{1}{r}\frac{d\varphi}{dr}, \quad \sigma_\theta = \frac{N_\theta}{h} = \frac{d^2\varphi}{dr^2}, \quad \tau_{r\theta} = 0 \tag{3.14}$$

The formulation employing the stress function in a static axisymmetric problem coincides with (3.9) where the axisymmetric versions of the operator $\nabla^4(\ldots)$ and the differential operator $L(w, \varphi)$ are

$$\nabla^4(\ldots) = \nabla^2(\ldots)\nabla^2(\ldots) = \left(\frac{d^2}{dr^2} + \frac{1}{r}\frac{d}{dr}\right)\left(\frac{d^2(\ldots)}{dr^2} + \frac{1}{r}\frac{d(\ldots)}{dr}\right)$$
$$= \frac{d^4(\ldots)}{dr^4} + \frac{2}{r}\frac{d^3(\ldots)}{dr^3} - \frac{1}{r^2}\frac{d^2(\ldots)}{dr^2} + \frac{1}{r^3}\frac{d(\ldots)}{dr}$$
$$= \frac{1}{r}\frac{d}{dr}\left\{r\frac{d}{dr}\left[\frac{1}{r}\frac{d}{dr}\left(r\frac{d(\ldots)}{dr}\right)\right]\right\} \tag{3.15}$$

and

$$L(w,\varphi) = \frac{1}{r}\frac{d\varphi}{dr}\frac{d^2w}{dr^2} + \frac{1}{r}\frac{dw}{dr}\frac{d^2\varphi}{dr^2} = \frac{1}{r}\frac{d}{dr}\left(\frac{dw}{dr}\frac{d\varphi}{dr}\right) \tag{3.16}$$

The linear axisymmetric strains in the middle plane and the changes of curvature and twist are obtained from (1.41), (1.42), respectively:

$$\begin{aligned}
\varepsilon_r^0 &= \frac{du_0}{dr} \\
\varepsilon_\theta^0 &= \frac{u_0}{r} \\
\varepsilon_{r\theta}^0 &= 0
\end{aligned} \tag{3.17}$$

and

$$\begin{aligned}
\kappa_r &= -\frac{d^2w}{dr^2} \\
\kappa_\theta &= -\frac{1}{r}\frac{dw}{dr} \\
\kappa_{r\theta} &= 0
\end{aligned} \tag{3.18}$$

It is observed that the radial displacement and transverse deflection in the middle-plane strains and in the changes of curvature and twist are uncoupled.

The bending stresses are available from (3.2) and (3.18):

$$\begin{aligned}
\sigma_r &= -\frac{Ez}{1-\nu^2}\left(\frac{d^2w}{dr^2} + \frac{\nu}{r}\frac{dw}{dr}\right) \\
\sigma_\theta &= -\frac{Ez}{1-\nu^2}\left(\nu\frac{d^2w}{dr^2} + \frac{1}{r}\frac{dw}{dr}\right) \\
\tau_{r\theta} &= 0
\end{aligned} \tag{3.19}$$

In-plane membrane stresses obtained by the substitution of (3.17) into (3.2) can easily be evaluated in terms of the in-plane radial displacement.

Note that in linear axisymmetric problems in-plane displacements and membrane stresses are uncoupled from transverse deflections and bending stresses. Accordingly, in cases of bending, buckling or transverse vibrations, it is only necessary to consider the axisymmetric version of the first equation (3.9). Furthermore, as follows from (3.14)

$$\sigma_\theta = \frac{d(r\sigma_r)}{dr} \tag{3.20}$$

If the linear problem is static, the axisymmetric version of the first equation (3.9) can be integrated yielding

$$D\frac{d\nabla^2 w}{dr} = p + \bar{N}_r\frac{dw}{dr} + \frac{C_0}{r} \tag{3.21}$$

$\bar{N}_r$ being an applied radial stress resultant that is constant throughout the plate as long as the problem is linear. In a solid circular plate, the singularity can be avoided only if $C_0 = 0$.

3.2.1 *Bending of a Solid Circular Plate Subject to a Uniform Pressure or to an Axisymmetric Pressure That Is a Function of the Radial Coordinate: Geometrically Linear Problem*

In a linear problem, equations of equilibrium (3.4) and the boundary conditions are uncoupled. Accordingly, in the absence of in-plane loads, in-plane displacements and membrane stresses are equal to zero, while the transverse deflection should be determined from the first equation (3.9) that becomes

$$D\nabla^4 w = p(r) \tag{3.22}$$

The complimentary solution of the homogeneous equation obtained from (3.22) is

$$w_h = C_1 \ln r + C_2 r^2 \ln r + C_3 r^2 + C_4 \tag{3.23}$$

where constants of integration C_i will be determined from the boundary conditions as shown below.

A particular integral of (3.22) can be derived by a successive integration of this equation. Omitting the intermediate steps, we present here the result of this integration (the validity of this result can be checked by substituting it into (3.22)):

$$w_p = \int \frac{1}{r} \int r \int \frac{1}{r} \int \frac{rp(r)}{D} dr dr dr dr \tag{3.24}$$

It is useful to outline here the results for an important case where the plate is subject to a uniform pressure, i.e. $p(r) = p_0$. Then, as follows from (3.23), (3.24) and (3.12),

$$w = C_1 \ln r + C_2 r^2 \ln r + C_3 r^2 + C_4 + \frac{p_0 r^4}{64D}$$

$$M_r = -D\left[-C_1 \frac{1-\nu}{r^2} + C_2 \left(2\left(1+\nu\right)\ln r + 3 + \nu\right) + 2C_3\left(1+\nu\right) + \frac{p_0 r^2}{16D}\left(3+\nu\right)\right]$$

$$M_\theta = -D\left[C_1\frac{1-\nu}{r^2} + C_2\left(2\left(1+\nu\right)\ln r + 1 + 3\nu\right) + 2C_3\left(1+\nu\right) + \frac{p_0 r^2}{16D}\left(1+3\nu\right)\right]$$

$$Q_r = -4D\left(C_2\frac{1}{r} + \frac{p_0 r}{8D}\right) \tag{3.25}$$

If the strength analysis employs the maximum principal stress criterion, the location of the maximum stress can be predicted in the general case. The maximum principal stress is the radial stress and its location coincides with that of the maximum radial stress couple that is found from the requirement

$$\frac{dM_r}{dr} = -D\left[\frac{2C_1\left(1-\nu\right)}{r^3} + 2C_2\left(1+\nu\right)\frac{1}{r} + \frac{p_0 r}{8D}\left(3+\nu\right)\right] = 0 \tag{3.26}$$

Once the constants of integration are specified from the boundary conditions, the radial location of the maximum stress σ_r can be determined from (3.26). Subsequently, this stress is found from (3.19). In the case where the von Mises strength criterion is employed, the procedure finding the location of failure is more involved, but it is still straightforward.

There are two boundary conditions that must be specified at each edge of an annular plate resulting in four equations for four constants of integration in (3.22). In the case of a solid circular plate that has only one edge, two missing conditions are contributed reflecting the requirement that displacements and stresses at the center of the plate $r = 0$ should be finite. The former requirement implies that $C_1 = 0$. In addition, the symmetry implies that the slope of the plate at the center should be equal to zero. The requirement to the slope and non-singular displacement and stresses at the plate center can be satisfied if $C_2 = 0$.

Example 3.1: Solid Circular Plate of Radius $r = a$ Subject to a Uniform Pressure p_0
This problem can be encountered in design of cargo hatches and covers of circular openings in bulkheads. Boundary conditions can vary from simply supported to clamped or elastically supported and elastically clamped plates. For example, in the latter case the torsional stiffness of the rim structure is finite but insufficient to completely prevent the rotation of the edge.

The deflection of the plate is represented by the corresponding simplification of (3.25):

$$w = C_3 r^2 + C_4 + \frac{p_0 r^4}{64D} \tag{a}$$

In a clamped plate where $w(a) = \frac{dw(a)}{dr} = 0$,

$$C_3 = -\frac{p_0 a^2}{32D}, \qquad C_4 = \frac{p_0 a^4}{64D}$$

$$w = \frac{p_0 \left(a^2 - r^2\right)}{64D} \tag{b}$$

The stresses in the plate are derived by substituting the deflection given by (a) into (3.19):

$$\sigma_r = \frac{3p_0 z}{4h^3}\left[(1+\nu)\,a^2 - (3+\nu)\,r^2\right]$$

$$\sigma_\theta = \frac{3p_0 z}{4h^3}\left[(1+\nu)\,a^2 - (1+3\nu)\,r^2\right] \tag{c}$$

The algebraically maximum radial and tangential stresses occur on the plate surfaces $\left(z = \pm\frac{h}{2}\right)$ at the center and along the boundary of the plate:

$$\sigma_r\,(r=0) = \sigma_\theta\,(r=0) = \frac{3\,(1+\nu)}{8}p_0\left(\frac{a}{h}\right)^2$$

$$\sigma_r\,(r=a) = -\frac{3}{4}p_0\left(\frac{a}{h}\right)^2$$

$$\sigma_\theta\,(r=a) = -\frac{3\nu}{4}p_0\left(\frac{a}{h}\right)^2 \tag{d}$$

The Poisson ratio of isotropic materials being smaller or equal to 0.5, it is obvious that the algebraically maximum radial stress is at the edge, while the maximum tangential stress is at the center.

If the strength analysis is conducted using the von Mises strength criterion, it is necessary to determine the radial location where the effective stress is maximum. In the plane stress axisymmetric problem, the von Mises criterion is

$$\sigma_{eff}^{e} = \sigma_r^2 - \sigma_r\sigma_\theta + \sigma_\theta^2 = \sigma_{all}^2 \tag{e}$$

σ_{all} being the allowable stress. It can easily be checked that the effective stress is maximum at

$$r = (1+\nu)\,a\sqrt{\frac{6}{(13+18\nu+13\nu^2)}} \tag{f}$$

For example, if the Poisson ratio is equal to 0.3, the maximum effective stress occurs at $r = 0.749a$.

If the plate is simply supported, boundary conditions are $w(a) = M_r(a) = 0$. Straightforward transformations yield

$$w = \frac{p_0 a^4}{64D}\left(\frac{r^4}{a^4} - 2\frac{3+\nu}{1+\nu}\frac{r^2}{a^2} + \frac{5+\nu}{1+\nu}\right)$$

$$\sigma_r = \frac{3p_0 z}{4h^3}(3+\nu)\left(a^2 - r^2\right)$$

$$\sigma_\theta = \frac{3p_0 z}{4h^3}\left[(3+\nu)a^2 - (1+3\nu)r^2\right] \tag{g}$$

The maximum stresses occur at the center of the plate being equal to

$$\sigma_r(0) = \sigma_\theta(0) = \frac{3(3+\nu)}{8}p_0\left(\frac{a}{h}\right)^2 \tag{h}$$

In applications, the edge of the plate is supported by a circular ring or stiffener providing finite resistance both to transverse deflections as well as to rotations. It is possible to account for both effects, i.e. both elastic support and elastic clamping, by satisfying the boundary conditions. In the present case these conditions are

$$V_r(a) = Q_r(a) = kw(a), \quad M_r(a) = \kappa\frac{dw(a)}{dr} \tag{i}$$

where k and κ are coefficients of elastic support and elastic clamping, respectively. The substitution of the radial stress couple and transverse stress resultant from (3.12) into (i) and using the solution (a) yields the values of constants of integration that can subsequently be applied to specify deflections and stresses.

3.2.2 *Bending of a Solid Circular Plate of Radius r = a Subject to a Concentrated Central Force P*

If a circular plate is subject to a load applied at the center, the problem is axisymmetric. There is no pressure distributed over the plate surface, i.e. the right side of the equation of equilibrium (3.22) is equal to zero. Accordingly, the solution is given by (3.23) where the singularity of deflections at the center is avoided by requiring that $C_1 = 0$. The constant of integration C_2 that was taken equal zero in Example 3.1 is retained in the solution.

Besides two boundary conditions along the edges of the plate, the additional condition introduced following Timoshenko and Woinowsky-Krieger (1959) is the equilibrium of a circular section of the plate of an arbitrary radius $r = r_1$ as reflected

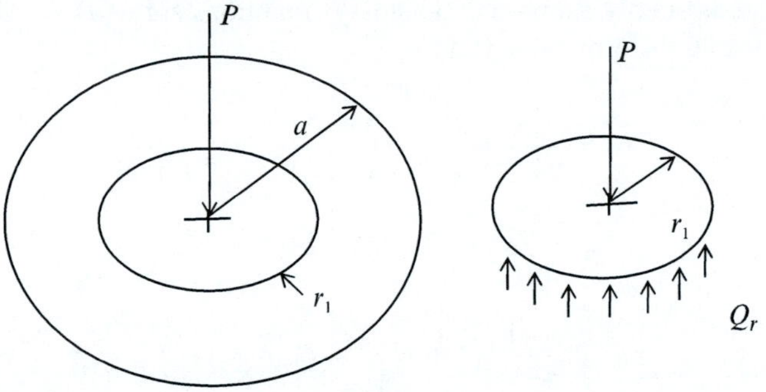

Fig. 3.3 Circular plate loaded at the center and the illustration of the equilibrium condition for a circle of radius r_1

in Fig. 3.3. It is evident that vertical forces acting on such section are in equilibrium if the transverse shear stress resultant balances the applied concentrated force

$$2\pi r_1 Q_r (r = r_1) + P = 0 \tag{3.27}$$

where $Q_r$can be evaluated in terms of unknown constants of integration by substituting (3.23) into (3.12). This procedure yields the constant of integration $C_2 = \frac{P}{8\pi D}$.

Combining the solution for C_2 with two boundary conditions along the edge of the plate we can obtain the remaining constants of integration, deflections and stresses. For example, if the plate is clamped along the boundary,

$$\begin{aligned} w &= \frac{P}{16\pi D}\left(2r^2 \ln \frac{r}{a} + a^2 - r^2\right) \\ \sigma_r &= \frac{3Pz}{\pi h^3}\left[(1+\nu)\ln \frac{a}{r} - 1\right] \\ \sigma_\theta &= \frac{3Pz}{\pi h^3}\left[(1+\nu)\ln \frac{a}{r} - \nu\right] \end{aligned} \tag{3.28}$$

It is evident that the stresses given by (3.28) are singular at the center of the plate. An approximate approach based on the analysis of Nadai (1925) was reproduced in the monograph of Ugural (1999). According to this approach, the maximum stresses given by (3.28) can be obtained using an "equivalent radius:"

$$\begin{aligned} r_{eq} &= \sqrt{1.6r_P^2 + h^2} - 0.675h \quad \textit{if} \quad r_P < 0.5h \\ r_{eq} &= r_P \quad \textit{if} \quad r_P \geqslant 0.5h \end{aligned} \tag{3.29}$$

where r_P is the radius of a circle loaded by the force P (in other words, we realize that even a "concentrated" force cannot physically be applied at one point). While this solution represents a historical interest, the availability of three-dimensional finite element modeling capable of an accurate account for the local state of stresses close to the center of the plate reduces its practical utility.

3.2.3 Annular Plate Subject to Loading Applied at the Inner Edge

This problem was discussed in detail by Timoshenko and Woinowsky-Krieger (1959). In particular, they considered two cases, i.e. the plate loaded by uniformly distributed bending stress couples M_1 and M_2 along both the inner and outer edges, respectively, as shown in Fig. 3.4, Case "a" and the plate subject to a uniformly distributed transverse shear stress resultant Q_1 (Fig. 3.4, Case "b"). In both problems the outer edge is simply supported, i.e. its deflections are equal to zero, while the inner edge is not supported. Other boundary conditions could also be considered, e.g., the rotations of the inner edge of the plate subject to a prescribed transverse shear stress resultant could be prevented. This and similar problems could be addressed using a similar approach to that considered below and using appropriate boundary conditions.

Similar to the previous problem (Sect. 3.2.2), the pressure is equal to zero and the solution of the equation of motion is given by (3.23) where all terms are retained since the singularity at the center does not present a problem in an annular plate.

The boundary conditions that have to be satisfied are

$$\begin{aligned} M_r\,(r=b) &= M_1, \quad M_r\,(r=a) = M_2 \\ Q_r\,(r=b) &= Q_1, \quad w\,(r=a) = 0 \end{aligned} \tag{3.30}$$

where in Case "a" $Q_1 = 0$ and in Case "b" $M_1 = M_2 = 0$.

Using the deflection given by (3.23) and the expressions for the bending couple and transverse shear stress resultant given in (3.12) we can derive the corresponding solutions for both cases. In particular, in Case "a",

$$\begin{aligned} w &= \frac{M_1 b^2 - M_2 a^2}{2\,(1+\nu)\,D\,(a^2-b^2)}\left(r^2-a^2\right) + \frac{(M_1-M_2)\,a^2b^2}{(1-\nu)\,D\,(a^2-b^2)}\ln\frac{r}{a} \\ M_r &= \frac{M_1 b^2 - M_2 a^2}{a^2-b^2} + \frac{(M_1-M_2)\,a^2b^2}{r^2\,(a^2-b^2)} \\ M_\theta &= \frac{M_2 a^2 - M_1 b^2}{a^2-b^2} - \frac{(M_1-M_2)\,a^2b^2}{r^2\,(a^2-b^2)} \end{aligned} \tag{3.31}$$

Case "b" and other problems related to annular plates are extensively discussed in the book of Timoshenko and Woinowsky-Krieger (1959). In particular, a superposition of available solutions enables us to model bending of an annular plate that is simply supported along the outer edge, free along the inner edge, and subject to a uniform pressure (Fig. 3.5). The results for this problem were obtained by a

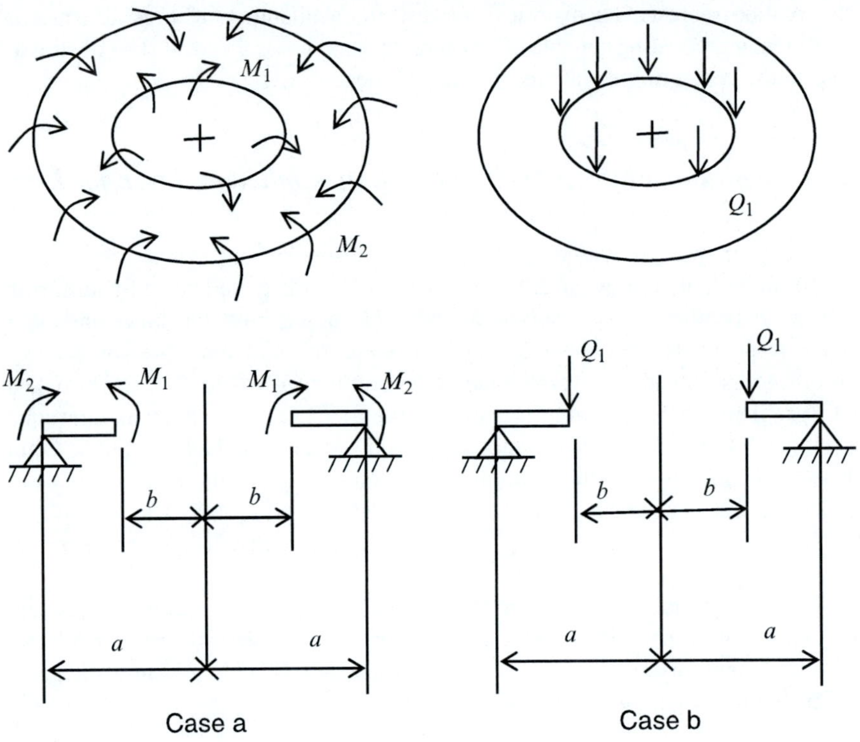

Fig. 3.4 Annular plate subject to uniformly distributed bending stress couples (Case a) and to uniformly distributed transverse shear stress resultants (Case b). The *outer* and *inner* radii are denoted by a and b, respectively

superposition of the solution for the solid simply supported plate and the solution for the section of this plate cut off by the inner radius (see details in Timoshenko and Woinowsky-Krieger 1959). The solution of this problem can also be obtained without superposition by substituting the expression for deflections of the plate subject to a uniform pressure (first equation 3.25) into the boundary conditions

$$\begin{aligned} M_r(b) &= Q_r(b) = 0, \\ w(a) &= M_r(a) = 0 \end{aligned} \tag{3.32}$$

The transverse shear stress resultant and stress couple in (3.32) have already been be evaluated in terms of constants of integration in (3.25). Thus, conditions (3.32) yield a system of four linear algebraic equations with respect to constants of integration.

Another problem that may represent an interest to designers of circular bulkheads is depicted in Fig. 3.6. The section of the bulkhead between reinforcements (3) is a sector plate (see a discussion on sector plates in Sect. 3.6). If the distance between the radial stiffeners is large, bending in the central part of the sector plate 3 can be

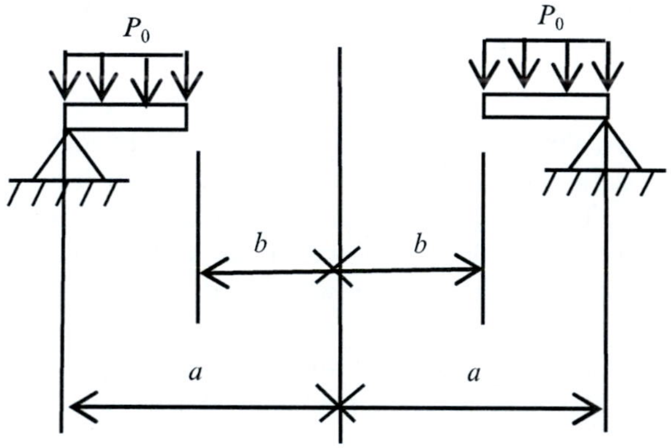

Fig. 3.5 Annular plate subject to a uniform pressure and simply supported along the outer edge

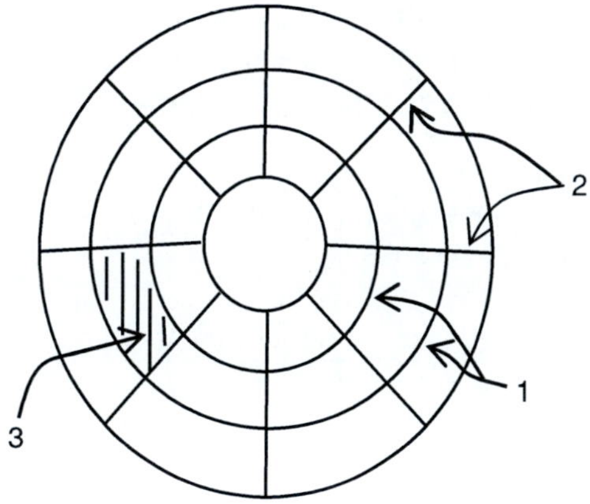

Fig. 3.6 Circular bulkhead reinforced by ring (*1*) and radial (*2*) stringers. Sector plate 3 is supported by rings and stringers

analyzed without accounting for their effect. Then such section can be considered as an annular plate. A typical load on bulkheads is represented by a hydrostatic (uniform) pressure. Ring stiffeners (1) usually provide conditions approaching simple support. Accordingly, the solution for such plate can be obtained using the solution (3.25) and specifying the constants of integration from the boundary conditions

$$
\begin{aligned}
w(b) &= M_r(b) = 0 \\
w(a) &= M_r(a) = 0
\end{aligned}
\tag{3.33}
$$

where $r = a$ and $r = b$ are the coordinates of ring stiffeners.

3.3 Geometrically Nonlinear Axisymmetric Bending Problem for a Solid Annular Plate

The previous paragraph outlines the approach to exact and relatively uncomplicated analytical solutions for both solid and annular circular plates. The validity of these solutions is limited by the allowable range of deflections. Indeed, the solutions developed using linear strain-displacement relationships become inaccurate at large deflections of the plate. As indicated elsewhere in this book, while the strict limit of the applicability of linear solutions is not available (such limit would depend on the allowable error for deformations or stresses using the linear theory), a rather typical practice is to recommend a nonlinear solution at deflections exceeding half-thickness of the plate.

The principal contribution of nonlinear deformations to the analysis is related to coupling between in-plane displacements and transverse deflections. This coupling complicates the analytical solution, so that a numerical analysis is often the preferred approach. Physically, coupling between displacements implies the emergence of in-plane membrane stress resultants in a plate experiencing bending, transverse deflections and bending stresses being in turn affected by these in-plane stress components.

The strain-displacement relationships (3.17) have to be modified in the case of a nonlinear axisymmetric deformation. While the changes of curvature (3.18) are not altered, the middle-plane strains are:

$$\begin{aligned}\varepsilon_r^0 &= \frac{du_0}{dr} + \frac{1}{2}\left(\frac{dw}{dr}\right)^2 \\ \varepsilon_\theta^0 &= \frac{u_0}{r} \\ \varepsilon_{r\theta}^0 &= 0\end{aligned} \tag{3.34}$$

Analytical methods that can be used for the nonlinear analysis of circular plates experiencing bending are outlined in the book of Timoshenko and Woinowsky-Krieger (1959). The exact solution of the nonlinear governing equations is usually impossible, but the solution can be based on approximate and numerical methods. In particular, the Rayleigh-Ritz method representing radial displacements and deflections in series with terms dependent on powers of the radial coordinate can be used for the analysis. The other approach is based on using the Galerkin procedure to integrate the equations of equilibrium. All methods of solution depend on the series representation of displacements and/or stress resultants, i.e. the accuracy is always dependent on the number of terms retained in these series.

The following approach to the solution of the problem of nonlinear bending of a solid circular plate clamped along the edge and subject to a uniform pressure utilizes the Rayleigh-Ritz method (Timoshenko and Woinowsky-Krieger 1959).

The expression for the strain energy of a circular plate experiencing an axisymmetric deformation is derived by substituting stress resultants and stress couples for an axisymmetric problem (3.12) into the formula for the strain energy (1.70) that is rewritten in the case of the axisymmetric plane problem in polar coordinates as follows:

$$U = \frac{1}{2} \iint_A \left(N_r \varepsilon_r^0 + N_\theta \varepsilon_\theta^0 + M_r \kappa_r + M_\theta \kappa_\theta \right) dA \tag{3.35}$$

The substitution of the middle plane strains and changes of curvature as given by (3.34) and (3.18) into (3.35) yields

$$\begin{aligned} U &= \frac{\pi E h}{1-\nu^2} \\ &\times \int_0^a \left[\left(\frac{\partial u_0}{\partial r} + \frac{1}{2}\left(\frac{\partial w}{\partial r}\right)^2 \right)^2 + \left(\frac{u_0}{r}\right)^2 + 2\nu \left(\frac{\partial u_0}{\partial r} + \frac{1}{2}\left(\frac{\partial w}{\partial r}\right)^2 \right) \left(\frac{u_0}{r}\right) \right] r dr \\ &+ D\pi \int_0^a \left[\left(\frac{\partial^2 w}{\partial r^2} \right)^2 + \frac{1}{r^2}\left(\frac{\partial w}{\partial r}\right)^2 + \frac{2\nu}{r} \frac{\partial w}{\partial r} \frac{\partial^2 w}{\partial r^2} \right] r dr \end{aligned} \tag{3.36}$$

The potential energy of the applied uniform pressure is evaluated from (1.68):

$$V = -2\pi \int_0^a p_0 w r dr \tag{3.37}$$

In a clamped plate with the boundary prevented from the radial displacement the boundary conditions are

$$w(a) = \frac{dw(a)}{dr} = 0, \quad u(a) = u(0) = 0 \tag{3.38}$$

where the last condition reflects the symmetry of radial displacements with respect to the plate center. Conditions (3.38) can be satisfied by representing the displacements in the form

$$\begin{aligned} u_0 &= r\,(a-r)\left(C_1 + C_2 r + C_3 r^2 + \cdots\right) \\ w &= C_0 \left(1 - \frac{r^2}{a^2} \right)^2 \end{aligned} \tag{3.39}$$

where $C_i \quad (i = 0,\ 1,\ 2,\ 3, \ldots)$ are unknown constants.

The substitution of series (3.39) into the Rayleigh-Ritz formulation $\frac{\partial(U+V_1)}{\partial C_i} = 0$ yields the system of nonlinear algebraic equations with respect to the unknown constants. In particular, the solution obtained retaining only terms with C_0, C_1 and C_2 in (3.39) and using the Poisson ratio equal to 0.3 yields

$$C_0 = w(0) = \frac{p_0 a^4}{64D} \frac{1}{1 + \underline{0.488\frac{C_0^2}{h^2}}},$$

$$C_1 = 1.185\frac{C_0^2}{a^3}, \qquad C_2 = -1.75\frac{C_0^2}{a^4} \tag{3.40}$$

Obviously, as reflected in (3.40), radial in-plane displacements are much smaller compared to the maximum deflection of the plate that occurs at the center. The underlined term in the expression for $w(0)$ represents the coupling between in-plane and transverse displacements. As can be checked by the inspection of the first equation (3.40), coupling, i.e. geometric nonlinearity, reduces deflections of the plate. For example, if the deflection at the center is equal to half-thickness of the plate, the stretching produced as a result of coupling reduces the deflection by 11%.

The stresses that can be found by (3.19) are also reduced as a result of the inclusion of geometric nonlinearity in the solution. This fact, previously referred to in Chap. 2, reflects on the conservative nature of linear solutions and explains why structures designed neglecting geometric nonlinearity are usually safe (though they are may be unnecessary heavy).

Coupling between in-plane displacements and transverse deflections that is present in nonlinear problems may cause a major difference in the solution. Accordingly, in-plane boundary conditions may become important in the bending problem. For example, if radial displacements at the edge are unconstrained, the formula for the deflection at the center of the plate becomes (Timoshenko and Woinowsky-Krieger 1959):

$$w(0) = \frac{p_0 a^4}{64D} \frac{1}{1 + \underline{0.146\frac{w_0^2}{h^2}}}, \tag{3.41}$$

The comparison of this equation with the first equation (3.40) yields the conclusion that the nonlinear effect (underlined term) weakened as a result of the change in the in-plane boundary condition. This is predictable since a constraint against radial displacements resulted in larger membrane stresses affecting the bending response (as reflected in the first equation (3.40)). In particular, the difference in the nonlinear and linear solutions corresponding to the bending deflection at the center of the plate equal to its half-thickness that was equal to 11% in the case of the in-plane constrained edge reduces to 3.5% if the edge is unconstrained.

3.4 Asymmetric Bending Problem for Circular Plates

Asymmetric problems in circular plates require a complicated analytical solution that is often avoided in favor of numerical methods of the analysis. Nevertheless, the analytical solution of the bending problem is available in the form that is applicable to an arbitrary distribution of transverse load applied to the plate as long as it is can be represented in the form of trigonometric series dependent on the circumferential coordinate (Mc Farland et al. 1972; Jawad 2004).

Consider bending of a solid circular plate by transverse pressure $p(r,\ \theta)$ that is a continuous function of both the radial as well as circumferential coordinates. The pressure is represented in Fourier series:

$$p(r,\theta) = f_0(r) + \sum_{m=1}^{M} [f_m(r)\cos m\theta + g_m(r)\sin m\theta] \tag{3.42}$$

where the functions of the radial coordinate are available from

$$f_0(r) = \frac{1}{\pi}\int_{-\pi}^{\pi} p(r,\theta)\, d\theta$$

$$f_m(r) = \frac{1}{\pi}\int_{-\pi}^{\pi} p(r,\theta)\cos m\theta\, d\theta$$

$$g_m(r) = \frac{1}{\pi}\int_{-\pi}^{\pi} p(r,\theta)\sin m\theta d\theta \tag{3.43}$$

The problem being static and linear, in-plane displacements and transverse bending deflections are uncoupled. In such case, we have to analyze the static version of the equation of motion (3.9), i.e. $D\nabla^4 w = p$ where the differential operator in the left side given by (3.10) and the pressure is a function of the radial and circumferential coordinates. The second equation (3.9) is the compatibility equation that does not affect the solution due to the absence of coupling between membrane and bending stresses.

The solution of the homogeneous equation obtained from the first equation (3.9) is sought in the Fourier series

$$w_h(r,\theta) = W_0'(r) + \sum_{m=1}^{M} \left[W'_m(r)\cos m\theta + W''_m(r)\sin m\theta\right] \tag{3.44}$$

The substitution of (3.44) into the homogeneous version of the equation of equilibrium (3.9) results in the equation

$$\sum_{i=0}^{M} S_i(r)\cos i\theta + \sum_{m=1}^{M} P_m(r)\sin m\theta = 0, \quad i = 0,\ m \tag{3.45}$$

where

$$S_i(r) = \frac{d^4W'_i}{dr^4} + \frac{2}{r}\frac{d^3W'_i}{dr^3} - \frac{1+2i^2}{r^2}\frac{d^2W'_i}{dr^2} + \frac{1+2i^2}{r^3}\frac{dW'_i}{dr} + \frac{i^2\left(i^2-4\right)}{r^4}W'_i$$

$$P_m(r) = \frac{d^4W''_m}{dr^4} + \frac{2}{r}\frac{d^3W''_m}{dr^3} - \frac{1+2m^2}{r^2}\frac{d^2W''_m}{dr^2} + \frac{1+2m^2}{r^3}\frac{dW''_m}{dr} + \frac{m^2\left(m^2-4\right)}{r^4}W''_m \tag{3.46}$$

Obviously, Eq. 3.45 can be satisfied only if all coefficients at trigonometric functions are equal to zero, i.e.

$$S_i(r) = 0, \quad P_m(r) = 0. \tag{3.47}$$

The solution of ordinary differential equations (3.47) is found representing the corresponding function of the radial coordinate in power series

$$W'_i(r) = \delta_s r^s, \quad W''_m(r) = \eta_s r^s \tag{3.48}$$

The substitution of (3.48) into (3.47) yields

$$\begin{aligned}
W'_0(r) &= A_0 + B_0 r^2 + C_0 \ln r + D_0 r^2 \ln r \\
W'_1(r) &= A_1 r + B_1 r^3 + C_1 r^{-1} + D_1 r \ln r \\
W'_i(r) &= A_i r^i + B_i r^{i+2} + C_i r^{-i} + D_i r^{2-i} \qquad \textit{for } i > 1 \\
W''_1(r) &= E_1 r + F_1 r^3 + G_1 r^{-1} + H_1 r \ln r \\
W''_m(r) &= E_m r^m + F_m r^{m+2} + G_m r^{-m} + H_m r^{2-m} \quad \textit{for } m > 1
\end{aligned} \tag{3.49}$$

where $A_0,\ A_1,\ A_i, \ldots, H_m$ are constants of integration that should be determined from the boundary conditions.

The particular integral corresponding to the load represented in the form (3.42) is sought in the form resembling (3.44) as

$$w_p(r,\theta) = S_0(r) + \sum_{m=1}^{M}\left[S_m(r)\cos m\theta + R_m(r)\sin m\theta\right] \tag{3.50}$$

The substitution of (3.42) and (3.50) into the first equation (3.9) and equating the coefficients at the same functions of the circumferential coordinate in the left and in the right sides of the resulting equation yields ordinary differential equations for functions $S_0(r)$, $S_m(r)$ and $R_m(r)$:

$$\frac{d^4S_0}{dr^4}+\frac{2}{r}\frac{d^3S_0}{dr^3}-\frac{1}{r^2}\frac{d^2S_0}{dr^2}+\frac{1}{r^3}\frac{dS_0}{dr}=\frac{f_0(r)}{D}$$

$$\frac{d^4S_m}{dr^4}+\frac{2}{r}\frac{d^3S_m}{dr^3}-\frac{1+2m^2}{r^2}\frac{d^2S_m}{dr^2}+\frac{1+2m^2}{r^3}\frac{dS_m}{dr}+\frac{m^2\left(m^2-4\right)}{r^4}S_m$$
$$=\frac{f_m(r)}{D}$$

$$\frac{d^4R_m}{dr^4}+\frac{2}{r}\frac{d^3R_m}{dr^3}-\frac{1+2m^2}{r^2}\frac{d^2R_m}{dr^2}+\frac{1+2m^2}{r^3}\frac{dR_m}{dr}+\frac{m^2\left(m^2-4\right)}{r^4}R_m$$
$$=\frac{g_m(r)}{D} \tag{3.51}$$

The solution of Eq. 3.51 can be specified dependent on the radial variations of the applied load, i.e. the functions in the right side of these equations. Subsequently, the solution

$$w(r,\theta)=W_0'(r)+\sum_{m=1}^{M}\left[W'_m(r)\cos m\theta+W''_m(r)\sin m\theta\right]+S_0(r)$$
$$+\sum_{m=1}^{M}\left[S_m(r)\cos m\theta+R_m(r)\sin m\theta\right] \tag{3.52}$$

must be subject to boundary conditions to specify the constants of integration. Boundary conditions have to employ the linearalized expression for the stress resultants and stress couples (3.3), including the contributions that were omitted in the axisymmetric formulation.

An example of the solution for a particular case where the applied pressure is a linear function of the radial coordinate is presented in the monograph of Jawad (2004). In this book the pressure applied to a simply supported plate was given by

$$p=f_1\frac{r}{a}\cos\theta \tag{3.53}$$

Following the procedure outlined above the deflection was obtained as

$$w=\frac{f_1}{192D}\left(\frac{r^5}{a}-\frac{2(5+\nu)}{3+\nu}ar^3+\frac{7+\nu}{3+\nu}a^3r\right)\cos\theta \tag{3.54}$$

Note that while the pressure in (3.53) is a linear function of the radial coordinate, the deflections and stresses that can be found using (3.54) are nonlinear functions of this coordinate.

While the solution shown in this paragraph is applicable for an arbitrary distribution of pressure applied to the plate as long as it can accurately be modeled by series (3.42), its limitations are evident. For example, a sector plate supported by concentric ring and radial stiffeners as shown in Fig. 3.6 and subject to a uniform pressure can be considered clamped along the radial stiffeners, as a result of symmetry of both geometry and load about each stiffener, i.e. the corresponding boundary conditions are $w = \frac{\partial w}{\partial \theta} = 0$. Obviously, such conditions cannot be satisfied by a series solution similar to that presented in this paragraph.

3.5 In-Plane Loading and Buckling of Circular Plates

Representative problems considered in this paragraph include bending of solid clamped plates subject to a combination of transverse pressure and radial compressive stress resultant, axisymmetric buckling of radially compressed simply supported and clamped solid circular plates, asymmetric buckling of clamped solid circular plates and buckling of annular plates.

3.5.1 Bending of a Solid Circular Plate Subject to Uniform Pressure and Compression

Consider the problem of bending of a solid circular plate subject to a combination of uniform transverse pressure p_0 and radial compression by the stress resultant $\bar{N}_r$ uniformly distributed along the edge $r = a$ (Nadai 1925). This problem is axisymmetric since the pressure is uniform.

Introducing the variable $\psi = -\frac{dw}{dr}$ Eq. 3.21 where $C_0 = 0$ to avoid singularity at the plate center becomes

$$\frac{d^2\psi}{dr^2} + \frac{1}{r}\frac{d\psi}{dr} + \left(\frac{n_r^2}{a^2} - \frac{1}{r^2}\right)\psi = -\frac{p_0}{D} \tag{3.55}$$

where $n_r^2 = \frac{\bar{N}_r a^2}{D}$ is a nondimensional applied stress resultant that is positive in compression.

The solution of (3.55) is (Timoshenko and Woinowsky-Krieger 1959):

$$\psi = C_1 J_1\left(\frac{n_r r}{a}\right) - \frac{p_0 r}{2\bar{N}_r}$$

$$w = \int \psi dr = C_1 \frac{a}{n_r} J_0\left(\frac{n_r r}{a}\right) - \frac{p_0 r^2}{4\bar{N}_r} + C_2 \tag{3.56}$$

where $J_1(\ldots)$ and $J_0(\ldots)$ are the Bessel functions of the first kind of the first and zero order, respectively, and C_1, C_2 are constants of integration.

If the plate is clamped around the boundary, the constants of integration C_1 and C_2 are specified from the conditions $w(r = a) = \psi(r = a) = 0$. Transformations yield

$$w = \frac{p_0 a^2}{2n_r^2 D}\left\{\frac{\left[J_0\left(\frac{n_r r}{a}\right) - J_0(n_r)\right]a^2}{n_r J_1(n_r)} - \frac{a^2 - r^2}{2}\right\} \tag{3.57}$$

It can be observed that as the compressive stress resultant increases deflections become larger and reach the infinite value if $J_1(n_r) = 0$. Of course, in reality, both physical and geometric nonlinearities have to be accounted for, i.e. the present solution becomes invalid at large deflections. It is obvious that similarly to the case of a combined transverse load and compression of rectangular plates, buckling load corresponds to $w = \infty$. Accordingly, the linear buckling stress resultant can be found from $J_1(n_r) = 0$. It can be shown that this condition yields

$$\bar{N}_{cr} = 14.68\frac{D}{a^2} \tag{3.58}$$

Denoting the nondimensional compressive load by $\tilde{N} = \frac{\bar{N}_r}{\bar{N}_{cr}}$, the deflection of the plate can approximately be represented as

$$w = \frac{w(p_0)}{1 - \tilde{N}} \tag{3.58a}$$

where $w(p_0)$ is the deflection of the plate subject to uniform pressure that is given by Eq. b in Example 3.1.

3.5.2 *Buckling of a Solid Circular Plate*

Although the solution for the buckling stress resultant of a clamped plate was shown in Sect. 3.5.1, it is useful to consider a "purely" buckling problem where the plate is subject to compression only. The problem of stability of clamped plates was first considered by Bryan (1890), while the simply supported plate was analyzed by Dinnik (1911).

The solution of the homogeneous equation (3.55) is

$$\psi = C_1 J_1\left(\frac{n_r r}{a}\right) + C_2 Y_1\left(\frac{n_r r}{a}\right) \tag{3.59}$$

where $Y_1\left(\frac{n_r r}{a}\right)$ is the Bessel function of the second kind of the first order.

The singularity caused by the function $Y_1\left(\frac{n_r r}{a}\right)$ at the center of a solid plate can be avoided only if $C_2 = 0$. If the solid plate is clamped along the boundary

$r = a$, the condition $\psi\,(r = a) = 0$ yields the buckling equation $J_1\,(n_r) = 0$ and the solution for the buckling stress resultant given by (3.57) that was shown in the previous section.

If the plate is simply supported, the corresponding boundary condition obtained using (3.12) is

$$M_r\,(r = a) = -D\left[\frac{d^2w(a)}{dr^2} + \frac{\nu}{r}\frac{dw(a)}{dr}\right] = D\left[\frac{d\psi(a)}{dr} + \frac{\nu}{r}\psi(a)\right] = 0 \quad (3.60)$$

The substitution of $\psi = C_1 J_1\left(\frac{n_r r}{a}\right)$ into this condition yields the value of the buckling stress resultant referred to above:

$$\bar{N}_{cr} = 4.20\frac{D}{a^2} \quad (3.61)$$

Thus, clamping increases the buckling load of the plate by nearly three and a half times.

The solution shown in this example was obtained by assumption that the mode shape of buckling of the plate is axisymmetric. An alternative buckling shape mode is asymmetric and the actually realized shape corresponds to the lowest buckling load (both axisymmetric and asymmetric cases should be analyzed and compared yielding the buckling load).

The asymmetric solution can be obtained by observing that as follows from (3.20), in linear buckling problems where in-plane stresses do not vary with radial coordinate $\sigma_r = \sigma_\theta$. In such case, the linear version of equations (3.9) yields

$$\nabla^4 w + \frac{\bar{N}_r}{D}\nabla^2 w = 0 \quad (3.62)$$

It can be verified by substitution that the solution of (3.62) corresponding to an asymmetric mode shape of instability with $m$nodal diameters and avoiding singularity at the plate center is

$$w = \left[C_1 J_m\left(\frac{n_r r}{a}\right) + C_2 r^m\right]\sin m\theta \quad (3.63)$$

$J_m\,(\ldots)$ being the Bessel function of the first kind and m-th order.

If the plate is clamped, boundary conditions specified above ($w\,(r = a) = \psi\,(r = a) = 0$) yield the following buckling equation:

$$\begin{bmatrix} \left[\dfrac{dJ_m\left(\frac{n_r r}{a}\right)}{d\left(\frac{n_r r}{a}\right)}\right]_{r=a} & m r^{m-1} \\ J_m\,(n_r) & r^m \end{bmatrix}\begin{Bmatrix} C_1 \\ C_2 \end{Bmatrix} = 0 \quad (3.64)$$

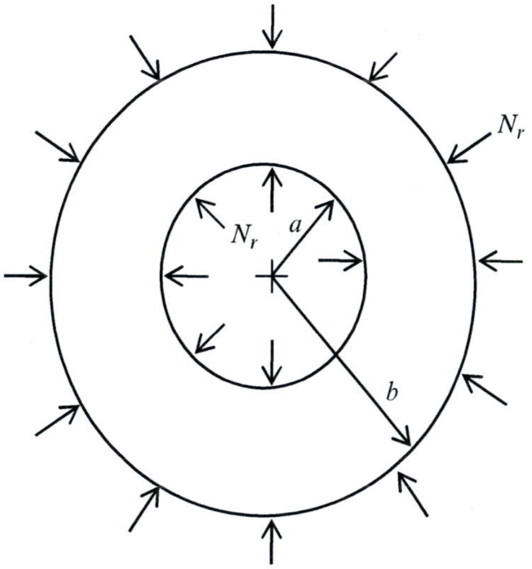

Fig. 3.7 Annular plate subject to radial compression along both edges

The nonzero solution of this equation obtained by requiring that the determinant of the matrix of coefficients is equal to zero results in $J_{m+1}(n_r) = 0$ (Vol'mir 1967). The buckling stress resultant corresponding to the smallest root that is found at m = 1 is equal to $\bar{N}_{cr} = 26.40\frac{D}{a^2}$. Obviously, axisymmetric buckling occurs at a lower applied stress resultant, i.e. asymmetric mode shape is not realized in the present problem.

3.5.3 Buckling of an Annular Plate

Compressed annular plates are found in circular bulkheads reinforced with ring stiffeners. Another example is a circular frame supporting a cylindrical shell subject to hydrostatic pressure.

Consider an annular plate loaded along the outer and inner edges by equal stress resultants $\bar{N}_r$ as shown in Fig. 3.7. In the case of an axisymmetric buckling the solution can be adopted from (3.59), complimented with the term that was neglected to avoid singularity in solid circular plates (see Eq. 3.21):

$$\psi = C_1 J_1\left(\frac{n_r r}{a}\right) + C_2 Y_1\left(\frac{n_r r}{a}\right) + \frac{C_3}{r} \tag{3.65}$$

Integrating (3.65), one obtains

$$w = \frac{a}{n_r}\left[C_1 J_0\left(\frac{n_r r}{a}\right) + C_2 Y_0\left(\frac{n_r r}{a}\right)\right] + C_3 \ln r + C_4 \tag{3.66}$$

On the other hand, if the mode shape of buckling is asymmetric, Eq. 3.63 is modified to account for terms neglected in the case of solid plates:

$$w = \left[C_5 J_m \left(\frac{n_r r}{a} \right) + C_6 r^m + C_7 Y_m \left(\frac{n_r r}{a} \right) + C_8 r^{-m} \right] \sin m\theta \tag{3.67}$$

Four constants of integration in either axisymmetric or asymmetric cases are determined from the boundary conditions. For example, Yamaki (1958) considered four cases:

(a) Clamping along both outer and inner edges;
(b) Simple support along both edges;
(c) Clamping along the outer edge and free deflection without rotations along the inner edge;
(d) Simple support along the outer edge and free deflection without rotations along the inner edge.

Other possible loading cases include annular plates subject to radial stresses applied along the inner edge and acing in the outward direction (Mansfield 1960, considered this problem for an infinite plate) or in the inward direction (Meissner 1933). Interestingly, buckling occurs in both these cases, even though the orientation of stresses in these two cases is opposite to each other. However, in both cases, one of the components of in-plane stresses in the polar coordinate system is compressive reflecting the Poisson effect. In particular, if the stresses act in the outward direction, radial stresses σ_r are compressive, while circumferential (tangential) stressesσ_θ are tensile. In this case, the mode shape of instability is axisymmetric. If radial stresses applied along the inner edge are oriented toward the center of the plate, the radial stresses are tensile, but the tangential stresses are compressive resulting in an asymmetric mode shape of instability with $m = 2$. Buckling of annular plates with the outer edge clamped and the inner edge free loaded by compressive radial stresses applied along the outer edge was considered by Majumdar (1968).

3.6 Bending of Plates of Non-rectangular and Non-circular Shapes

Plates of found in applications may have triangular, skewed, trapezoidal, oval and elliptical shapes. Most of these plates are analyzed by numerical methods using commercially available software. However, there are several exceptions where an accurate analytical solution, often tracing back to the beginning of the twentieth century, is available.

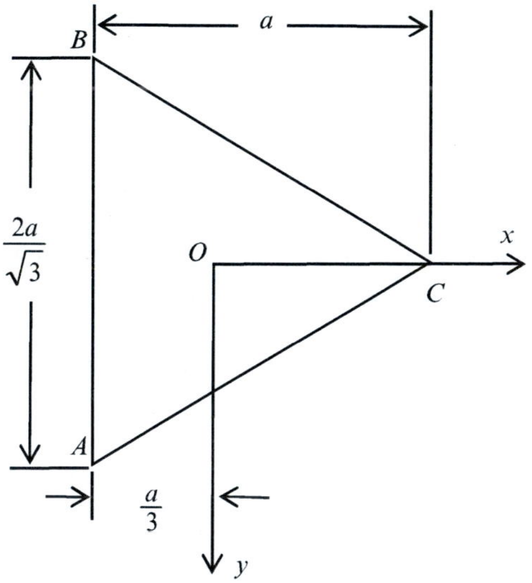

Fig. 3.8 Equilateral triangular plate

3.6.1 Bending of Equilateral Triangular Plates Subject to Uniform Pressure

An example where the problem of bending of triangular plates is encountered in found isogrid plates where the skin is supported by intersecting ribs forming a triangular pattern. Uniform pressure represents typical load acting on such plates.

Consider a triangular plate shown in Fig. 3.8 where all angles are equal to 60° and all edges have equal length (equilateral triangular plate). The edges of the plate are simply supported. Equations of the boundaries of the triangular plate shown in Fig. 3.8 are

$$
\begin{aligned}
AB: &\quad x + \frac{a}{3} = 0 \\
AC: &\quad \frac{x}{\sqrt{3}} + y - \frac{2a}{3\sqrt{3}} = 0 \\
BC: &\quad \frac{x}{\sqrt{3}} - y - \frac{2a}{3\sqrt{3}} = 0
\end{aligned}
\tag{3.68}
$$

The problem of simply supported equilateral plates subject to stress couples uniformly distributed along all edges was discussed by Timoshenko and Woinowsky-Krieger (1959). A more relevant for practical applications situation where the load is represented by uniform pressure was also discussed in this book. In this case, deflections in the form

$$w = \frac{p_0}{64aD}\left[x^3 - 3xy^2 - a\left(x^2 + y^2\right) + \frac{4a^3}{27}\right]\left(\frac{4}{9}a^2 - x^2 - y^2\right) \tag{3.69}$$

satisfy both boundary conditions as well as the equation of equilibrium. Subsequently, the stresses in the plate can be calculated using constitutive equations in the rectangular coordinate system. The largest bending stresses occur along the lines bisecting the angles of the triangle where in-plane shear stress is equal to zero due to symmetry. In particular, if the Poisson ratio of the material is equal to 0.3, these stresses are:

$$\sigma_x^{\max} = \pm 0.1488 p_0 \left(\frac{a}{h}\right)^2 \quad (x = -0.062a)$$

$$\sigma_y^{\max} = \pm 0.1554 p_0 \left(\frac{a}{h}\right)^2 \quad (x = 0.129a) \tag{3.70}$$

3.6.2 *Bending of Isosceles Triangular Plates*

The problem where a simply supported isosceles triangular plate OBC is subject to a concentrated force as shown in Fig. 3.9 can be solved by the method of images introduced by Nadai (1925). The method is based on introducing an imaginary symmetric plate shown by dashed lines in the figure resulting in a square plate. A fictitious force equal to the applied force but acting in the opposite direction is applied at the point that mirrors the point of application of the load relative to the diagonal BC of the square plate. As a result of antisymmetry of forces F and F' with respect to the diagonal BC, the deflections along this diagonal are equal to zero. This is a boundary condition for the triangular plate OBC. Subsequently, deflections of the plate due to forces F and F' are obtained by superposition of deflections due to each of these forces that are available from Example 2.3. These deflections that satisfy boundary conditions along edges OB and OC are:

$$w = w(F) + w\left(F'\right) = \frac{4Fa^2}{\pi^4 D}\sum_{m=1}^{M}\sum_{n=1}^{N}\frac{1-(1)^{m+n}}{(m^2+n^2)^2}\sin\alpha_m\xi\,\sin\alpha_n\eta\,\sin\alpha_m x\,\sin\alpha_n y \tag{3.71}$$

The solution of the problem where a simply supported right triangular plate is subject to uniform pressure can be obtained from (3.71) by substituting $F = p_0 d\xi\, d\eta$ into this equation and integrating over the surface of the plate. The result is

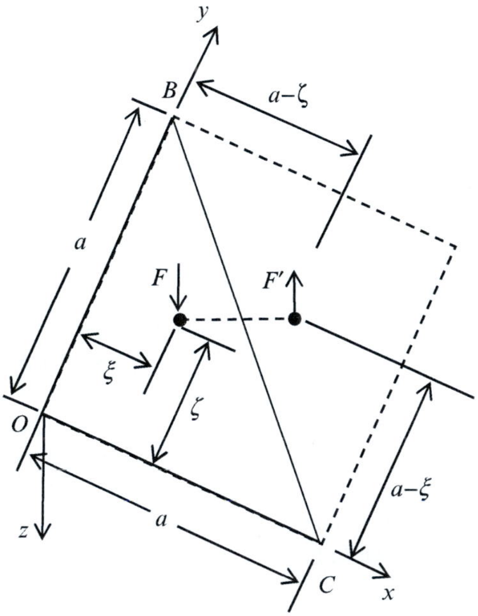

Fig. 3.9 Isosceles triangular plate

$$w = \frac{16 p_0 a^4}{\pi^6 D}\left[\begin{array}{l}\sum\limits_{m=1,3,5,\ldots}\sum\limits_{n=2,4,6,\ldots}\dfrac{n}{m\left(n^2-m^2\right)\left(m^2+n^2\right)^2}\sin\alpha_m x\sin\alpha_m y+\\ \sum\limits_{m=2,4,6,\ldots}\sum\limits_{n=1,3,5,\ldots}\dfrac{m}{n\left(m^2-n^2\right)\left(m^2+n^2\right)^2}\sin\alpha_m x\sin\alpha_m y\end{array}\right] \tag{3.72}$$

3.6.3 Bending and Buckling of Skew Plates

Skew plates (Fig. 3.10) represent practical interest for airplane wings and other applications. Unfortunately, the analysis of such plates is complicated since the equation of equilibrium transferred into the oblique coordinate system defies an exact solution. The results are usually obtained numerically. Representative procedures can be found in the monograph of Szillard (2004) and papers of Rao and Farran (1986), York (1996), Huyton and York (2001, 2006). For example, Huyton and York (2001) presented design curves for buckling strength of thin skew plates of various aspect ratios, skew angle and rotational restraint of edges.

An example of a numerical bending analysis of clamped skew plates experiencing large deformations as a result of uniform pressure or a concentrated force is presented in Fig. 3.11 (Duan and Mahendran 2003). Load-deflection curves are

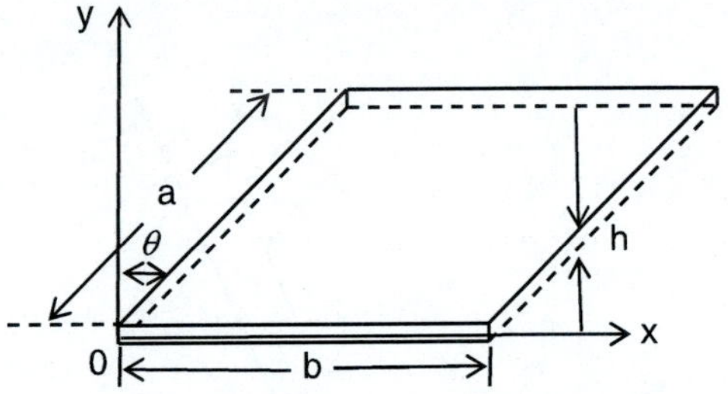

Fig. 3.10 Skew plate

presented in this figure for various aspect ratios (a and b being half-lengths of skew and longitudinal edges, respectively). The nondimensional maximum deflection normalized with respect of the plate thickness is shown as a function of uniform pressure (in the paper, q denotes the uniform pressure) and a concentrated force P. The effect of the aspect ratio on bending is evident in this figure. Furthermore, geometric nonlinearity becomes noticeable at the deflection exceeding half-thickness of the plate. The skew angle is another parameter that affects the response of skewed plates (this effect is not shown in Fig. 3.11).

3.6.4 Response of Elliptical and Super-Elliptical Plates

Elliptical plates are used as hatch covers where the absence of corners is preferable due to the requirement to reduce stress concentration. The cover is sometimes bolted to the rim around the hatch providing conditions that approach clamping. In the case where clamping cannot be assured it is often justified to use the assumption of a simply supported edge. The load applied to these plates is almost invariably represented by a uniform pressure.

Consider a so-called super-elliptical plate the boundary equation

$$\left(\frac{x}{a}\right)^{2k} + \left(\frac{y}{b}\right)^{2k} = 1 \tag{3.73}$$

where a and b are the major and minor semi-axes and k is a real number. Static and dynamic analyses of such plates have been published by a number of authors (Leissa 1967; Wang et al. 1994; Altekin and Altay 2008; Ceribasi and Altay 2009). However, with the exception of the case $k = 1$ corresponding to an elliptical plate, the solutions are numerical. For example, the Rayleigh-Ritz method that does not require satisfaction of natural (static) boundary conditions can be applied representing the deflection of a simply supported plate by

$$w = \left[\left(\frac{x}{a}\right)^{2n} + \left(\frac{y}{b}\right)^{2n} - 1\right]\left[C_{00} + C_{20}\left(\frac{x}{a}\right)^{2} + C_{02}\left(\frac{y}{b}\right)^{2}\right] \tag{3.74}$$

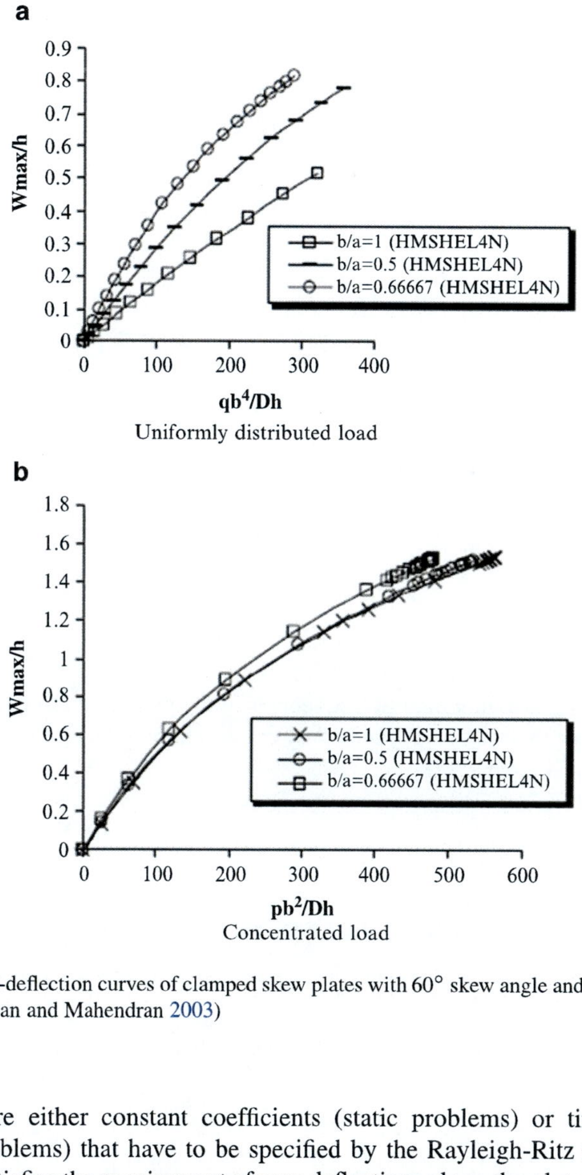

Fig. 3.11 Load-deflection curves of clamped skew plates with 60° skew angle and different aspect ratios (From Duan and Mahendran 2003)

where C_{ij} are either constant coefficients (static problems) or time functions (dynamic problems) that have to be specified by the Rayleigh-Ritz method. This expression satisfies the requirement of zero deflections along the plate boundary but violates the condition of zero bending stress couples in the planes perpendicular to the boundary. If the plate is clamped, the previous formula for deflections can be modified by taking a square of the first term in the right side and satisfying all boundary conditions. Other expressions for the deflection were used to solve dynamic problems for super-elastic plates with a variable thickness and/or to

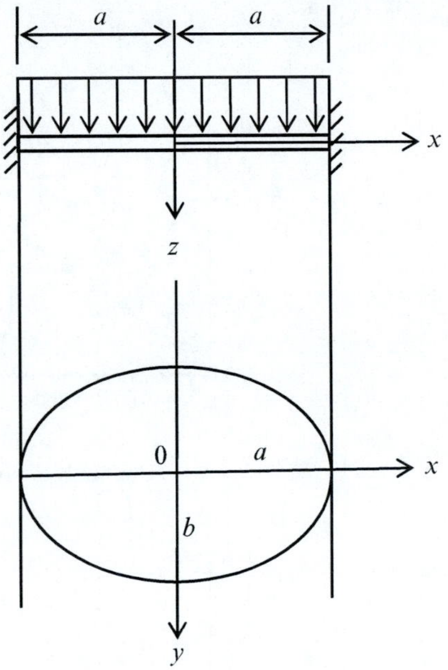

Fig. 3.12 Elliptical plate clamped around the boundary and subject to a uniform pressure

accurately evaluate higher natural frequencies of the plate (e.g., Ceribasi and Altay 2009) as well as to analyze such boundary conditions as point-supports of the edge (e.g., Altekin and Altay 2008).

If $k = 1$ in (3.73), the geometry is reduced to classical elliptical plates extensively investigated in literature (e.g., Timoshenko and Woinowsky-Krieger 1959). We will consider the case of the plate subject to uniform pressure (Fig. 3.12). If the plate is clamped around the boundary, the corresponding boundary conditions $w = \frac{\partial w}{\partial n} = 0$ where n is a normal to the edge are identically satisfied by representing the deflection in the form

$$w = W\left(1 - \frac{x^2}{a^2} - \frac{y^2}{b^2}\right)^2 \tag{3.75}$$

The constant W is found by substituting (3.75) into (2.1):

$$W = \frac{p_0}{8D} \frac{a^4 b^4}{3a^4 + 2a^2 b^2 + 3b^4} \tag{3.76}$$

The corresponding maximum stresses are available through the substitution of (3.75) into the expressions for changes of curvature and twist (1.29) and the subsequent substitution of (1.29) into the plane-stress version of the constitutive relations (2.9a). The highest bending stresses occur on the plate surfaces at the center of the plate and at the ends of the major and minor axes of the ellipse:

$$
\begin{aligned}
&x = y = 0: && \sigma_x^{\max} = \pm\frac{24DW}{h^2}\left(\frac{1}{a^2}+\frac{\nu}{b^2}\right), && \sigma_y^{\max} = \pm\frac{24DW}{h^2}\left(\frac{1}{b^2}+\frac{\nu}{a^2}\right) \\
&x = \pm a,\ y = 0: && \sigma_x^{\max} = \mp\frac{48DW}{a^2h^2}, && \sigma_y^{\max} = \mp\frac{48\nu DW}{a^2h^2} \\
&x = 0,\ y = \pm b: && \sigma_x^{\max} = \mp\frac{48\nu DW}{b^2h^2}, && \sigma_y^{\max} = \mp\frac{48DW}{b^2h^2}
\end{aligned}
\qquad (3.77)
$$

At all locations listed above the in-plane shear stress is equal to zero. The change in the sign of the stress from the center of the plate to boundaries on the same plate surface reflected in Eq. 3.77 occurs since the stress couples applied to the plate by the clamped boundary are reactive. Accordingly, they have the opposite direction as compared to the bending stress couples at the center of the plate (the situation is similar to a distribution of moments in a clamped beam subject to pressure). Note that if $a = b$ the present solution reduces to the results for circular plates. The results for clamped elliptical plates subject to a linearly varying pressure and for simply supported elliptical plates undergoing uniform pressure are presented by Timoshenko and Woinowsky-Krieger (1959).

3.6.5 Sector Plate Subject to Bending

An example of a sector plate is shown in Fig. 3.6 where a circular bulkhead is reinforced by ring and radial stiffeners. If the sector angle is small, the analysis should account for the influence of straight edges on bending of the plate and employ an asymmetric approach. Besides numerical procedures usually utilizing finite element or finite difference methods, the analytical solution is available as is demonstrated in this section.

The boundary conditions along the straight edges of the sector plate shown in Fig. 3.13 can be modeled as clamping or simple support. The former condition is realized if the plate represents a part of the circular bulkhead divided into identical sector plates by radial stiffeners (Fig. 3.6) and loaded by a hydrostatic pressure. In such case clamping is effectively enforced through the symmetry of the adjacent sector plates with respect to the radial stiffener as well as because uniform pressure is also symmetric with respect to the stiffener. Simple support along the straight edges is the boundary condition for a single sector plate or in case where the symmetry of pressure or geometry with respect to the radial stiffener is violated. In this section we consider a sector plate simply supported along straight edges and subject to a uniform pressure p_0.

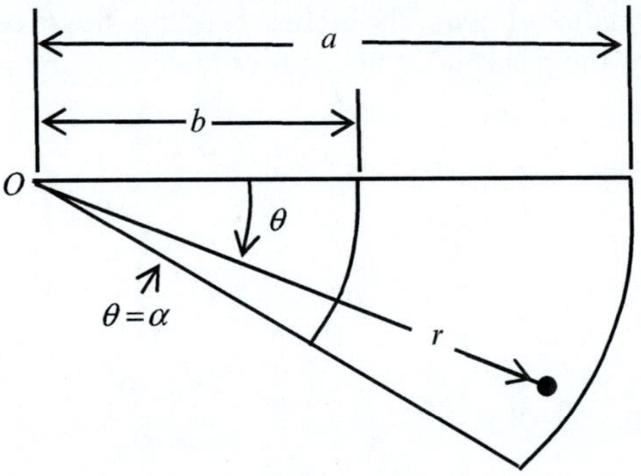

Fig. 3.13 Sector plate. Straight boundaries: $\theta = 0,\ \theta = \alpha$. Curved boundaries: $r = a,\ r = b$

The conditions of simple support along the straight edges $\theta = 0$ and $\theta = \alpha$ are satisfied if

$$w = 0, \quad \frac{\partial^2 w}{\partial \theta^2} = 0 \tag{3.78}$$

The pressure can be represented in Fourier series as follows:

$$p = \frac{4p_0}{\pi} \sum_{n=1,3,\ldots}^{\infty} \frac{1}{n} \sin \frac{n\pi\theta}{\alpha} \tag{3.79}$$

Of course, the series in (3.79) and subsequent solution are truncated, i.e. the "infinite" number of terms should not be understood literally.

The equation of equilibrium for asymmetric bending available from the first equation (3.9), i.e.$D\nabla^4 w = p$, and boundary conditions (3.78) are satisfied if the particular integral of the equation of equilibrium is represented in Fourier series (Nadai 1925)

$$w_p = \frac{4p_0 r^4}{\pi D} \sum_{n=1,3,\ldots}^{\infty} \frac{1}{n\,(16 - n^2\pi^2/\alpha^2)\,(4 - n^2\pi^2/\alpha^2)} \sin \frac{n\pi\theta}{\alpha} \tag{3.80}$$

The general solution of the homogeneous equation of equilibrium that also satisfies boundary conditions (3.78) is

$$w_h = \sum_{n=1}^{\infty} \left(A_n^* r^{n\pi/\alpha} + B_n^* r^{-n\pi/\alpha} + C_n^* r^{2+n\pi/\alpha} + D_n^* r^{2-n\pi/\alpha}\right) \sin \frac{n\pi\theta}{\alpha} \tag{3.81}$$

where n are odd numbers and constants of integration A_n^* through D_n^* are determined from boundary conditions along the curved edges.

The solution can now be obtained by substituting the deflection $w = w_h + w_p$ into the boundary conditions along the edges $r = b$ and $r = a$. There are two conditions along each of these edges that can be used to specify four constants of integration for each integer n.

3.7 Design Philosophy and Recommendations

Plates of a non-rectangular shape are found in engineering applications, although they are less typical than rectangular counterparts. Among these plates, circular plates have been most intensively investigated. Besides their practical applications, the reason for the interest to circular plates is the convenience of using a polar coordinate system employed to characterize their geometry. Other plate shapes are seldom characterized in rectangular or polar coordinate systems, although there are several exceptions, such as elliptical and triangular configurations discussed in the chapter.

While the shape of the plate may prevent an analytical solution, this is seldom a limitation due to the availability of numerical procedures. On the other hand, technological considerations sometimes impose constraints on the shape of plates used in applications. These limitations are due to the convenience of the manufacture as well as the need to avoid sharp angles formed by the plate boundaries that cause stress concentrations.

The number of exact analytical solutions for non-rectangular plates is limited. In particular, numerous axisymmetric problems for circular and annular plates have a closed form solution. The solutions for the case of large axisymmetric deformations of such plates are also available, but such problems are usually treated numerically. Analytical solutions are also available for asymmetrically deformed circular and sector plates characterized by a geometrically linear theory. Other plates that can be analyzed using convenient analytical solutions include elliptical plates as well as equilateral and isosceles triangular plates with boundary conditions specified in the chapter.

Practical recommendations regarding the distinction between geometrically linear and nonlinear formulations for circular plates and plates of other shapes are identical to those for rectangular plates. Designs based on a geometrically linear analysis become unnecessary conservative if deflections exceed half-thickness of the plate. In case where linear solutions predict larger deflections, it is appropriate to use a nonlinear analysis for the evaluation of stresses and deformations.

It is useful to mention here that besides the shape of the plate dictated by geometry of the structure, plates of irregular shapes may be encountered as a result of using non-straight stringers to optimize the structural performance. For example, if the orientation of the stringer follows the path of the largest principal stresses in the plate, the sections of the plate between such stringer and other support structures may not be rectangular. Such situations may arise as a result of the desire to reduce the weight of the structure, particularly in the aerospace industry.

References

Altekin, M., & Altay, G. (2008). Static analysis of point-supported super-elliptical plates. *Archives of Applied Mechanics, 78*, 259–266.

Bryan, G. H. (1890). On the stability of a plate under thrusts in its own plane with applications to the "buckling" of the sides of a ship. *Proceedings of the London Mathematical Society, 22*, 55–67.

Ceribasi, S., & Altay, G. (2009). Free vibration of super elliptical plates with constant and variable thickness by Ritz method. *Journal of Sound and Vibration, 319*, 668–680.

Dinnik, A. N. (1911). *On stability of a compressed circular plate*. Proceedings of the Kiev Polytechnic Institute, Kiev, (In Russian).

Duan, M., & Mahendran, M. (2003). Large deflection analyses of skew plates using hybrid/mixed finite element method. *Computers and Structures, 81*, 1415–1424.

Huyton, P. B., & York, C. B. (2001). Buckling of compression loaded skew plates with rotational edge restraint. *ASCE Journal of Aeronautical Engineering, 14*, 92–101.

Huyton, P. B., & York, C. B. (2006). *Buckling of skew plates with planform taper*. Paper AIAA-2006-1572.

Jawad, M. H. (2004). *Design of plate and shell structures*. New York: ASME Press.

Leissa, A. W. (1967). Vibration of a simply-supported elliptical plate. *Journal of Sound and Vibration, 6*, 145–148.

Majumdar, S. (1968). *Buckling of thin annular plates due to radial compressive loading*, Thesis for the degree of aeronautical engineer, California Institute of Technology, Pasadena, CA.

Mansfield, E. H. (1960). On the buckling of an annular plate. *Quarterly Journal of Mechanics and Applied Mathematics, 13*, 16–23.

McFarland, D. E., Smith, B. L., & Bernhart, W. D. (1972). *Analysis of plates*. New York: Macmillan.

Meissner, E. (1933). Ueber das Knicken kreisformiger Scheiben. *Schweizische Bauzeitung, 101*, 87–89.

Nadai, A. (1925). *Die elastische platten*. Berlin, Germany: Springer.

Rao, H. V. S. G., & Farran, H. J. (1986). Macro-element analysis of skewed and triangular orthotropic thin plates with beam boundaries. *Computers & Structures, 22*, 399–404.

Reddy, J. N. (1999). *Theory and analysis of elastic plates*. Philadelphia: Taylor & Francis.

Reddy, J. N. (2007). *Theory and analysis of elastic plates and shells*. Boca Raton, FL: CRC Press/Taylor & Francis Group.

Szillard, R. (2004). *Theory and application of plate analysis: Classical, numerical and engineering methods*. New York: Wiley.

Timoshenko, S., & Woinowsky-Krieger, S. (1959). *Theory of plates and shells*. New York: McGraw-Hill.

Ugural, A. C. (1999). *Stresses in plates and shells* (2nd ed.). New York: McGraw-Hill (same information is available in the third edition published in 2009).

Vol'mir, A. S. (1967). *Stability of deformable systems*. Moscow: Nauka Publishing House (in Russian).

Wang, C. M., Wang, L., & Liew, K. M. (1994). Vibration and buckling of superelliptical plates. *Journal of Sound and Vibration, 171*, 301–314.

Yamaki, N. (1958). Buckling of a thin annular plate under uniform compression. *Journal of Applied Mechanics, 25*, 267–273.

York, C. B. (1996). Influence of continuity and aspect-ratio on the buckling of skew plates and plate assemblies. *International Journal of Solids and Structures, 33*, 2133–2159.

Chapter 4
Dynamic Problems in Isotropic Plates

Plates found in applications are often subject to dynamic loads. These loads can be directly applied to the plate (i.e., wave impact, wind gusts, blast overpressure, impact by birds or other objects, etc.). In numerous applications, dynamic loads are applied to the plate by unbalanced rotating machinery supported by the plate or through the kinematic excitation by beams that support both the plate and the engine. In all these problems, the structural integrity of the plate has to be analyzed to prevent immediate failure due to excessive dynamic stresses or fatigue damage as a result of continuous large-amplitude vibrations. The present chapter provides an insight into vibrations of isotropic plates, including free and forced vibrations, response to non-harmonic dynamic loads, large-amplitude vibrations and dynamic instability. Dynamic problems of composite plates that can be investigated using an extension of analytical and numerical tools employed for the analysis of their isotropic counterparts are outside the scope of this book. An exception applicable to the analysis of composite plates is vibration of plates reinforced with stringers whose constitutive equations and equations of motion resemble those of composite plates (Sect. 4.5).

4.1 Typical Problems

The subject of dynamics of plates refers to multiple and diverse problems that have to be addressed by a designer. To better understand the subject we begin with the classification of relevant problems.

Dynamic problems can be subdivided into the following classes:

1. Free vibrations;
2. Forced vibrations due to a harmonic excitation;
3. Forced non-periodic response to a non-harmonic load;
4. Forced response to in-plane (parametric) dynamic load.

V. Birman, *Plate Structures*, Solid Mechanics and Its Applications 178,
DOI 10.1007/978-94-007-1715-2_4, © Springer Science+Business Media B.V. 2011

The solutions of all these classes of problems depend on the motion being characterized by a geometrically linear or nonlinear theory. It is rather typical to consider the motion linear if the maximum dynamic deflection of the plate does not exceed its half-thickness, while nonlinear solutions are applied in case of larger deformations. In addition, damping can have a significant effect on the response of the plate. If the response is considered by assumption that damping is negligible, the error is usually small as long as the plate is not in the narrow range of frequencies corresponding to resonance conditions discussed below. Note that plates are continuous systems whose vibrations share many features with vibrations of beams, shells and other structures. A brief outline of the classes of problems listed above is provided in this paragraph.

Free vibrations refer to the motion of the plate that occurs if it is displaced from static equilibrium by an initial displacement, rotation or an impulse resulting in initial velocity. The motion following the removal of the initial excitation is referred to as "free vibrations" since the plate vibrates in the absence of external driving force or moment. The frequencies of free vibrations are called natural frequencies (the lowest natural frequency is referred to as the fundamental frequency). The number of natural frequencies equals the number of degrees of freedom of the system. Plates are characterized by a distributed mass over the domain occupied in space, rather than a system of concentrated masses. Accordingly, the number of degrees of freedom of the plate is infinite. While the natural frequencies are determined from the equations of motion, the amplitudes of free motion are specified from the initial conditions. Damping results in a decrease of natural frequencies, although in plates such decrease is usually small and can be neglected. The only significant effect of damping in case of free vibrations of plates is a reduction and eventual suppression of motion with time. This occurs since while additional energy is not supplied to the plate experiencing free vibrations, the available energy is "spent" overcoming the resistance due by damping.

The motion of plates subject to a harmonic-in-time driving load represents a superposition of free vibrations and forced response. The former are usually quickly eliminated through damping, even if its effect on forced motion is negligible. A typical frequency-amplitude relationship of a plate is schematically shown in Fig. 4.1. The response is shown for a spectrum of driving frequencies in the range overlapping the lowest three natural frequencies of the plate. Note that the amplitude of motion becomes infinite when the frequency of the applied load (driving frequency) is equal to one of the natural frequencies. This condition called "resonance" explains the reason for the free vibration analysis where a typical goal is to design the system (plate) with natural frequencies that are remote from anticipated driving frequencies. Consider an example, where the same beams support an unbalanced engine and the plate. In such case, the plate natural frequencies should not coincide with the frequencies of dynamic loads transmitted by the engine to the beams since these are also the frequencies of kinematic excitation applied by the same beams to the plate.

While amplitudes of motion corresponding to the resonance in Fig. 4.1 are infinite, in reality they are limited due to unavoidable damping present in the system.

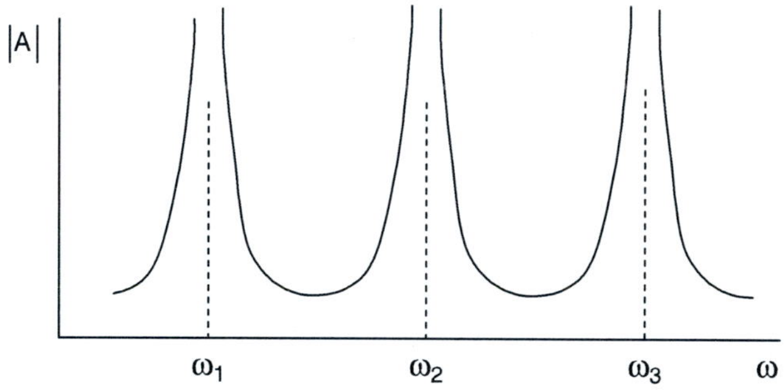

Fig. 4.1 Forced vibrations of a continuous system (plate) without accounting for damping. The absolute value of the amplitude is shown as a function of the driving frequency (frequency of the applied load), ω_i are the natural frequencies of the system

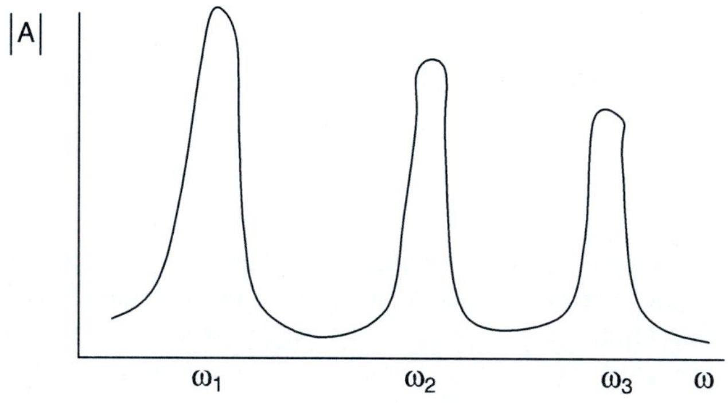

Fig. 4.2 Forced vibrations of a continuous system (plate) accounting for damping

The effect of damping on the forced motion is reflected in Fig. 4.2. The most significant effect is on the motion at resonance where damping results in limited amplitudes. This effect becomes more pronounced at higher resonance frequencies. Accordingly, resonances occurring at the fundamental frequency of motion and at other low natural frequencies are often considered the "most damaging." The other consequence of damping is a small "shift" of the resonance frequencies to smaller values compared to those evaluated without damping (in typical plate structures such shift is usually negligible).

The designer should be concerned with two damage scenarios. The plate may collapse due to excessive amplitude of dynamic stresses. This mode of failure can be predicted based on the analysis of forced motion (e.g., Sect. 4.3). The other possible mode of failure is associated with continuous vibrations that do not immediately result in the collapse, but rather generate fatigue damage that may occur after millions of cycles of motion. Fatigue analysis has to account for coupling of the vibration and fatigue problems (e.g., the stiffness of a plate with fatigue

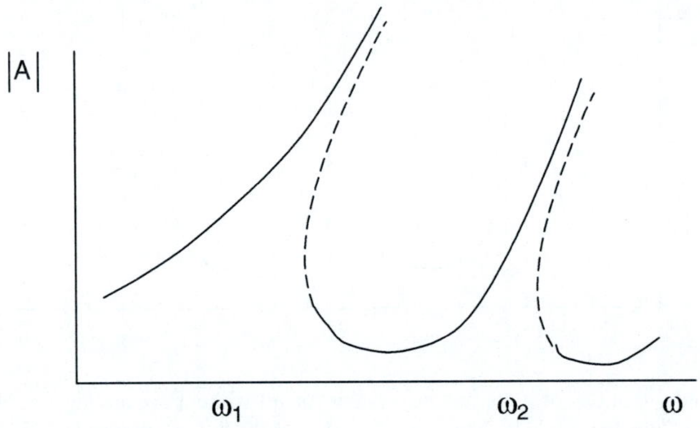

Fig. 4.3 Geometrically nonlinear frequency-amplitude relationship of system (plate) with a hardening nonlinearity

cracks is reduced affecting the amplitudes of motion and fatigue damage during the subsequent life).

Non-periodic response of a plate to such loads as impact or blast overpressure is analyzed using standard methods of the theory of vibration combined with the theory of plates (Sect. 4.4). The solution is relatively straightforward if the motion is linear since in such case equations of motion can often be reduced to uncoupled ordinary differential equations where time serves as an argument. Every uncoupled equation in such formulation characterizes a single mode of motion. The solution of uncoupled equations is available by initial value methods or by the convolution integral.

Both free and forced vibrations of the plate can be affected by geometrically nonlinear effects that become essential at large amplitudes of motion. Plates possess a hardening nonlinear response meaning that forced vibration frequency-amplitude relationships "bend" to the right (to larger frequencies) on the frequency-amplitude plane (Fig. 4.3). The response is schematically shown in Fig. 4.3 for driving frequencies in the range of the lowest two natural frequencies of the plate. In the presence of damping the amplitude of motion is limited. The "middle" brunches identified by broken curves in Fig. 4.3 are not realized since the corresponding motion is unstable. Examples of geometrically nonlinear problems of isotopic plates are discussed in Sect. 4.6.

In addition to problems where dynamic load is applied to the surface of the plate, these structures sometimes operate under in-plane dynamic forces. If such forces are supplemental to transverse loading, they affect the motion of the plate. In the situation where in-plane dynamic forces represent the only load applied to the plate they may result in parametric or dynamic instability characterized by large-amplitude transverse vibrations. An example of such problem is illustrated in Sect. 4.7.

The chapter also includes a paragraph on forced vibrations of reinforced plates (Sect. 4.5) since such problems are routinely encountered by designers. The analysis of such vibrations should be concerned with possible resonance of both the entire plate with the stringers as well as resonance of sections of the plate between the stringers and resonance of the stringers. The reason is that each of these structures, i.e. the reinforced plate, sections of the plate between the stringers and the stringers themselves, have their own natural frequencies. Accordingly, undesirable large-amplitude vibrations occur if the frequency of driving forces coincides with any of these natural frequencies.

4.2 Free Vibrations of Rectangular Isotropic Plates

Consider an isotropic rectangular plate with geometry corresponding to that in Fig. 2.11. The plate is subject to in-plane static stress resultants $\bar{N}_x$ and $\bar{N}_y$. Free linear vibrations of the plate are analyzed using the equation of equilibrium (2.38) where following D'Alembert's principle pressure is replaced with the inertial term:

$$D\nabla^4 w = -\hat{m}\frac{\partial^2 w}{\partial t^2} + \bar{N}_x \frac{\partial^2 w}{\partial x^2} + \bar{N}_y \frac{\partial^2 w}{\partial y^2} \tag{4.1}$$

Usually in dynamic problems boundary conditions are not explicitly affected by time. However, designers should be aware that this statement is correct only if boundary structures providing support to the plate are "rigid," i.e. they do not participate in motion. An example where vibrations of boundary structures interact with and affect dynamics of the plate would be the case where the plate was supported by flexible beams. In such case it is necessary to consider the motion of the entire plate-beam system.

In this paragraph we consider the case of a simply supported plate where boundary conditions correspond to (2.2). The motion of the plate is represented by series (2.5) where the amplitude of each harmonic is a function of time:

$$w = \sum_{m=1}^{M}\sum_{n=1}^{N} W_{mn}(t) \sin \alpha_m x \sin \beta_n y \tag{4.2}$$

Series (4.2) satisfy the boundary conditions. The functions of time are harmonic, so that

$$W_{mn}(t) = A_{mn} \sin \omega_{mn} t + B_{mn} \cos \omega_{mn} t \tag{4.3}$$

where A_{mn} and B_{mn} are constants and ω_{mn} are frequencies corresponding to motion with m and n halfwaves in the x- and y-directions, respectively.

Upon the substitution of (4.2) into (4.1) the equations of motion for each harmonic are uncoupled:

$$\left[\alpha_m^4 + 2\alpha_m^2\beta_n^2 + \beta_n^4 - \frac{\hat{m}}{D}\omega_{mn}^2 + \frac{\alpha_m^2}{D}\bar{N}_x + \frac{\beta_n^2}{D}\bar{N}_y\right] W_{mn}(t) = 0 \tag{4.4}$$

The natural frequency corresponding to the $mn-th$ mode of motion is obtained from (4.4). In the absence of in-plane stress resultants it is given by

$$\omega_{mn} = \pi^2 \sqrt{\frac{D}{\hat{m}}}\left(\frac{m^2}{a^2} + \frac{n^2}{b^2}\right)^2 \tag{4.5}$$

If in-plane stress resultants are present, the expression for the natural frequency becomes

$$\omega_{mn} = \omega_{mn}\left(\bar{N}_x = \bar{N}_y = 0\right)\sqrt{1 - \frac{\bar{N}_x}{\bar{N}_{x,cr}\,(m,n)} - \frac{\bar{N}_y}{\bar{N}_{y,cr}\,(m,n)}} \tag{4.6}$$

where the frequency $\omega_{mn}\left(\bar{N}_x=\bar{N}_y=0\right)$ is given by (4.5) and the terms $\bar{N}_{x,cr}\,(m,n)$ and $\bar{N}_{y,cr}\,(m,n)$ are critical loads of the plate subject to compression along the x- and y-axes, respectively (see Sect. 2.6).

The fundamental (lowest) frequency of the plate without in-plane loading is available from (4.5) using $m = n = 1$. As is evident from (4.5) this frequency always corresponds to a single half-wave along both pairs of parallel edges, irrespectively of the plate aspect ratio.

It can be observed from (4.6) that the natural frequencies are reduced if the plate is subject to compressive in-plane loads. In the contrary, tension results in larger natural frequencies. The natural frequency becomes equal to zero when applied loads reach a critical combination, i.e. the term under the square root in the right side of (4.6) becomes equal to zero.

Boundary conditions have a profound effect on natural frequencies of plates. This is reflected in Table 4.1 where the lowest five frequencies of square plates are compared for several representative boundary conditions (the Poisson ratio in all cases was equal to 0.3).

The analysis of free vibration motion requires us to specify constants A_{mn} and B_{mn} in (4.3) that depend on initial conditions. Consider the case where both the initial deflection and initial velocity of the plate are prescribed being functions of the x- and y-coordinates. Then they can be represented in the form resembling series (4.2):

$$w\,(x,y,t=0) = \sum_{m=1}^{M}\sum_{n=1}^{N} W_{mn}(0)\sin\alpha_m x \sin\beta_n y$$

$$\dot{w}\,(x,y,t=0) = \sum_{m=1}^{M}\sum_{n=1}^{N} \dot{W}_{mn}(0)\sin\alpha_m x \sin\beta_n y \tag{4.7}$$

Table 4.1 Natural frequencies of square isotropic plates with various boundary conditions (Based on data from Leissa (1973))

Boundary conditions	k for fundamental frequency	k (second mode)	k (third mode)	k (fourth mode)	k (fifth mode)
SSSS	19.74	49.35	78.96	98.70	128.30
CFCF	22.27	26.53	43.66	61.47	67.55
CSCS	28.95	54.74	69.33	94.58	102.21
CCCC	36.00	73.41	108.27	131.64	132.24
CCFF	6.94	24.03	26.68	47.79	63.04
CFFF	3.49	8.52	21.43	27.33	31.11
FFFF	13.49	19.79	24.43	35.02	61.53

$\omega = k\sqrt{\frac{D}{\hat{m}a^4}}$ S = simply supported edge, C = clamped edge, F = free edge

where the dot identifies the derivative with respect to time, i.e. $\dot{(\ldots)} = \frac{\partial(\ldots)}{\partial t}$, and $W_{mn}(0)$ and $\dot{W}_{mn}(0)$ are available from

$$W_{mn}(0) = \frac{4}{ab}\int_0^b\int_0^a w(x,\ y,\ 0)\sin\alpha_m x\sin\beta_n y dx dy$$

$$\dot{W}_{mn}(0) = \frac{4}{ab}\int_0^b\int_0^a \dot{w}(x,\ y,\ 0)\sin\alpha_m x\sin\beta_n y dx dy \tag{4.8}$$

Constants A_{mn} and B_{mn} are now obtained using (4.3) and (4.8):

$$B_{mn} = W_{mn}(0),\quad A_{mn} = \frac{\dot{W}_{mn}}{\omega_{mn}} \tag{4.9}$$

4.3 Forced Harmonic Vibrations of Rectangular Isotropic Plates

Forced vibrations of rectangular isotropic panels can be studied by modifying the equation of motion to account for the driving load. This load is usually applied in the form of a distributed pressure, concentrated dynamic loads or kinematic excitation due to vibrations of supporting structures. In the former case, equation of motion (4.1) becomes (static in-plane loads are not included)

$$D\nabla^4 w = -\hat{m}\frac{\partial^2 w}{\partial t^2} + p(x, y, t) \tag{4.10}$$

Boundary conditions are not affected by the load applied to the plate, except for the case where the plate is driven by the motion of its supports that is discussed below or if the support structures are flexible implying the necessity to analyze a joint plate-support dynamics.

The effect of damping on vibrations of plates is often accounted for using the viscous damping model. While a detailed discussion of damping is outside the scope of this book, a useful reference to the phenomenon can be found in Nashif et al. (1985). In the case of viscous damping, its effect is reflected through the presence of the corresponding term in the equation of motion. Accordingly, the left side of (4.10) would contain the viscous elastic term $c\frac{\partial w}{\partial t}$, c being the damping coefficient. In the subsequent solutions damping is excluded from the consideration, though we implicitly account for its effect suppressing free motion of the plate.

Let the applied pressure be represented in double Fourier series

$$p(x, y, t) = \sum_{m=1}^{M}\sum_{n=1}^{N} p_{mn}(t) \sin\alpha_m x \sin\beta_n y \tag{4.11}$$

The substitution of these series and the motion modeled by (4.2) into (4.10) yields uncoupled equations for the amplitudes of harmonics of a simply supported plate:

$$\frac{\hat{m}}{D}\ddot{W}_{mn} + \left(\alpha_m^2 + \beta_n^2\right)^2 W_{mn}(t) = \frac{p_{mn}(t)}{D} \tag{4.12}$$

Equation 4.12 is valid for the analysis of motion of simply supported plates with negligible damping experiencing linear vibrations. The solution depends on the function $p_{mn}(t)$ and it can be obtained by standard methods of the theory of vibration of single-degree of freedom systems.

Consider the particular case where the load is harmonic, i.e.

$$p_{mn}(t) = p_{mn} \sin\omega t \tag{4.13}$$

ω being the driving frequency. The time-dependent term $W_{mn}(t)$ includes the free vibration contribution (4.3) representing the solution of the homogeneous equation obtained from (4.12). A particular integral of the nonhomogeneous equation (4.12) representing the forced response is combined with (4.3) resulting in

$$W_{mn}(t) = A_{mn}\sin\omega_{mn}t + B_{mn}\cos\omega_{mn}t + \frac{p_{mn}}{D\left(\alpha_m^2 + \beta_n^2\right)^2 - \hat{m}\omega^2}\sin\omega t \tag{4.14}$$

As noted above, free vibrations become negligible after a short transient period as a result of energy dissipation. On the other hand, forced vibrations can continue indefinitely as long as energy is supplied to the plate by the driving load.

Accordingly, the free vibration contribution in (4.14) is usually neglected and the resulting motion referred to as "steady state vibrations" can be represented as

$$W_{mn}(t) = \frac{p_{mn}}{\hat{m}\left(\omega_{mn}^2 - \omega^2\right)} \sin \omega t \tag{4.15}$$

where the natural frequency ω_{mn} is introduced through (4.5).

It is observed from (4.14) or (4.15) that if the driving frequency is equal to the natural frequency of the corresponding mode, the amplitude of motion is infinite, i.e. the plate vibrates in resonance (see Fig. 4.1). Accounting for damping limits the amplitudes as is schematically shown in Fig. 4.2. This observation is valid for all boundary conditions. In the presence of static in-plane loads, the form of Eq. 4.15 is not altered, but the natural frequency is now given by (4.6).

4.3.1 *Kinematic Excitation of the Plate*

Consider now an important practical case where forced vibrations of the plate are driven by the motion of beams or other structures supporting its edges. If this motion is harmonic and torsional stiffness of the supporting structure is negligible, dynamic boundary conditions are represented by a generalization of conditions (2.2):

$$x = 0,\ \ x = a: \quad w = W \sin \omega t, \quad M_x = -D\left(\frac{\partial^2 w}{\partial x^2} + \nu \frac{\partial^2 w}{\partial y^2}\right) = 0$$

$$y = 0,\ \ x = b: \quad w = W \sin \omega t, \quad M_y = -D\left(\frac{\partial^2 w}{\partial y^2} + \nu \frac{\partial^2 w}{\partial x^2}\right) = 0 \tag{4.16}$$

where the amplitude of the edge motion W is known.

The equation of motion of the plate is (4.1) since no driving loads are applied to its surface. The steady-state solution can be sought in the form

$$w = \left[\sum_{m=1}^{M}\sum_{n=1}^{N} \bar{W}_{mn} \sin \alpha_m x \sin \beta_n y + W\right] \sin \omega t \tag{4.17}$$

Equation 4.17 identically satisfy boundary conditions (4.16).

The substitution of (4.17) into (4.1) where in-plane loads are absent yields

$$\sum_{m=1}^{M}\sum_{n=1}^{N}\left[D\left(\alpha_m^2 + \beta_n^2\right)^2 - \hat{m}\omega^2\right] \bar{W}_{mn} \sin \alpha_m x \sin \beta_n y = \hat{m}\omega^2 W \tag{4.18}$$

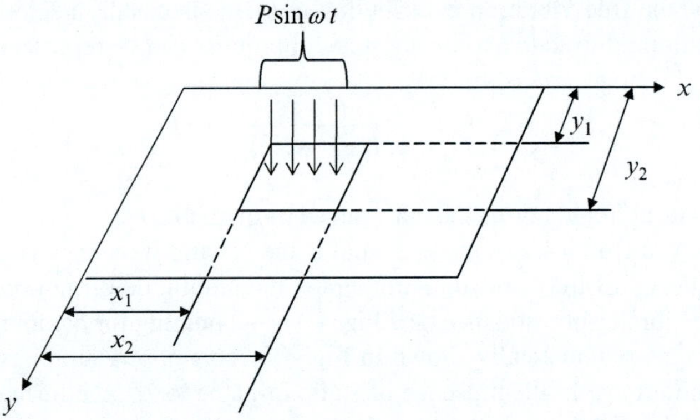

Fig. 4.4 Rectangular plate subject to a load distributed over the segment $x_1 \leqslant x \leqslant x_2,\ y_1 \leqslant y \leqslant y_2$

The solution for amplitudes $\bar{W}_{mn}$ is available by representing the term in the right side of (4.18) in double Fourier series

$$W = \sum_{m=1}^{M} \sum_{n=1}^{N} \hat{W}_{mn} \sin \alpha_m x \sin \beta_n y \tag{4.19}$$

where

$$\hat{W}_{mn} = \frac{16}{mn\pi^2} W \quad m, n = odd \tag{4.20}$$

Even values of m or n yield $\hat{W}_{mn} = 0$.

The substitution of (4.19) into (4.18) and equating the coefficients at the same trigonometric functions in the left and right sides of this equation result in

$$\bar{W}_{mn} = \frac{\hat{m}\omega^2 \hat{W}_{mn}}{D\left(\alpha_m^2 + \beta_n^2\right)^2 - \hat{m}\omega^2} \tag{4.21}$$

4.3.2 Energy Method for the Analysis of Plate Vibrations

Energy methods can successfully be applied to the analysis of both free and forced vibrations of plates. In this paragraph we illustrate the application of the Rayleigh-Ritz method to the analysis of forced vibrations of a simply supported plate subject to a harmonic load $P(t) = P \sin \omega t$ uniformly distributed over the segment $x_1 \leqslant x \leqslant x_2,\ \ y_1 \leqslant y \leqslant y_2$ as shown in Fig. 4.4.

The potential energy of the plate, including the strain energy given by (2.33) and the energy of the applied load given by (1.68) is

$$\Pi = \frac{D}{2}\int_0^b\int_0^a \left\{\left(\frac{\partial^2 w}{\partial x^2}+\frac{\partial^2 w}{\partial y^2}\right)^2 - 2(1-\nu)\left[\frac{\partial^2 w}{\partial x^2}\frac{\partial^2 w}{\partial y^2} - \left(\frac{\partial^2 w}{\partial x \partial y}\right)^2\right]\right\} dxdy$$
$$- \frac{P(t)}{(x_2-x_1)(y_2-y_1)}\int_{y_1}^{y_2}\int_{x_1}^{x_2} w dxdy \tag{4.22}$$

The kinetic energy of the plate is given by equation (1.69) that can be simplified since in-plane displacements and inertia are usually negligible in the linear problem concerned with transverse (lateral) vibrations. Accordingly,

$$K = \frac{1}{2}\hat{m}\int_0^b\int_0^a \left(\frac{\partial w}{\partial t}\right)^2 dxdy \tag{4.23}$$

The solution is obtained by the Rayleigh-Ritz method. The steady-state motion of the plate is modeled by series

$$w = \sum_{m=1}^{M}\sum_{n=1}^{N} W_{mn} \sin\alpha_m x \sin\beta_n y \sin\omega t \tag{4.24}$$

that satisfy all boundary conditions.

The Rayleigh-Ritz method implies the application of equation (1.65) that becomes

$$\frac{\partial(\Pi - K)}{\partial W_{mn}} = 0 \tag{4.25}$$

The substitution of (4.24) into (4.22) and (4.23) and the subsequent application of (4.25) yield uncoupled equations for each amplitude W_{mn}:

$$W_{mn} = \frac{4}{ab}\frac{f_{mn}}{D\left(\alpha_m^2+\beta_n^2\right)^2 - \hat{m}\omega^2} \tag{4.26}$$

where

$$f_{mn} = \frac{P}{(x_2-x_1)(y_2-y_1)}\int_{y_1}^{y_2}\int_{x_1}^{x_2} \sin\alpha_m x \sin\beta_n y dxdy \tag{4.27}$$

4.4 Non-periodic Response (Representative Example of Blast Loading)

The response of the plate to a dynamic non-periodic load is analyzed reducing the partial differential equations of motion where the response depends on both spatial coordinates and time to ordinary differential equations where time is the only argument. Subsequently, the response is found using the convolution integral or numerical initial value methods well known in the theory of vibration. In this paragraph we consider a representative non-periodic dynamic problem, i.e. the response of a plate to explosive blast overpressure.

A plate subject to blast loading experiences overpressure that is applied almost instantaneously and varies on the time scale of milliseconds. An example of blast overpressure measurements recorded during the test of a steel plate is shown in Fig. 4.5.

An analytical model for the air blast overpressure was developed in the eighties (e.g., Gupta et al. 1987) and extensively used in structural studies (Birman and Bert 1987; Librescu and Nosier 1990; Genin and Birman 2009; Nguyen et al. 2011). The overpressure is represented as a function of time by the modified Friedlander exponential decay equation:

$$p_{blast}(t) = p_0\left(1 - \frac{t}{t_p}\right)\exp\left(-\frac{a' t}{t_p}\right) \tag{4.28}$$

where p_0 is the peak overpressure, t_p is a positive phase duration of the pulse and a' is an empirical decay parameter.

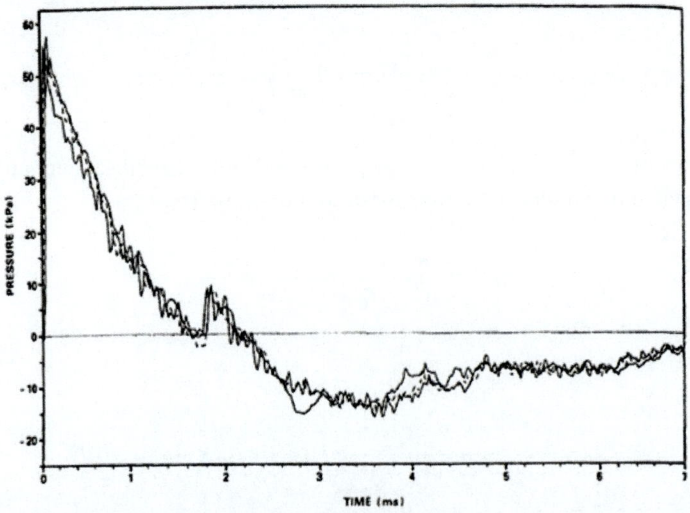

Fig. 4.5 Record of time history of the air blast overpressure (From Houlston et al. 1985)

The blast overpressure is usually uniformly distributed over the surface of the plate so that it can be represented in series (4.11) where

$$p_{mn}(t) = \frac{16}{mn\pi^2} p_{blast}(t) \tag{4.29}$$

where m and n are odd numbers. The motion of the plate that is simply supported around the perimeter is given by (4.2).

The nondimensional form of the uncoupled modal equation obtained as a result of the substitution of (4.2) and (4.11) into (4.10) is (Birman and Bert 1987):

$$\frac{d^2\tilde{W}_{mn}}{d\tau^2} + \bar{\omega}_{mn}^2 \tilde{W}_{mn} = \tilde{p}_{mn}(\tau) \tag{4.30}$$

where

$$\tilde{W}_{mn} = \frac{W_{mn}}{h}, \quad \tau = \omega_{11} t,$$
$$\bar{\omega}_{mn} = \frac{\omega_{mn}}{\omega_{11}}, \quad \tilde{p}_{mn} = \frac{p_{mn}}{\hat{m} h \omega_{11}^2} \tag{4.31}$$

The solution obtained for the case where the plate was at rest at the instant of blast, i.e. the initial deflection and velocity are equal to zero, is

$$\tilde{W}_{mn}(\tau) = \frac{16\bar{p}}{mn\pi^2\bar{\omega}_{mn}} \frac{1}{\left(\frac{a'}{\tau_p}\right)^2 + \bar{\omega}_{mn}^2} \times$$
$$\left\{ \begin{array}{l} \left(1 - \frac{\tau}{\tau_p}\right)\bar{\omega}_{mn} \exp\left(-\frac{a'\tau}{\tau_p}\right) + \frac{a'}{\tau_p} \sin\bar{\omega}_{mn}\tau - \bar{\omega}_{mn}\cos\bar{\omega}_{mn}\tau - \\ \frac{1}{\left[\left(\frac{a'}{\tau_p}\right)^2 + \bar{\omega}_{mn}^2\right]\tau_p} \left[\begin{array}{r} 2\frac{a'}{\tau_p}\bar{\omega}_{mn} \exp\left(-\frac{a'\tau}{\tau_p}\right) + \left(\left(\frac{a'}{\tau_p}\right)^2 - \bar{\omega}_{mn}^2\right)\sin\bar{\omega}_{mn}\tau \\ -2\frac{a'}{\tau_p}\bar{\omega}_{mn}\cos\bar{\omega}_{mn}\tau \end{array} \right] \end{array} \right\} \tag{4.32}$$

In (4.32),

$$\bar{p} = \frac{p_0}{\hat{m} h \omega_{11}^2}, \quad \tau_p = \omega_{11} t_p \tag{4.33}$$

Equation 4.32 is valid both for isotropic as well as for symmetrically laminated composite plates as long as we use appropriate natural frequencies. An example of the response of a cross-ply symmetrically laminated plate to blast load is shown in Fig. 4.6. A square 8-ply glass/epoxy cross-ply laminate of thickness $h = 0.008$ m and in-plane dimension $a = 0.5$ m was considered in this example. The layers

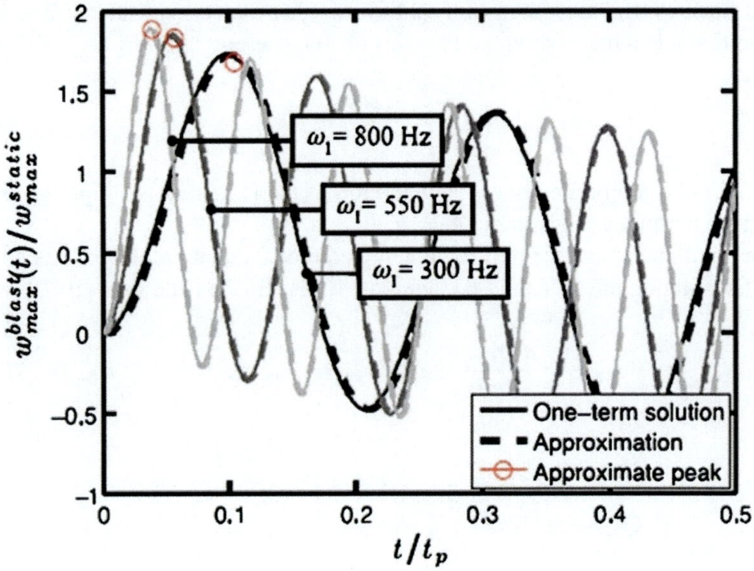

Fig. 4.6 Nondimensional deflection of a plate subject to blast loading as a function of the fundamental frequency denoted by ω_1. "Approximation" refers to the solution shown in Eq. 4.34 (From Genin and Birman 2009)

included glass fibers and spherical glass particles added to the epoxy matrix to enhance the stiffness of the laminate. The fundamental frequency of the laminate with a variable content of particles and fibers, ω_{11}, varied in the range from 250 to 800 Hz. For this natural frequency range, and using the blast considered by Librescu and Nosier (1990), i.e. $t_p = 0.1$ s, $a' = 2$, several oscillations occur in the panel during the positive pressure phase of the blast duration, t_p (Fig. 4.6). As can be observed in this figure, the peak deflection in the plates with a higher fundamental frequency was larger than in the plates that had lower fundamental frequencies.

The approximate solution for the peak deflection at the center of the plate during the positive phase of the pulse was derived in the form (Genin and Birman 2009)

$$w_{\max}^{blast} = w_{\max}^{static}\left(2 - \frac{\pi}{\omega_{11}t_p}\left(1 + a'\right)\right) \tag{4.34}$$

where the static deflection resulting from the application of the quasi-static peak pressure is

$$w_{\max}^{static} = \frac{16p_0}{\pi^2 \hat{m}\omega_{11}^2} \tag{4.35}$$

4.5 Vibrations of Reinforced Plates

Vibration problems of reinforced plates have been extensively considered due to a wide range of applications of such structures. The motion of a reinforced plate occurs both at the global level, i.e. the structure vibrating as one unit including the skin and stringers, and at the local level. The latter involves the motion of unreinforced sections of the plate between the stringers (see for example, sections of the plate in Fig. 2.18). The interaction between global and local vibrations that takes place in case of nonlinear vibrations is accounted for in finite element and other numerical solutions. The analytical solution for nonlinear vibrations accounting for mode interactions requires too many simplifications to be an accurate tool. In this paragraph we concentrate on global geometrically linear vibrations of reinforced plates that are uncoupled with local vibration modes.

We begin with a simple case where the stringers are symmetric about the plate middle plane, although the utility of such design is limited. Then the global motion of the plate reinforced by stringers can be analyzed by adding the inertia term to equation (2.66):

$$\left[\bar{D}_{11}\frac{\partial^4}{\partial x^4}+2D\frac{\partial^4}{\partial x^2\partial y^2}+\bar{D}_{22}\frac{\partial^4}{\partial y^4}\right]w-\bar{N}_x\frac{\partial^2 w}{\partial x^2}-\bar{N}_y\frac{\partial^2 w}{\partial y^2}=p\left(x,y,t\right)-\hat{m}\frac{\partial^2 w}{\partial t^2} \tag{4.36a}$$

In the case where the plate is simply supported along all edges the analytical solution is available representing pressure and deflections in the form (4.11) and (4.2), respectively. This yields the system of uncoupled equations of motion for every harmonic in these series:

$$\hat{m}\ddot{W}_{mn}+\left(\bar{D}_{11}\alpha_m^4+2D\alpha_m^2\beta_n^2+\bar{D}_{22}\beta_n^4+\bar{N}_x\alpha_m^2+\bar{N}_y\beta_n^2\right)W_{mn}(t)=p_{mn}(t) \tag{4.36b}$$

The solution of problems of free vibrations, forced harmonic vibrations and forced non-harmonic motion are available using (4.36b). In particular, the amplitude of motion caused by harmonic excitation (4.13) is

$$\bar{W}_{mn}=\frac{p_{mn}}{\bar{D}_{11}\alpha_m^4+2D\alpha_m^2\beta_n^2+\bar{D}_{22}\beta_n^4+\bar{N}_x\alpha_m^2+\bar{N}_y\beta_n^2-\hat{m}\omega^2} \tag{4.36c}$$

Plates with stringers on one surface are more typical in industry. Then equation (4.36a) is not applicable and it is necessary to account for coupling between bending and in-plane displacements. For example, in the case of forced vibrations of the plate reinforced by stringers in the x- and y-directions equations of motion available as an extension of (2.65) accounting for inertial terms become

$$A_{11}u_{,xx} + A_{66}u_{,yy} + (A_{12} + A_{66})\, v_{,xy} - B_{11}w_{,xxx} - (B_{12} + 2B_{66})\, w_{,xyy} = \hat{m}u_{,tt}$$
$$(A_{12} + A_{66})\, u_{,xy} + A_{66}v_{,xx} + A_{22}v_{,yy} - (B_{12} + 2B_{66})\, w_{,xxy} - B_{22}w_{,yyy} = \hat{m}v_{,tt}$$
$$D_{11}w_{,xxxx} + 2\,(D_{12} + 2D_{66})\, w_{,xxyy} + D_{22}w_{,yyyy} - B_{11}u_{,xxx}$$
$$- (B_{12} + 2B_{66})\left(u_{,xyy} + v_{,xxy}\right) - B_{22}v_{,yyy} - \bar{N}_x w_{,xx} - \bar{N}_y w_{,yy} = p\,(x, y, t) - \hat{m}w_{,tt} \tag{4.37}$$

The mass of the plate per unit surface area is

$$\hat{m} = m_p + \delta\,(y - y_s)\,\rho_s A_s + \delta\,(x - x_r)\,\rho_r A_r$$

where m_p is the mass of the panel without stringers per unit surface area, ρ_s is the mass density of the material of the stringer oriented in the x-direction, and A_s is the cross sectional area of this stringer. The mass density and cross sectional area of the stringers oriented in the y-direction are ρ_r and A_r, respectively.

Consider the case where the applied uniformly distributed harmonic pressure is represented by (4.11) and (4.13). The analytical solution is available that satisfies the conditions of simple support along all boundaries, i.e. equations (2.69), in the form

$$u = \sum_{m=1}^{M}\sum_{n=1}^{N} U_{mn} \cos\alpha_m x \sin\beta_n y \sin\omega t$$
$$v = \sum_{m=1}^{M}\sum_{n=1}^{N} V_{mn} \sin\alpha_m x \cos\beta_n y \sin\omega t$$
$$w = \sum_{m=1}^{M}\sum_{n=1}^{N} W_{mn} \sin\alpha_m x \cos\beta_n y \sin\omega t \tag{4.38}$$

The substitution of (4.38) into (4.37) yields systems of uncoupled equations for each mode of motion (i.e. for each harmonic in series (4.38)):

$$\begin{bmatrix} S_{11mn} - \hat{m}\omega^2 & S_{12mn} & S_{13mn} \\ S_{12mn} & S_{22mn} - \hat{m}\omega^2 & S_{23mn} \\ S_{13mn} & S_{23mn} & S_{33mn} - \hat{m}\omega^2 \end{bmatrix} \begin{Bmatrix} U_{mn} \\ V_{mn} \\ W_{mn} \end{Bmatrix} = \begin{Bmatrix} 0 \\ 0 \\ p_{mn} \end{Bmatrix} \tag{4.39}$$

where S_{ijmn} are given by (2.72).

Equations 4.39 can be applied to solve both free and forced vibration problems. In the former case, the vector in the right side is equal to zero. Accordingly, the requirement to non-zero amplitudes of free vibration implies that the determinant of the system of homogeneous equations obtained from (4.39) must be equal to zero, i.e. $\det[S] = 0$. Three natural frequencies obtained from the solution of $\det[S] = 0$ include one lower value and two values that are much higher. The lower value corresponds to the frequency of predominantly transverse vibrations, while two higher frequencies are associated with predominantly in-plane free vibrations. This reflects the fact that in-plane stiffness of the plate is much higher than its transverse stiffness. Usually, the effect of the higher frequencies on the frequency of predominantly transverse vibrations can be neglected, so that this frequency can be found from the correspondingly simplified condition:

$$\begin{vmatrix} S_{11mn} & S_{12mn} & S_{13mn} \\ S_{12mn} & S_{22mn} & S_{23mn} \\ S_{13mn} & S_{23mn} & S_{33mn} - \hat{m}\omega^2 \end{vmatrix} = 0 \tag{4.40}$$

Additional natural frequencies correspond to vibrations of panels of the plate between the stringers. Coupling between these additional frequencies and the frequencies of the reinforced plate is negligible as long as the problem is geometrically linear. Also, the natural frequencies of the stringers (with adjacent sections of skin) have to be considered, though these frequencies are typically much higher than those of either the reinforced plate or plate sections between the stringers.

Example 4.1 Vibrations of Stringer-Reinforced Functionally Graded Plates Functionally graded composite materials (FGM) are formed of two or more phases (matrix, particles and fibers) that vary throughout the domain occupied by the structure with the goal of optimizing its performance (Birman and Byrd 2007). Examples of such materials are particulate composites with a piece-wise or continuous variation of the volume fraction of particles and fiber-reinforced composites where the volume fraction or orientation of fibers depends on the location.

The effect of stringers on the natural frequencies is reflected in Fig. 4.7 (Birman and Byrd 2008). The panels were manufactured from titanium boride/titanium (TiB/Ti). Functionally graded panels had seven 0.254 mm thick layers with varying volume fraction of the constituent materials; the properties of the layers are listed in Table 4.2. The panels were reinforced by a system of equally-spaced blade stringers oriented along the x-axis. The thickness of titanium blade stringers located on the titanium-rich surface of the plate was equal to the thickness of the panel (1.778 mm). Accordingly, the panels were pseudo-isotropic, but the theory for isotropic panels had to be extended to the functionally-graded case as shown in the paper (this extension can be comprehended upon reading Sect. 5.6). The horizontal and vertical axes in Fig. 4.7 correspond to the ratio of the fundamental frequency of the reinforced plate to that of the plate without stringers and the ratio of the height of the stringer to the thickness of the panel, respectively.

As follows from Fig. 4.7, increasing the fundamental frequency to a desirable level can easily be achieved through the use of relatively light stringers. This reflects a dramatic improvement in the stiffness of plates available using stringers.

Besides harmonic vibrations, reinforced plates are sometimes subject to non-harmonic loads, such as blast. Similar to forced vibrations, it is necessary to account for "global" dynamic response of the entire plate with reinforcement as well as local motion between the stringers. The solution of this problem is outside the scope of

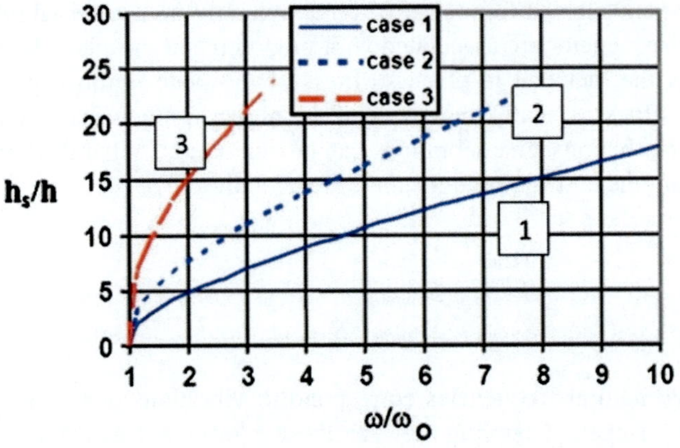

Fig. 4.7 Blade stringer height as a function of desired fundamental frequency (ω_0 corresponding to m = n = 1). The thickness of the plate is $h = 1.778mm$. The stringer spacing is equal to 0.1667 m. Case 1: $a = 0.5m, b = 1.5m, \omega_0 = 141.00\frac{1}{s}$. Case 2: $a = b = 1.0m, \omega_0 = 63.46\frac{1}{s}$. Case 3: $a = 1.0m, b = 0.5m, \omega_0 = 158.60\frac{1}{s}$ (From Birman and Byrd 2008)

Table 4.2 Properties of the layers of the functionally graded plate

Layer number	$\rho\ \left(\frac{kg}{m^3}\right)$	$E\ (GPa)$	ν
1	4570.0	106.9	0.34
2	4568.5	117.0	0.31
3	4567.0	133.0	0.28
4	4565.5	159.0	0.25
5	4564.0	193.0	0.22
6	4562.5	237.0	0.19
7	4561.0	274.0	0.17

the book, but it is instructive to refer to available experimental and numerical (FEA) studies (e.g., Houlton et al. 1985; Houlton and DesRochers 1987).

4.6 Large-Amplitude Vibrations

Geometrically nonlinear effects that usually become essential at the amplitude of vibrations equal to or exceeding half-thickness of the plate can be analyzed using equations of motion or by energy methods. Examples of early analytical solutions can be found in the monographs of Chia (1980) or Vol'mir (1972). However, due to difficulties involved in satisfying the boundary conditions as well as a result of coupling of modes of motion, such solutions are seldom applied in industry. The majority of contemporary solutions rely on the numerical analysis. In this paragraph we review a representative problem for a simply supported rectangular plate experiencing free vibrations illustrating qualitative effects of large vibrations on dynamic characteristics of the plate and the limitations of the analytical approach.

The nonlinear equation of motion and the compatibility equation for a plate in the Cartesian coordinate system are (1.96) and (1.99), respectively. Limiting the analysis of an isotropic plate to a single-degree-of-freedom approach, and concentrating on the fundamental frequency and mode of motion, the deflection is sought in the form

$$w\,(x, y, t) = W(t)\sin\alpha_1 x \sin\beta_1 y \tag{4.41}$$

Equation 4.41 may adequately represent the motion of the plate vibrating with moderate amplitudes in the vicinity to the fundamental frequency.

The substitution of (4.41) into (1.99) yields

$$\frac{1}{E}\nabla^4\varphi = \frac{\pi^4}{2a^2b^2}W^2(t)\,(\cos 2\alpha_1 x + \cos 2\beta_1 y) \tag{4.42}$$

The solution of (4.42) including the complementary solution of the homogeneous equation and a particular integral is

$$\varphi = \frac{E}{32}W^2(t)\left[\frac{a^2}{b^2}\cos 2\alpha_1 x + \frac{b^2}{a^2}\cos 2\beta_1 y\right] + \frac{\bar{N}_x y^2}{2h} + \frac{\bar{N}_y x^2}{2h} \tag{4.43}$$

where the applied in-plane stress resultants $\bar{N}_x$ and $\bar{N}_y$are positive in tension.

Evidently, the shear stress is absent throughout the plate. The boundary conditions for in-plane stress resultants $\bar{N}_x$ and $\bar{N}_y$ can be satisfied only in the integral sense as was indicated in Sect. 2.11, i.e.

$$x = 0,\ x = a:\quad h\int_0^b \frac{\partial^2\varphi}{\partial y^2}dy = \bar{N}_x b$$

$$y = 0,\ y = b:\quad h\int_0^a \frac{\partial^2\varphi}{\partial x^2}dx = \bar{N}_y a \tag{4.44}$$

Obviously, such integral satisfaction of in-plane boundary conditions may yield inaccurate results. Besides, adding terms to the mode shape of deflections that is necessary due to coupling of modes at large deformations results in a complicated mathematical formulation. This is a drawback of the analytical solution.

The exact integration of equation of motion (1.96) is impossible, but the solution is usually obtained using the Galerkin procedure (e.g., Vol'mir 1972; Chia 1980). In the present formulation this procedure implies

$$\int_0^a\int_0^b \left(\frac{D}{h}\nabla^4 w - \frac{\partial^2\varphi}{\partial x^2}\frac{\partial^2 w}{\partial y^2} - \frac{\partial^2\varphi}{\partial y^2}\frac{\partial^2 w}{\partial x^2} + 2\frac{\partial^2\varphi}{\partial x\partial y}\frac{\partial^2 w}{\partial x\partial y} + \frac{\hat{m}}{h}\frac{\partial^2 w}{\partial t^2}\right)$$

$$\times \sin\alpha_1 x \sin\beta_1 y dx dy = 0 \tag{4.45}$$

The substitution of (4.41) and (4.43) into (4.45) and integration result in Duffing's equation:

$$\Psi(t) = \frac{d^2W}{dt^2} + \omega_{11}^2 W\left(1 + K\frac{W^2}{h^2}\right) = 0 \tag{4.46}$$

where the coefficient K depends on in-plane boundary conditions, material properties and geometry of the plate. In particular, if the edges of the plate are not constrained against in-plane displacements, i.e. $\bar{N}_x = \bar{N}_y = 0$,

$$K = \frac{3\left(1-\nu^2\right)\left(1+\frac{a^4}{b^4}\right)}{4\left(1+\frac{a^2}{b^2}\right)^2} \tag{4.47}$$

ν being Poisson's ratio of the plate material.

Let us assume that the motion is harmonic, i.e. $W = \tilde{W}h\sin\omega t$. Then the Duffing equation (4.46) can be integrated over the period of motion:

$$\int_0^{\frac{2\pi}{\omega}} \Psi(t)dt = 0 \tag{4.48}$$

The integration yields a relationship between the natural frequency, the amplitude of nonlinear vibrations and the fundamental frequency of linear vibrations ($\omega_{11} \equiv \omega_0$):

$$\omega = \omega_0\left(1 + \frac{3}{4}K\tilde{W}^2\right) \tag{4.49}$$

Due to limitations of a single-degree-of-freedom analytical solution equation (4.49) and similar results can be used only as the first approximation. However, a fundamental observation valid in nonlinear dynamics of plate structures can be made based on the analysis of this equation. It is obvious that plates represent nonlinear systems with a hardening nonlinearity, i.e. increasing amplitudes of motion correspond to larger natural frequencies.

Nonlinear frequency-amplitude relationships encompass the free-vibration "skeleton" curves. A schematic illustration of such relationship is shown in Fig. 4.8. The ratio $\frac{\omega}{\omega_{11}}$ along the horizontal axis represents the driving frequency normalized with respect to the fundamental frequency of the plate. The curve "f" originating from $\frac{\omega}{\omega_{11}} = 1$ at zero amplitude represents free vibrations of the plate. Other curves depict the forced response for different amplitudes of the harmonic pressure. The "middle" brunches of the forced motion curves are unstable (as indicated by broken curves) and they are not realized. If the frequency increases, the response corresponds to the "upper" brunch, but at a certain value of the driving frequency

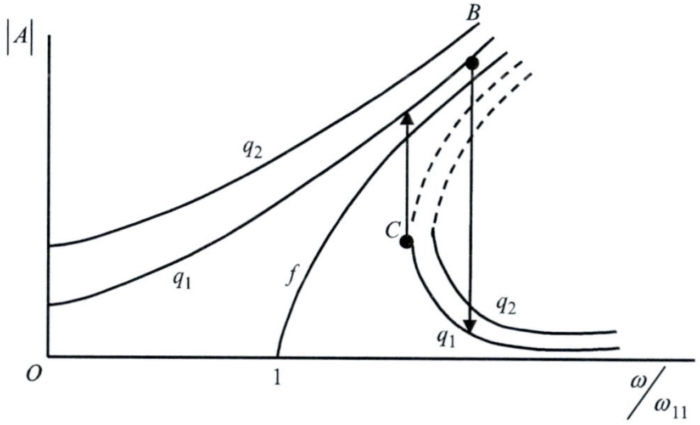

Fig. 4.8 Schematic amplitude vs. nondimensional frequency relationship for forced nonlinear vibrations of a plate subject to a harmonic uniformly distributed pressure; "f" = free nonlinear vibrations; q_1 and q_2 = forced vibrations ($q_2 > q_1$). Points B and C correspond to snap through between the upper and lower branches of the relationships for q_1

this response becomes unstable, the amplitude abruptly decreases (point B), and at larger frequencies vibrations corresponds to the "lower" brunch. In the contrary, if the frequency of the dynamic pressure decreases from large values, the response is represented by the lower brunch of the graphs. At the point where the tangent to this brunch is perpendicular to the horizontal axis (point C), snap-through to the upper brunch occurs and at lower frequencies the response corresponds to the upper brunch. Note that such behavior is typical for all vibrating systems with a hardening nonlinearity.

4.7 Dynamic Instability of Plates

In numerous engineering applications plates are subject to dynamic in-plane loads. As a result of dynamically applied impulsive loads, the plate can buckle; such phenomenon is usually referred to as dynamic buckling (e.g., Birman 1989). On the other hand, if in-plane loads are harmonic functions of time, the plate may experience transverse vibrations with gradually increasing amplitudes that are often called dynamic or parametric instability. The phenomenon of dynamic instability that we consider in this paragraph depends on the relationship between the driving and natural frequencies of the plate as well as on the amplitude of driving in-plane stress resultants. While in-plane vibrations are present at all combinations of the amplitude and frequency of the driving load, dynamic instability occurs only within certain regions on the driving frequency-driving amplitude plane. The boundaries of the regions of instability on this plane represent harmonic and periodic solutions of the equations of motion.

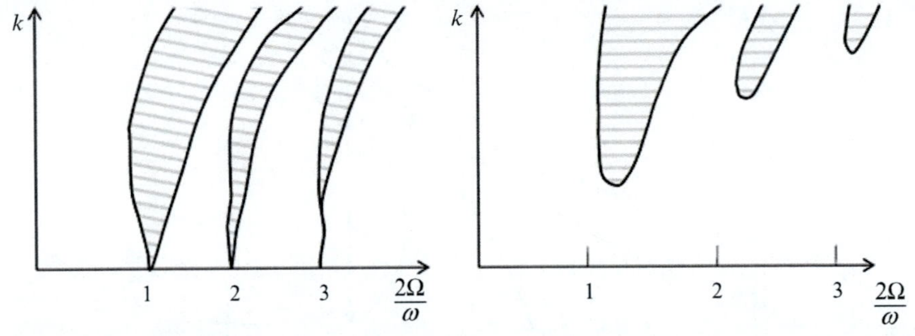

Fig. 4.9 Dynamic instability of plates subject to in-plane harmonic-in-time loads. The regions of dynamic instability are shaded. The response of an undamped plate is shown in figure "on the left", the qualitative effect of damping is reflected in figure "on the right"

Numerous studies devoted to dynamic instability of plates were reviewed in the classical monograph of Bolotin (1956). A review of the subject was published by Xie (2006). The qualitative response of rectangular plates to harmonic-in-time in-plane loading is depicted in Fig. 4.9. The vertical axis reflects the amplitude of dynamic in-plane load (the coefficient k is defined below in equation (4.54)), while the ratio along the horizontal axis represents a doubled fundamental frequency of the plate loaded by in-plane static loads divided by the frequency of the applied load. The amplitude of motion increases with time in the regions of instability schematically shown in this figure. The "widest" region that is called the principal instability region has the "origin" at the zero amplitude of the load and the driving frequency equal to twice the fundamental frequency of the plate. Note that the motion within the regions of instability is theoretically infinite, unless we account for the effect of damping. In the presence of damping there is a minimum amplitude of the driving load required to trigger dynamic instability, while if damping is neglected, dynamic instability can be caused by a load of a very small amplitude (the difference between dynamic instability of undamped and damped plates is evident from the comparison of Fig. 4.9). The amplitude of the in-plane load required to cause dynamic instability of damped systems increases for higher instability regions as compared to the principal region (Fig. 4.9). Accounting for damping results in limited amplitudes of motion within instability regions, even in the linear formulation.

The present solution is based on pioneering work of Bolotin on dynamic stability (Bolotin 1954). Consider a rectangular plate of size $a \times b$ subject to a dynamic in-plane stress resultant applied at the edge x = a

$$\bar{N}_x = N_0 + N_t \cos \omega t \tag{4.50}$$

The plate is simply supported, while in-plane boundary conditions are as follows:

- All edges remain straight during the motion;

- The edges x = 0, y = 0 and y = b cannot move in the plane of the plate;
- The edge x = a is not restrained against the motion in the plane of the plate.

Considering the case where the frequency of the dynamic load is close to twice the fundamental frequency of the plate we limit the analysis representing the motion by the first harmonic of series (4.2) where $m = n = 1$.

The substitution of (4.50) and the first harmonic of (4.2) into the system of equations of motion in terms of the deflection and stress function and transformations similar to those in Sect. 4.6 yield the following equation of motion (Bolotin 1954):

$$\frac{d^2\tilde{W}_{11}}{dt^2} + \omega_{11}^2\left(1 - \frac{N_0 + N_t \cos\omega t}{N_{cr}}\right)\tilde{W}_{11} + \eta\tilde{W}_{11}^3 = 0 \tag{4.51}$$

where the nondimensional deflection is defined as in (4.31) and $N_{cr} = N_{x,cr}$ is the buckling load of the plate corresponding to the case of a uniaxial compression. The fundamental frequency is specified in the absence of in-plane loads, i.e. $\omega_{11} = \omega_{11}\left(\bar{N}_x = \bar{N}_y = 0\right)$. The motion described by (4.51) belongs to the class referred to as parametric vibrations (vibrations excited by in-plane dynamic loads).

The coefficient at the nonlinear term is given by

$$\eta = \frac{0.75\left(1-\nu^2\right)\left(1+3\lambda^4\right)}{\left(1+\lambda^2\right)^2}\omega_{11}^2 \tag{4.52}$$

$\lambda = \frac{a}{b}$ being the aspect ratio.

Introducing the squared nondimensional fundamental frequency of the plate subject to static in-plane compression

$$\Omega^2 = \omega_{11}^2\left(1 - \frac{N_0}{N_{cr}}\right) \tag{4.53}$$

and the coefficient reflecting the magnitude of dynamic loading

$$k = \frac{N_t}{N_{cr} - N_0} \tag{4.54}$$

the equation of motion (4.51) is reduced to a nonlinear version of the Mathieu equation:

$$\frac{d^2\tilde{W}_{mn}}{dt^2} + \Omega^2\left(1 - k\cos\omega t\right)\tilde{W}_{mn} + \eta\tilde{W}_{mn}^3 = 0 \tag{4.55}$$

The linear version of equation (4.55) is referred to as the Mathieu equation. Different combinations of the driving frequency and the amplitude of the dynamic load correspond to stable or unstable motion of the plate. The motion within stable regions is oscillatory, though not regularly periodic. In the regions of dynamic instability the amplitude increases with time (though this increase can be controlled by damping and geometrically nonlinear effects). The combinations (regions) of stable and unstable motion on the driving frequency-driving load amplitude plane (so-called Ince-Strutt diagram) are separated by periodic solutions of the equation of motion.

In this paragraph the analysis is limited to the motion of the plate with a negligible damping. Then introducing a nondimensional time scale$\tau = \Omega t$, the periodic solution in the vicinity of the frequency $\omega = 2\Omega$ can be sought in the from

$$\tilde{W}_{mn} = F_1 \cos\frac{\omega\tau}{2\Omega} + F_2 \sin\frac{\omega\tau}{2\Omega} \tag{4.56}$$

The substitution of (4.56) into (4.55), transformations and neglecting superharmonics of the third order (terms proportional to $\frac{3\omega\tau}{2\Omega}$ that have little effect on motion in the vicinity of the nondimensional time $\frac{\omega\tau}{2\Omega}$ corresponding to the origin of the principal instability region) lead to the following equation:

$$\left(1 - \frac{\omega^2}{4\Omega^2} - \frac{k}{2} + \frac{3}{4}\delta\left(F_1^2 + F_2^2\right)\right) F_1 \cos\frac{\omega\tau}{2\Omega} + \left(1 - \frac{\omega^2}{4\Omega^2} + \frac{k}{2} + \frac{3}{4}\delta\left(F_1^2 + F_2^2\right)\right) F_2 \sin\frac{\omega\tau}{2\Omega} = 0 \tag{4.57}$$

where $\delta = \frac{\eta}{\Omega^2}$.

Obviously, Eq. 4.57 can be satisfied at all time instances only if the terms in the brackets are equal to zero. This yields two nonlinear equations with respect to the amplitude $F = \sqrt{F_1^2 + F_2^2}$:

$$1 - \frac{\omega^2}{4\Omega^2} \pm \frac{k}{2} + \frac{3}{4}\delta F^2 = 0 \tag{4.58}$$

The boundaries of the principal instability region specified from the linear version of (4.58) are

$$\frac{\omega^2}{4\Omega^2} = 1 \pm \frac{k}{2} \quad or \quad \frac{\omega}{2\Omega} \approx 1 \pm \frac{k}{4} \tag{4.59}$$

The boundaries of the principal region of instability according to (4.59) are shown in Fig. 4.10 by vertical lines. In the linear problem, the amplitudes within the interval of driving frequencies defined by (4.59) gradually increase to infinity.

In the presence of the nonlinear term, the boundaries of the principal instability region are determined from (4.58):

$$F = \sqrt{\frac{4}{3\delta}\left(\frac{\omega^2}{4\Omega^2} - 1 \pm \frac{k}{2}\right)} \tag{4.60}$$

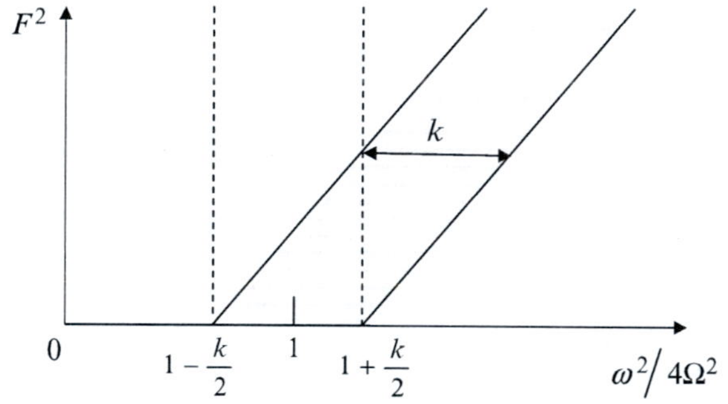

Fig. 4.10 The boundaries of the principal instability region of the plate. *Broken vertical lines*: linear problem; *solid inclined lines*: nonlinear problem

The relationship between the squared nondimensional driving frequency and the squared amplitude of motion given by (4.60) is schematically shown in Fig. 4.10. If damping is present, the unstable motion with large amplitudes occurs at the combinations of frequencies and amplitudes within solid straight lines in Fig. 4.10. Note that the width of the principal instability region corresponds to the coefficient k that reflects the amplitude of the applied in-plane load (see Eq. 4.54). In reality, the amplitudes remain limited due to the dissipation of energy, so that immediate failure is unlikely in realistic structures. However, fatigue damage remains a serious consideration. Accordingly, designers should avoid the situation where the plate can experience dynamic instability.

4.8 Design Philosophy and Recommendations

Design of plates subject to dynamic loading should address two possible modes of failure. The plate can collapse in case where dynamic stresses exceed the failure stress of the material. An example of such problem would be an explosive blast destroying the plate. In other cases dynamic stresses remain within allowable limits and do not cause instantaneous failure, but rather result in fatigue damage. Both situations, i.e. instantaneous and fatigue failure, can be prevented employing the analysis that predicts dynamic deflections and stresses in the plate.

Excessive deflections are seldom the problem by themselves, i.e. they can often be tolerated as long as the stresses remain within allowable limits. However, similar to the problem of static bending, the knowledge of these deflections is necessary to identify the problem as either geometrically linear or nonlinear. In addition to deflections and stresses, accelerations have to be limited in some applications. Such limitations on accelerations are often enforced if structural vibrations can affect people (e.g., pilots and passengers of military and civilian airplanes) since inertial forces and moments are proportional to accelerations.

The knowledge of natural frequencies is important to avoid resonances of either transverse or parametric vibrations (the latter resonance is associated with dynamic instability). The analysis predicts infinite amplitudes of forced vibrations in the geometrically linear problem for plates (and for other structures) experiencing resonance as long as damping is neglected. The presence of damping limits the amplitude of motion, particularly at resonance, though it may still reach unacceptable values resulting in excessive dynamic stresses.

The most dangerous resonance of transverse vibrations occurs when the driving frequency coincides with the fundamental (lowest) natural frequency of the plate. Resonances with higher natural frequencies are relatively less dangerous as the result of a larger effect of damping.

Static in-plane compressive loads reduce natural frequencies of the plate. At the limit, when the compressive load is equal to the buckling value, the fundamental frequency is reduced to zero (as a matter of fact, one of the methods of buckling analysis specifies the buckling load from this condition). In the contrary, tension results in higher natural frequencies. The resonance of plates loaded by static in-plane loads occurs at driving frequencies that are different from those evaluated without accounting for such loads due to a difference in the natural frequencies. Accordingly, the designer should be aware of possible in-plane static loads since they may significantly affect the response of the plate in dynamic environments.

Reinforcing the plate with stringers is an effective method of reducing its dynamic response. The analysis of dynamic characteristics of reinforced plates should account for the possibility of three resonance cases:

- Resonance of the entire reinforced plate, including both skins and stringers;
- Resonance of sections of the plate between the stringers;
- Resonance of stringers with the adjacent skin.

If any of these resonances takes place, excessive vibrations may cause fatigue damage or an immediate dynamic collapse.

If the plate subject to a harmonic in-plane load undergoes in-plane vibrations that have small amplitude, such plate is considered dynamically stable. However, at certain combinations of applied load amplitudes and driving frequencies, the plate becomes dynamically unstable developing significant transverse vibrations (dynamic instability). The combinations of amplitudes and frequencies of in-plane driving loads separating regions of stable and unstable vibrations correspond to periodic motion of the plate. The motion of the plate within the regions of dynamic instability is limited due to the effects of damping and geometric nonlinearity. Nevertheless, dynamic instability should be avoided since large-amplitude vibrations may result either in immediate failure of the plate or in fatigue damage.

References

Birman, V. (1989). Problems of dynamic buckling of antisymmetric rectangular laminates. *Composite Structures, 12*, 1–15.

Birman, V., & Bert, C. W. (1987). Behavior of laminated plates subjected to conventional blast. *International Journal of Impact Engineering, 6*, 145–155.

Birman, V., & Byrd, L. W. (2007). Modeling and analysis of functionally graded materials and structures. *Applied Mechanics Reviews, 60*, 195–216.

Birman, V., & Byrd, L. W. (2008). Stability and natural frequencies of functionally graded stringer-reinforced plates. *Composites: Part B, 39*, 816–825.

Bolotin, V. V. (1954). *Selected nonlinear problems of dynamic stability of plates, communications of the Academy of Science of the USSR*. Department of Technical Sciences. (No. 10, pp. 47–59).

Bolotin, V. V. (1956). *Dynamic stability of elastic systems*. Moscow: Gostechizdat (English translation: Holden Day, San Francisco, 1964).

Chia, C. Y. (1980). *Nonlinear analysis of plates*. New York: McGraw-Hill.

Genin, G. M., & Birman, V. (2009). Micromechanics and structural response of functionally graded particulate-matrix, fiber-reinforced composites. *International Journal of Solids and Structures, 46*, 2136–2150.

Gupta, A. D., Gregory, F. H., Bitting, R. L., & Bhattacharya, S. (1987). Dynamic response of an explosively loaded hinged rectangular plate. *Computers & Structures, 26*, 339–344.

Houlston, R., Slater, J. E., Pegg, N., & DesRochers, C. G. (1985). On analysis of structural response of ship panels subject to air blast loading. *Computers & Structures, 21*, 273–289.

Houlton, R., & DesRochers, C. G. (1987). Nonlinear structural response of ship panels subjected to air blast loading. *Computers & Structures, 26*, 1–15.

Leissa, A. W. (1973). The free vibration of rectangular plates. *Journal of Sound and Vibration, 31*, 257–293.

Librescu, L., & Nosier, A. (1990). Response of shear deformable elastic laminated composite flat panels to sonic boom and explosive blast loading. *AIAA Journal, 28*, 345–352.

Nashif, A. D., Jones, D. I. G., & Henderson, J. P. (1985). *Vibration damping*. New York: Wiley.

Nguyen, C. H., Butukuri, R. R., Chandrashekhara, K. & Birman, V. (2011). Dynamics and buckling of sandwich panels with stepped facings. *International Journal of Structural Stability and Dynamics* (in press).

Vol'mir, A. C. (1972). *Nonlinear dynamics of plates and shells*. Moscow: Nauka Publishers.

Xie, W.-C. (2006). *Dynamic stability of structures*. New York: Cambridge University Press.

Chapter 5
Mechanics of Composite Plates

Composite plates are increasingly used in engineering applications where they provide enhanced strength and stiffness without incurring additional weight compared to metallic counterparts or alternatively, enable a designer to reduce the weight, without sacrificing strength and stiffness. This is reflected in Fig. 5.1 where both specific strength as well as specific stiffness of composite materials are shown to be remarkable superior to typical alloys (specific strength and specific stiffness are strength per unit weight and elastic modulus per unit weight, respectively). Potential advantages of composite plates are related to the opportunity to "tailor" the response by orienting stiff high-strength fibers in the direction of maximum stresses. In other words, such plates utilize the principal difference of composites from isotropic materials, i.e. their anisotropic direction-dependent properties. A further enhancement in the response may be achieved by using sandwich plates where a light and relatively compliant core joins two stiff opposite facings and whose concept can be traced back to the classical I-beam.

5.1 Basic Concepts of Thin Laminated Plates

Each layer (lamina) in a composite laminated plate includes fibers embedded within the matrix. A number of fiber arrangements can be considered, including unidirectional, woven or discontinuous (short) fibers. Short fibers can be either aligned along a preferential direction or randomly oriented in the plane of the lamina (thicker plates with a 3-D random fiber distribution are also used in applications). While structures with a random orientation of fibers are isotropic, materials with unidirectional, woven or oriented discontinuous fibers have direction-dependent anisotropic properties. In this chapter we concentrate on laminated plates (laminates) composed of laminae with unidirectional fibers.

A laminate consisting of laminae that are perfectly bonded forming the composite structure is shown in Fig. 5.2. Furthermore, Fig. 5.3 illustrates a unidirectional lamina and its principal coordinate system 1-2-3 with the axes oriented along and

V. Birman, *Plate Structures*, Solid Mechanics and Its Applications 178,
DOI 10.1007/978-94-007-1715-2_5, © Springer Science+Business Media B.V. 2011

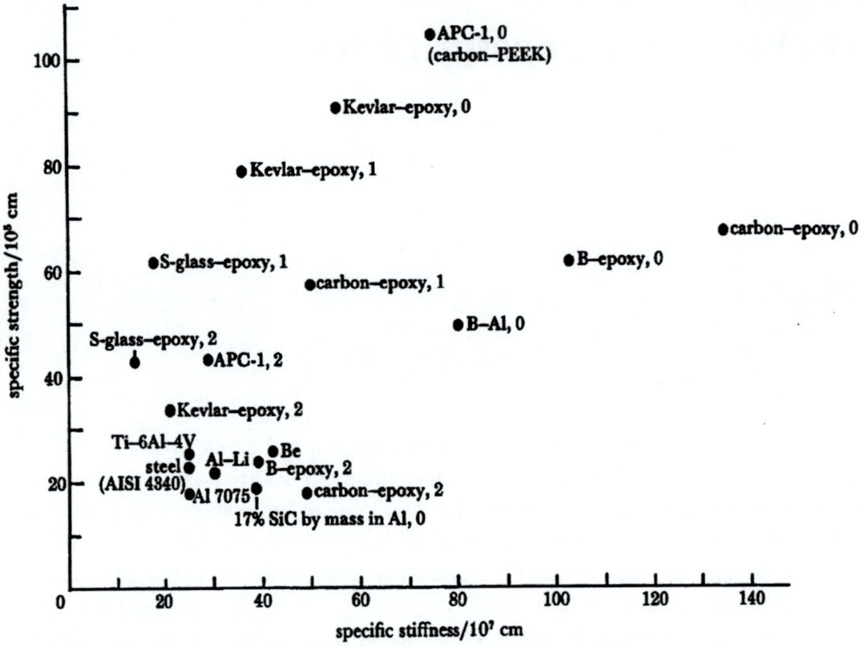

Fig. 5.1 Effectiveness of composite materials compared to alloys illustrated through the comparison of specific strength and specific stiffness (Kelly 1987). *1*: the fiber volume fraction of 50% at 0° layers, 40% at +/−45° and 10% at 90° (composite laminates). *2*: balanced laminated composites with equal proportion of 45°, 90° and 135° layers. *0*: unidirectionally laminated composites with volume fraction of fibers between 40% and 60%

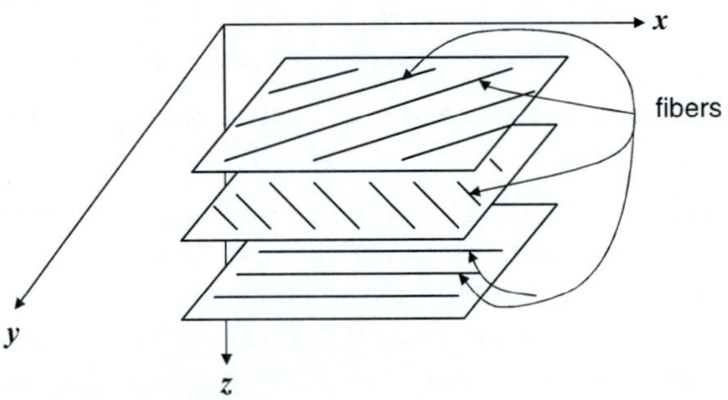

Fig. 5.2 Schematic illustration of a laminated plate. The fibers in each lamina are oriented at a different angle to the reference coordinate system

perpendicular to the fiber direction. The reference coordinate system that is also shown in Fig. 5.3 corresponds to the coordinate system employed in the analysis of the entire structure (e.g., Cartesian coordinate system used to analyze rectangular plates). The angle between the direction of fibers in the individual lamina and the axis of the laminate shown in this figure is called the lamination angle of the lamina.

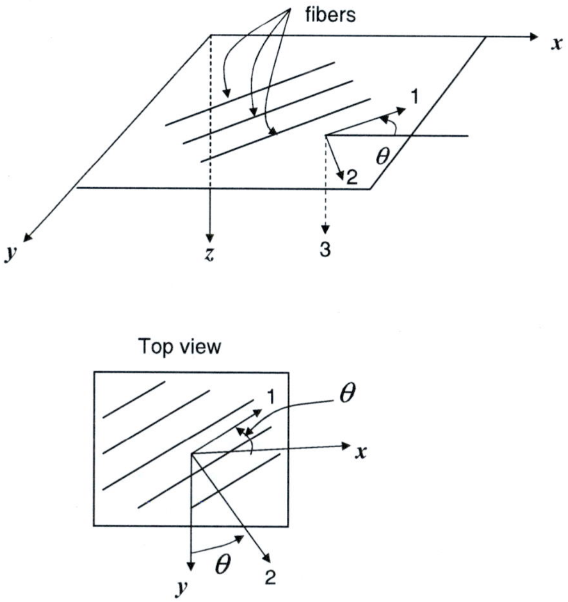

Fig. 5.3 Lamina with unidirectional fibers. The lamina principal coordinate system is associated with the axes 1, 2 and 3. The reference coordinate system is associated with the directions x, y and z

The mathematical foundations of the theory of thin composite plates differ from those of thin isotropic plates only in the constitutive relations. The kinematic equations and the equations of equilibrium in terms of stress couples and stress resultants remain without change whether the material is anisotropic or isotropic. Similarly, static boundary conditions in terms of stress resultants and stress couples and kinematic boundary conditions are not affected by the material of the plate. However, composite materials being anisotropic, the strength criteria used in the analysis of laminated plates differ from those applicable to isotropic plates.

The assumptions involved in the theory of thin laminated plates are consistent with those for thin plates from isotropic materials. However, laminated plates being constructed of a number of layers, perfect bonding between the layers becomes an additional and important requirement. If bonding between the layers is violated, the so-called delamination crack begins to propagate violating the integrity of the structure. Such crack is difficult to detect without expensive testing; eventually, it "unzips" the plate resulting in catastrophic failure.

The constitutive relations for a lamina are affected by the choice of the coordinate system. At the lamina level, the natural choice of in-plane coordinates is shown in Fig. 5.3 where the 1-axis is oriented in the fiber direction, while the in-plane 2-axis and the out-of-plane 3-axis are perpendicular to the fibers. The strain-stress relationships in this coordinate system are

$$\begin{Bmatrix} \varepsilon_1 \\ \varepsilon_2 \\ \varepsilon_3 \\ \gamma_{23} \\ \gamma_{31} \\ \gamma_{12} \end{Bmatrix} = \begin{bmatrix} 1/E_1 & -\upsilon_{21}/E_2 & -\upsilon_{31}/E_3 & 0 & 0 & 0 \\ -\upsilon_{12}/E_1 & 1/E_2 & -\upsilon_{32}/E_3 & 0 & 0 & 0 \\ -\upsilon_{13}/E_1 & -\upsilon_{23}/E_2 & 1/E_3 & 0 & 0 & 0 \\ 0 & 0 & 0 & 1/G_{23} & 0 & 0 \\ 0 & 0 & 0 & 0 & 1/G_{31} & 0 \\ 0 & 0 & 0 & 0 & 0 & 1/G_{12} \end{bmatrix} \begin{Bmatrix} \sigma_1 \\ \sigma_2 \\ \sigma_3 \\ \tau_{23} \\ \tau_{31} \\ \tau_{12} \end{Bmatrix} \tag{5.1}$$

where E_i are the moduli of elasticity in the respective directions, G_{ij} are the shear moduli in the ij-planes, and ν_{ij} are the Poisson ratios (ν_{ij} is the strain in the j-direction produced by a unit strain in the i-direction).

The material characterized by Eq. 5.1 is called specially orthotropic (see Eqs. 1.17 that represent an inverse of (5.1)); such material has three planes of property symmetry. There are 12 engineering constants that characterize the response of such material, but only 9 of these constants are independent since the matrix of compliance coefficients in the right side of (5.1) is symmetric, so that $\frac{\nu_{ij}}{E_i} = \frac{\nu_{ji}}{E_j}$. In the reference coordinate system x-y-z forming an angle with the principal material system 1-2-3 that is not equal to 0° or 90° the material response is referred to as generally orthotropic. The constitutive relations for such material are discussed in the next paragraph.

The analysis of thin composite plates being undertaken by assumption of plane stress, i.e. neglecting the stresses in the z-direction, Eq. 5.1 are simplified accordingly. The inverse of the remaining three equations that relate in-plane stresses and strains is

$$\begin{Bmatrix} \sigma_1 \\ \sigma_2 \\ \tau_{12} \end{Bmatrix} = \begin{bmatrix} Q_{11} & Q_{12} & 0 \\ Q_{21} & Q_{22} & 0 \\ 0 & 0 & Q_{66} \end{bmatrix} \begin{Bmatrix} \varepsilon_1 \\ \varepsilon_2 \\ \gamma_{12} \end{Bmatrix} \tag{5.2}$$

where Q_{ij} are so-called reduced stiffnesses expressed in terms of engineering constants by

$$\begin{aligned} Q_{11} &= \frac{E_1}{1-\upsilon_{12}\upsilon_{21}} \\ Q_{12} &= \frac{\upsilon_{12}E_2}{1-\upsilon_{12}\upsilon_{21}} = Q_{21} \\ Q_{22} &= \frac{E_2}{1-\upsilon_{12}\upsilon_{21}} \\ Q_{66} &= G_{12} \end{aligned} \tag{5.3}$$

The values of engineering constants for a composite lamina combining fibers and matrix depend on the volume occupied by each constituent phase. A large relative volume of fibers defined as the ratio of the volume occupied by the fibers to the volume of the lamina (it is referred to as "fiber volume fraction" and denoted

Table 5.1 Engineering constants of unidirectionally laminated composite materials

Material	E_1, GPa	E_2, GPa	G_{12}, GPa	ν_{12}	V_f
Kevlar 49/Epoxy	75.80	5.50	2.07	0.34	0.60
Carbon/epoxy AS4/3501-6	142.00	10.30	7.20	0.27	0.60
E-glass/epoxy Scotchply® 1002	38.6	8.27	4.14	0.26	0.45
E-glass/vinylester	24.4	6.87	2.89	0.32	0.30

Table 5.2 Strengths of unidirectionally laminated composite materials

Material	$s_1^{(+)}$, MPa	$s_1^{(-)}$, MPa	$s_2^{(+)}$, MPa	$s_2^{(-)}$, MPa	s_{12}, MPa
Kevlar 49/Epoxy	1,380.0	586.0	34.5	138.0	44.1
Carbon/epoxy AS4/3501-6	1,830.0	1096.0	57.0	228.0	71.0
E-glass/epoxy Scotchply® 1002	1,103.0	621.0	27.6	138.0	82.7
E-glass/vinylester	584.0	803.0	43.0	187.0	64.0

by V_f) results in higher stiffness and strength, particularly in the fiber direction. The volume fraction of fibers is naturally limited by the packing limitation. For example, if the aligned identical fibers of a circular cross section are packed within the composite material, the maximum possible volume fraction (such that fibers "touch" each other but do not overlap) is easily found to be $\frac{\pi}{4} = 0.785$. The volume fraction of typical composites in applications varies between 0.30 and 0.65. Typical values of engineering constants for representative composite materials are presented in Table 5.1 (Barbero 1998; Gibson 2007).

As is evident from Table 5.1, the stiffness of a lamina in the fiber direction is significantly higher than that in the direction transverse to the fibers. A similar observation could be made for the strengths of a lamina where we distinguish between the tensile and compressive strengths in the fiber and transverse directions. Accordingly, five characteristic strength values for a composite lamina include the tensile and compressive longitudinal strengths in the fiber direction $\left(s_1^{(+)}, s_1^{(-)}\right)$, the tensile and compressive strengths in the transverse direction $\left(s_2^{(+)}, s_2^{(-)}\right)$ and the in-plane shear strength (s_{12}). The values of these strengths for materials presented in Table 5.1 are reproduced in Table 5.2 (Barbero 1998; Gibson 2007).

As reflected in Table 5.2, unidirectionally laminated composites are most vulnerable to transverse tension. Although the compressive transverse strength is higher than that in tension, it is still low relative to the strengths in the fiber direction. The explanation is that while fibers take the lion share of the load in the axial direction, their contribution to load sharing in the transverse direction is smaller, while the strength of the matrix material is invariably low compared to that of the fibers.

The relatively low transverse stiffness and strength of unidirectionally laminated composite laminae lead to the design concept of a laminate. A laminate represents a combination of bonded laminae with the angles of lamination chosen to provide the required performance. It is necessary to design laminate that possesses appropriate strength and stiffness in the major loading direction, while retaining sufficient properties in the transverse direction (a unidirectional material is usually too vulnerable to transverse loading to present a viable design).

Laminates are analyzed using the coordinate system that generally does not coincide with the principal coordinate system of individual layers. The origin of the laminate coordinate system is located at its middle plane while in-plane axes are oriented similarly to the orientation of such axes in isotropic plates. In particular, for rectangular plates considered in this chapter, the orientation of in-plane axes coincides with that of the edges. The coordinate systems of individual laminae form an angle with the laminate coordinate system. This lamination angle (θ) is used to specify the lamina orientation. The rule of sign for the angle is irrelevant for the subsequent analysis; it is usually assumed that the angle is positive if the rotation from the x-axis to the 1-axis observed looking in the positive z-direction is in the counterclockwise direction (Fig. 5.3). Laminates are identified by the angles of lamination of the corresponding layers counted from one surface $\left(z = -\frac{h}{2}\right)$ to the opposite surface $\left(z = \frac{h}{2}\right)$. For example, a four-layered laminate with the angles of lamination +45°, −45°, −45°, +45° will be identified as [45°/−45°/−45°/45°]. A laminate formed by 8 layers symmetric about the middle plane with the layers on one side of the plane oriented at 0°, 30°, −30°, +90° is identified as $[0°/30°/-30°/90°]_s$, etc. A lamina oriented at an angle to the laminate coordinate system is called a generally orthotropic lamina; such lamina has different constitutive relations in the laminate coordinate system from the specially orthotropic relations (5.1). Laminates where all laminae are oriented in either x or y directions are called "cross-ply" (e.g., an 8-layer plate $[0°/90°/90°/0°]_s$). If a laminate has at least one layer of a different orientation, forming an angle with x and y axes, it is called "angle-ply."

As will be discussed below, the best design is usually achieved forming laminates that are symmetric about the middle plane. If the load is predominantly applied in one direction, it is preferable to orient most layers in this direction. For example, if the bending load produces the maximum stresses in the x-direction, the laminate $[0°/30°/-30°/90°]_s$ may be acceptable since the outer laminae oriented in the load direction are located at the maximum distance from the middle plane, providing high stiffness and strength in this direction. The thickness of a typical lamina is usually limited to a fraction of a millimeter. Composite laminates consist of numerous laminae, so that the examples of laminates in this paragraph are given only to elucidate notations and they do not refer to laminates found in practice.

5.2 Governing Equations for Thin Composite Plate

The formulation of the constitutive relations for thin laminated plates involves the following steps. First, we have to modify the stress-strain equations (5.2) to account for the transformation of coordinates in the case of a generally orthotropic lamina.

This transformation enables us to consider strains and stresses in the lamina in the coordinate system of the laminate irrespectively of the lamination angle. As is shown in textbooks on composite materials (e.g., Gibson 2007), the transformed stress-strain relationships are

$$\begin{Bmatrix} \sigma_x \\ \sigma_y \\ \tau_{xy} \end{Bmatrix} = \begin{bmatrix} \bar{Q}_{11} & \bar{Q}_{12} & \bar{Q}_{16} \\ \bar{Q}_{12} & \bar{Q}_{22} & \bar{Q}_{26} \\ \bar{Q}_{16} & \bar{Q}_{26} & \bar{Q}_{66} \end{bmatrix} \begin{Bmatrix} \varepsilon_x \\ \varepsilon_y \\ \gamma_{xy} \end{Bmatrix} \tag{5.4}$$

where transformed reduced stiffnesses are expressed in terms of reduced stiffnesses as follows

$$\begin{aligned}
\bar{Q}_{11} &= Q_{11}\cos^4\theta + Q_{22}\sin^4\theta + 2\left(Q_{12} + 2Q_{66}\right)\sin^2\theta\cos^2\theta \\
\bar{Q}_{12} &= \left(Q_{11} + Q_{22} - 4Q_{66}\right)\sin^2\theta\cos^2\theta + Q_{12}\left(\cos^4\theta + \sin^4\theta\right) \\
\bar{Q}_{22} &= Q_{11}\sin^4\theta + Q_{22}\cos^4\theta + 2\left(Q_{12} + 2Q_{66}\right)\sin^2\theta\cos^2\theta \\
\bar{Q}_{16} &= \left(Q_{11} - Q_{12} - 2Q_{66}\right)\cos^3\theta\sin\theta - \left(Q_{22} - Q_{12} - 2Q_{66}\right)\cos\theta\sin^3\theta \\
\bar{Q}_{26} &= \left(Q_{11} - Q_{12} - 2Q_{66}\right)\cos\theta\sin^3\theta - \left(Q_{22} - Q_{12} - 2Q_{66}\right)\cos^3\theta\sin\theta \\
\bar{Q}_{66} &= \left(Q_{11} + Q_{22} - 2Q_{12} - 2Q_{66}\right)\sin^2\theta\cos^2\theta + Q_{66}\left(\sin^4\theta + \cos^4\theta\right)
\end{aligned} \tag{5.5}$$

Note that the difference in notation between transformed reduced and reduced stiffnesses is often muted in technical documentation (usually, the reference is simply to "reduced stiffnesses"), i.e. the designer should be alert to what exactly the reference to "reduced stiffness" implies in the particular problem.

It is important to establish relationships between stresses in the lamina coordinate system (1-2-3) and the laminate system (x-y-z) since we will utilize them to check the strength:

$$\begin{Bmatrix} \sigma_1 \\ \sigma_2 \\ \tau_{12} \end{Bmatrix} = \begin{bmatrix} \cos^2\theta & \sin^2\theta & 2\cos\theta\sin\theta \\ \sin^2\theta & \cos^2\theta & -2\cos\theta\sin\theta \\ -\cos\theta\sin\theta & \cos\theta\sin\theta & \cos^2\theta - \sin^2\theta \end{bmatrix} \begin{Bmatrix} \sigma_x \\ \sigma_y \\ \tau_{xy} \end{Bmatrix} \tag{5.6}$$

Extrapolating the concept of stress resultants and stress couples introduced for isotropic plates, we can integrate the layer-wise stresses and their moments about the middle plane. Relationships (5.4) referring to the stresses in the k-th layer and representing the strains in terms of middle plane components and the changes of its curvature and twist are:

$$\begin{Bmatrix} \sigma_x \\ \sigma_y \\ \tau_{xy} \end{Bmatrix}_k = \begin{bmatrix} \bar{Q}_{11} & \bar{Q}_{12} & \bar{Q}_{16} \\ \bar{Q}_{12} & \bar{Q}_{22} & \bar{Q}_{26} \\ \bar{Q}_{16} & \bar{Q}_{26} & \bar{Q}_{66} \end{bmatrix}_k \begin{Bmatrix} \varepsilon_x^0 + z\kappa_x \\ \varepsilon_y^0 + z\kappa_y \\ \gamma_{xy}^0 + z\kappa_{xy} \end{Bmatrix} \tag{5.7}$$

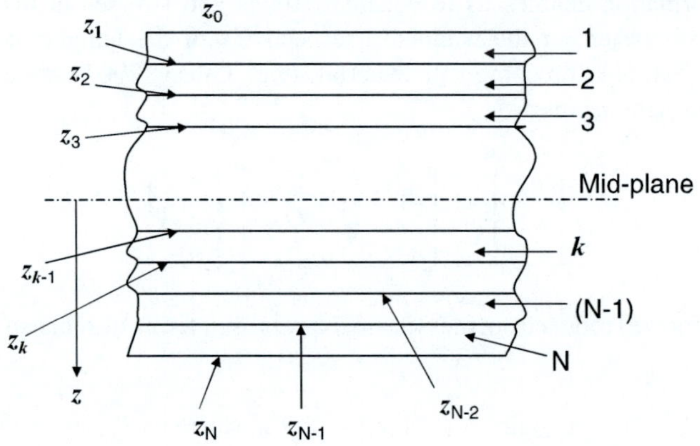

Fig. 5.4 Layers of a laminated plate. The numbers of layers vary from 1 to N

The integration according to (1.57) yields

$$\begin{Bmatrix} N_x \\ N_y \\ N_{xy} \\ M_x \\ M_y \\ M_{xy} \end{Bmatrix} = \begin{bmatrix} A_{11} & A_{12} & A_{16} & B_{11} & B_{12} & B_{16} \\ A_{12} & A_{22} & A_{26} & B_{12} & B_{22} & B_{26} \\ A_{16} & A_{26} & A_{66} & B_{16} & B_{26} & B_{66} \\ B_{11} & B_{12} & B_{16} & D_{11} & D_{12} & D_{16} \\ B_{12} & B_{22} & B_{26} & D_{12} & D_{22} & D_{26} \\ B_{16} & B_{26} & B_{66} & D_{16} & D_{26} & D_{66} \end{bmatrix} \begin{Bmatrix} \varepsilon_x^0 \\ \varepsilon_y^0 \\ \gamma_{xy}^0 \\ \kappa_x \\ \kappa_y \\ \kappa_{xy} \end{Bmatrix} \tag{5.8}$$

where the stiffness terms include so-called extensional A_{ij}, coupling B_{ij} and bending D_{ij} stiffnesses:

$$\begin{aligned} A_{ij} &= \int_{-t/2}^{t/2} (\bar{Q}_{ij})_k \, dz = \sum_{k=1}^{N} (\bar{Q}_{ij})_k \, (z_k - z_{k-1}) \\ B_{ij} &= \int_{-t/2}^{t/2} (\bar{Q}_{ij})_k \, z \, dz = \frac{1}{2} \sum_{k=1}^{N} (\bar{Q}_{ij})_k \, (z_k^2 - z_{k-1}^2) \\ D_{ij} &= \int_{-t/2}^{t/2} (\bar{Q}_{ij})_k \, z^2 \, dz = \frac{1}{3} \sum_{k=1}^{N} (\bar{Q}_{ij})_k \, (z_k^3 - z_{k-1}^3) \end{aligned} \tag{5.9}$$

The coordinates of the interfaces of the layers referred to in (5.9) are shown in Fig. 5.4.

The terminology of composite materials refers to symmetric and antisymmetric laminates. Both these classes of laminates have equal number of layers on each side of the middle plane. The laminate is symmetric if each couple of layers at equal and opposite distance from the middle plane has the same thickness and the same lamination angle, while in antisymmetric laminates the angles of lamination of each couple of layers are opposite, although these layers are otherwise identical. Accordingly, [45°/−45°/−45°/45°] would be an example of a symmetric laminate, while the laminate [−45°/45°/−45°/45°] is antisymmetric. It is easily shown from the second equation (5.9) that each couple of layers in a symmetric laminate produces a net-zero contribution to the coupling stiffness, so that in these laminates $B_{ij} = 0$. This fact has important implications on composite structures due to the following observations presented here without a proof, for the sake of brevity:

1. Nonzero coupling stiffness in asymmetrically laminated plates results in larger deflections and stresses as compared to the case where the same layers are symmetrically arranged about the middle plane. This is generally the rule as long as symmetrically laminated outer layers are oriented in the plane of the maximum bending stress couple. For example, if a four-layer plate is subject to the bending stress couple M_x, the symmetric lamination $[0°/90°/90°/0°]$ is a much better design than either an alternative symmetrically laminated plate $[90°/0°/0°/90°]$ or an antisymmetric laminate $[0°/90°/0°/90°]$.
2. Nonzero coupling stiffness in asymmetric or antisymmetric laminates results in a smaller buckling load and a lower fundamental frequency that those in a laminate where otherwise identical layers are symmetric about the middle plane. This reduction in the eigenvalues is always undesirable in the case of the buckling load and usually detrimental in the case of the fundamental (lowest) frequency of the structure.

Accordingly, most composite plates (as well as beams and shells) found in applications are symmetric about the middle plane. One exception is found in the structures subject to thermal loading applied at one of the surfaces. In such case, it may be beneficial to design the structure that is asymmetric about the middle plane at room temperature. However, due to a nonuniform temperature distribution through the thickness, the properties of various layers degrade at a different rate and the structure may actually become "symmetric" or nearly symmetric under operational high-temperature conditions. We do not consider this case in the present chapter and limit ourselves to symmetric configurations avoiding a further reference to antisymmetric and asymmetric structures.

Additional simplifications of Eqs. 5.8 are possible in symmetric laminates. In particular, it can be shown that extensional stiffness terms $A_{16} = A_{26} = 0$. A numerical analysis of multilayered symmetric angle-ply laminates illustrates that as the number of layers increases, the bending stiffness terms D_{16} and D_{26} become small compared to other bending terms (e.g., Barbero 1998). Accordingly, if the number of laminae exceeds 12 (with a typical laminae thickness this corresponds to the laminate thickness of only about 1.5 mm), it is possible to neglect these terms. In this case Eq. 5.8 are simplified:

$$\begin{Bmatrix} N_x \\ N_y \\ N_{xy} \\ M_x \\ M_y \\ M_{xy} \end{Bmatrix} = \begin{bmatrix} A_{11} & A_{12} & 0 & 0 & 0 & 0 \\ A_{12} & A_{22} & 0 & 0 & 0 & 0 \\ 0 & 0 & A_{66} & 0 & 0 & 0 \\ 0 & 0 & 0 & D_{11} & D_{12} & 0 \\ 0 & 0 & 0 & D_{12} & D_{22} & 0 \\ 0 & 0 & 0 & 0 & 0 & D_{66} \end{bmatrix} \begin{Bmatrix} \varepsilon_x^0 \\ \varepsilon_y^0 \\ \gamma_{xy}^0 \\ \kappa_x \\ \kappa_y \\ \kappa_{xy} \end{Bmatrix} \tag{5.10}$$

A popular type of laminates are symmetric cross-ply structures composed of the layers oriented at 0° and 90° relative to the x-axis (the number of layers in each direction may vary). As follows from (5.5), in such structures $\bar{Q}_{16} = \bar{Q}_{26} = 0$. Accordingly, $A_{16} = A_{26} = D_{16} = D_{26} = 0$, so that (5.10) represent the constitutive law for the symmetric cross-ply laminates, even if the number of layers is small. Note that Eq. 5.10 are also applicable to the analysis of specially orthotropic plates.

Linear equations of equilibrium for symmetric laminates in terms of displacements are obtained by the substitution of constitutive relations (5.10) and the linearized strain-displacement relationships (1.28), (1.29) into the equations written in terms of stress resultants and stress couples (1.84). This results in coupled equations for in-plane displacements obtained from the first two equations of equilibrium and the uncoupled equation for the transverse deflection shown here for a laminate subject to transverse pressure $p(x, y)$:

$$\begin{gathered} A_{11}u^0{}_{,xx} + A_{66}u^0{}_{,yy} + (A_{12} + A_{66})\, v^0{}_{,xy} = 0 \\ (A_{12} + A_{66})\, u^0{}_{,xy} + A_{22}u^0{}_{,yy} + A_{66}v^0{}_{,xx} = 0 \end{gathered} \tag{5.11a}$$

and

$$D_{11}w_{,xxxx} + 2(D_{12} + 2D_{66})\, w_{,xxyy} + D_{22}w_{,yyyy} = p(x, y) \tag{5.11b}$$

where $(\ldots)_{,i} = \frac{\partial(\ldots)}{\partial i}$, $i = x, y$. For comparison, if all stiffness terms were present, the equilibrium equations would be coupled (e.g., Gibson 2007).

The analytical solution for the general case where all stiffness terms are present is impossible. Closed form solutions in double Fourier series exist for the following particular cases, besides the case of symmetric laminates discussed in Sect. 5.4 (Jones 1999):

1. Antisymmetric cross-ply laminates ($A_{16} = A_{26} = B_{12} = B_{16} = B_{26} = B_{66} = D_{16} = D_{26} = 0$). The solution exists for simply supported plates with the boundary conditions $x = 0, a: \; w = M_x = v = N_x = 0, \; y = 0, b: \; w = M_y = u = N_y = 0$.
2. Antisymmetric angle-ply laminates ($A_{16} = A_{26} = B_{11} = B_{12} = B_{22} = B_{66} = D_{16} = D_{26} = 0$). The solution exists for the simply supported plates with the boundary conditions $x = 0, a: \; w = M_x = u = N_{xy} = 0, \; y = 0, b: \; w = M_y = v = N_{xy} = 0$.

However, as noted above, antisymmetric laminates are avoided in typical design practices, so that available solutions for such laminates are useful only as benchmark results. In angle-ply symmetric laminates with a small number of layers the equilibrium equation (5.11b) is complicated by the presence of additional terms in the left side $4\left(D_{16}w,_{xxxy} + D_{26}w,_{xyyy}\right)$ that make the Fourier analysis impossible. These terms become negligible in multilayered symmetric structures, so that the plate can be analyzed using (5.11b).

It is noted that besides equilibrium equations (5.11a and 5.11b), the boundary conditions are also uncoupled for multilayered symmetric laminates. This follows from the constitutive relations (5.10) where the first three equations relate in-plane stress resultants to in-plane displacements in the middle plane, while the last three equations refer stress couples to the transverse deflection. Accordingly, the first three equations lead to in-plane static boundary conditions that are used with (5.11a), while the last three equations provide out-of-plane static conditions analyzed with (5.11b). In case where the plate is subject to pressure applied at one of the surfaces, i.e. $p(x, y)$, the bending problem is uncoupled from the problem of in-plane middle plane displacements. The latter displacements being unaffected by the applied loads are equal to zero, i.e. $u_0 = v_0 = 0$.

If displacements are large, the problem becomes geometrically nonlinear and in-plane displacements and transverse deflections are coupled. This is immediately evident if the nonlinear strain-displacement relationships (1.28) and (1.29) are substituted into constitutive relations (5.8) or (5.10). In the general anisotropic case (5.8) both stress resultants as well as stress couples are affected by all three displacement components u_0, v_0 and w. Therefore, coupling of these components occurs both through the equations of equilibrium in terms of displacements and via static boundary conditions. In the case of a multilayered symmetric laminate, constitutive equations (5.10) result in the stress resultants that depend on $u^0,\ v^0$ and w, while the stress couples are independent of in-plane displacements.

An alternative formulation of the nonlinear problem is obtained using the stress function (1.95) that identically satisfies the first two equations of equilibrium (1.84). The inverse of the first three constitutive relations (5.10) for multilayered symmetrically laminated plates is

$$\begin{Bmatrix} \varepsilon_x^0 \\ \varepsilon_y^0 \\ \gamma_{xy}^0 \end{Bmatrix} = \begin{bmatrix} a_{11} & a_{12} & 0 \\ a_{12} & a_{22} & 0 \\ 0 & 0 & a_{66} \end{bmatrix} \begin{Bmatrix} N_x \\ N_y \\ N_{xy} \end{Bmatrix} = \begin{bmatrix} A_{11} & A_{12} & 0 \\ A_{12} & A_{22} & 0 \\ 0 & 0 & A_{66} \end{bmatrix}^{-1} \begin{Bmatrix} N_x \\ N_y \\ N_{xy} \end{Bmatrix} \tag{5.12}$$

where the compliance coefficients a_{ij} are easily available in terms of the extensional stiffnesses A_{ij} by the inverse of the matrix

$$\left[a_{ij}\right] = \left[A_{ij}\right]^{-1} \tag{5.13}$$

The substitution of (5.12) into the strain compatibility equation (1.97) and the substitution of both (5.12) and the last three equations (5.10) into the third equilibrium equation (1.84) yields the system of two nonlinear differential equations with respect to the stress function and the deflection:

$$a_{22}\varphi,_{xxxx} + (2a_{12} + a_{66})\,\varphi,_{xxyy} + a_{11}\varphi,_{yyyy} = \left(w,_{xy}\right)^2 - w,_{xx}w,_{yy}$$

$$\begin{aligned} &D_{11}w,_{xxxx} + 2\left(D_{12} + 2D_{66}\right)w,_{xxyy} + D_{22}w,_{yyyy} \\ &\quad = p\left(x, y\right) + h\left(\varphi,_{yy}w,_{xx} - 2\varphi,_{xy}w,_{xy} + \varphi,_{xx}w,_{yy}\right) \end{aligned} \tag{5.14}$$

The solution employing the stress function formulation can be obtained in a number of cases (e.g., Yang et al. 2006). Unfortunately, it is often difficult to satisfy in-plane boundary conditions and numerous studies achieved this task only in the integral sense (e.g., Liew et al. 2003) satisfying the conditions in average over the length of the edge, rather than at every point.

Finally, it is useful to discuss how temperature affects the formulation of the problem (the effect of moisture on the formulation is similar, see for example, Gibson 2007). The effects of thermal loading on composite structures include:

1. Effect on the strain-displacement relationships (see Sect. 1.9). This effect leads to thermally-induced terms in the constitutive relations as discussed in Chap. 6.
2. Effect on the material properties. This involves the effect on the moduli of elasticity and Poisson ratios as well as temperature-affected strengths of the material, its thermal conductivities, etc. Experimental data for the entire list of properties affected by temperature is usually difficult to find. However, a partial analysis accounting only for thermal effects on some of the properties may lead to dangerous results. For example, the analysis that takes into account a degradation in the moduli of elasticity but disregards decreases in the strength associated with the local temperature may lead to catastrophic results.
3. Effect of temperature on the solution of the heat transfer problem. This problem has to be addressed to evaluate a temperature distribution throughout the structure.

5.3 Strength Criteria for Laminated Composites

Deflections found from the analysis are usually valuable as a check whether the solution can be obtained using the linear theory. Similar to isotropic plates, the stress check is needed to evaluate the strength of the structure, while deflections usually serve only to validate the theory adopted for the analysis. Contrary to isotropic plates, the stress analysis of laminates is more complicated requiring a layerwise strength check. Moreover, it is impossible to predict where failure takes place, i.e. although the strains are linearly distributed throughout the thickness, the combination of stresses in a layer located in the interior of the plate may result

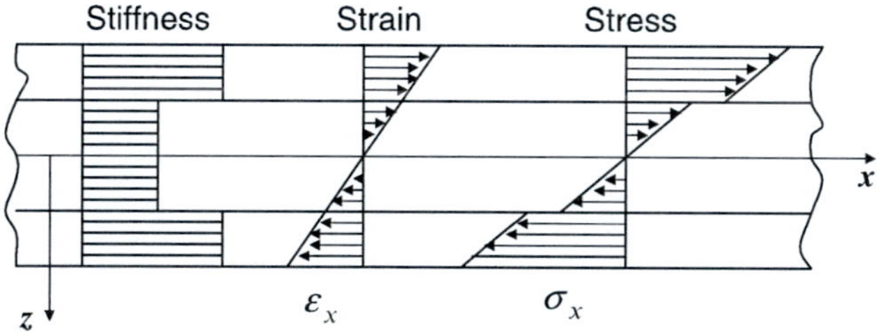

Fig. 5.5 Example of a distribution of stiffness, strain and stress in a laminated plate

in failure in this layer, prior to any damage in the outermost layers. The stresses are piecewise functions of the z-coordinate changing from layer to layer. This is reflected in Fig. 5.5 where we schematically illustrate a possible scenario of a distribution of strains and stresses in a four-layered composite laminate subject to bending. In the case shown in Fig. 5.5 the stresses in the outer layers exceed those in the inner layers. However, if the stiffness distribution is shown in Fig. 5.5 was altered so that the stiffness of the inner layers greatly exceeded that of the outer ones, the stresses in the former layers could exceed their counterparts in the latter layers. Note that larger stresses in a layer do not automatically imply that this layer fails first. This is because the strength may also vary from layer to layer, so that the strength analyst should avoid rush judgments influenced by the experience of design of isotropic structures.

The analysis begins with the evaluation of displacements and strains in the layers. Subsequently, the stresses in each layer are found from (5.7). The failure criterion should now be applied to each layer to check its strength. This requires us to transform the stresses in every layer to its principal coordinate system using (5.6). There is a large number of strength criteria for laminated orthotropic laminae. This is not surprising considering that even in the case of isotropic materials a designer can choose between numerous yield criteria (yielding is often associated with the loss of strength). The situation in composite materials is even more complex than in isotropic counterparts since there are a number of possible failure modes, including failure of fibers and matrix, debonding of fibers from matrix and delamination between the layers. Moreover, failure can be studied at the homogeneous material level or at the micromechanical level, distinguishing between various modes of failure.

There are a number of reviews of failure theories for composite materials (e.g., Christensen 1997; Puck and Schurmann 2002; Hinton et al. 2002). We do not attempt to review these theories in the framework of this book and simply outline several popular methods used for the analysis of composite plates in numerous applications. The plate is assumed in the state of plane stress except for Hashin's and Christensen's criteria.

The maximum principal stress criterion provides the envelope of local stresses in the lamina principal axes:

$$\begin{aligned} & s_1^{(-)} < \sigma_1 < s_1^{(+)} \\ & s_2^{(-)} < \sigma_2 < s_2^{(+)} \\ & |\tau_{12}| < s_{12} \end{aligned} \tag{5.15}$$

The values of strength for several representative composite materials referred to in (5.15) are found in Table 5.2. The lamina fails if any of inequalities (5.15) is violated.

Both the Tsai-Hill and the Tsai-Wu strength criteria are derived from the quadratic polynomial failure criteria (Jones 1999; Christensen 2005). In particular, the former (Tsai-Hill) criterion reads

$$\frac{\sigma_1^2}{s_1^2} - \frac{\sigma_1\sigma_2}{s_1^2} + \frac{\sigma_2^2}{s_2^2} + \frac{\tau_{12}^2}{s_{12}^2} = 1 \tag{5.16}$$

The choice of tensile or compressive strengths $s_1,\ s_2$ in (5.16) is dictated by the sign of the stresses $\sigma_1,\ \sigma_2$, i.e. a tensile stress in either direction means that we should use the tensile strength in the same direction, etc.

In the case of plane stress the Tsai-Wu strength criterion has the form (Christensen 2005):

$$\begin{aligned} & \left(\frac{1}{s_1^{(+)}} - \frac{1}{s_1^{(-)}}\right)\sigma_1 + \left(\frac{1}{s_2^{(+)}} - \frac{1}{s_2^{(-)}}\right)\sigma_2 + \frac{\sigma_1^2}{s_1^{(+)}s_1^{(-)}} + \frac{\sigma_2^2}{s_2^{(+)}s_2^{(-)}} \\ & + 2F_{12}\sigma_1\sigma_2 + \frac{\tau_{12}^2}{s_{12}^2} = 1 \end{aligned} \tag{5.17}$$

The coefficient F_{12} has to be specified from a biaxial test. The requirement that the material should not fail under hydrostatic pressure yields

$$F_{12} = \frac{1}{4s_{23}^2} - \frac{1}{s_2^{(+)}s_2^{(-)}} - \frac{1}{4s_1^{(+)}s_1^{(-)}} \tag{5.18}$$

where s_{23} is the shear strength in the plane perpendicular to the fibers. The effect of F_{12} on the strength being relatively small, it was suggested that the corresponding term can be disregarded (Narayanaswami and Adelman 1977).

Two criteria shown below account for different modes of failure dependent on the local stresses. It is convenient to illustrate these criteria for a three-dimensional state of stress, understanding that they can always be reduced to the particular case of plane stress. The strength criterion of Hashin (1980) distinguishes between failure modes dominated by tensile or compressive failure of fibers and matrix The criterion consists of four conditions shown here for a transversely isotropic lamina:

Tensile fiber breakage ($\sigma_1 > 0$)

$$\left(\frac{\sigma_1}{s_1^{(+)}}\right)^2 + \frac{1}{s_{12}^2}\left(\tau_{12}^2 + \tau_{13}^2\right) = 1 \tag{5.19a}$$

Compressive fiber failure ($\sigma_1 < 0$)

$$\frac{-\sigma_1}{s_1^{(-)}} = 1 \tag{5.19b}$$

Tensile matrix failure ($\sigma_2 + \sigma_3 > 0$)

$$\frac{(\sigma_2 + \sigma_3)^2}{s_2^{(+)}} + \frac{\tau_{12}^2 + \tau_{13}^2 + \tau_{23}^2 - \sigma_2\sigma_3}{s_{12}^2} = 1 \tag{5.19c}$$

Compressive matrix failure ($\sigma_2 + \sigma_3 < 0$)

$$\frac{1}{s_2^{(-)}}\left[\left(\frac{s_2^{(-)}}{2s_{23}}\right)^2 - 1\right](\sigma_2 + \sigma_3) + \frac{(\sigma_2 + \sigma_3)^2}{4s_{23}^2} + \frac{\tau_{12}^2 + \tau_{13}^2}{s_{12}^2} + \frac{\tau_{23}^2 - \sigma_2\sigma_3}{s_{23}^2} = 1 \tag{5.19d}$$

The criterion of Christensen (1997) shown here for a three-dimensional state of stress also distinguishes between fiber and matrix dominated failure modes.

Fiber failure mode:

$$\left(\frac{1}{s_1^{(+)}} - \frac{1}{s_1^{(-)}}\right)\sigma_1 + \frac{\sigma_1^2}{s_1^{(+)}s_1^{(-)}} - \frac{1}{4}\left(\frac{1}{s_1^{(+)}} + \frac{1}{s_1^{(-)}}\right)^2 \sigma_1(\sigma_2 + \sigma_3) = 1 \tag{5.20a}$$

Matrix failure mode:

$$\left(\frac{1}{s_2^{(+)}} - \frac{1}{s_2^{(-)}}\right)(\sigma_2 + \sigma_3) + \frac{1}{s_2^{(+)}s_2^{(-)}}\left[(\sigma_2 - \sigma_3)^2 + 4\tau_{23}^2\right] + \frac{\left(\tau_{12}^2 + \tau_{13}^2\right)}{s_{12}^2} = 1 \tag{5.20b}$$

In addition to the criteria outlined above and concerned with failure of a lamina, delamination criteria have been developed to predict the conditions for "unzipping" of laminates due to fracture along laminae interfaces (e.g., Tong and Steven 1999).

In the conclusion of this paragraph it is emphasized that in composite laminates failure of one layer does not automatically imply failure of the entire laminate. While in many applications failure of even a single layer is unacceptable, progressive damage can be monitored allowing for a sequential failure of laminae. This methodology is explained in literature on composite materials and structures (e.g., Jones 1998).

5.4 Representative Bending Problems for a Thin Composite Plate

There are several problems concerned with bending of composite plates where the exact solution is available. Some of these problems can serve as benchmark solutions verifying available numerical models. Moreover, the simplest among these problems is Navier's solution for simply supported specially orthotropic plates that is applicable to many real structures.

5.4.1 Bending of a Simply Supported Specially Orthotropic Plate Subject to Transverse Pressure (Navier's Solution)

The discussion on the boundary conditions presented in Chap. 2 remains valid for composite plates. Accordingly, the assumption that the edges are simply supported is often acceptable, either as a conservative simplification in situations where an accurate evaluation of boundary conditions is too cumbersome or if supporting structures represent open-profile beams with high stiffness against bending and low torsional stiffness. The applicability of the specially orthotropic mathematical formulation to symmetrically laminated cross-ply or symmetric multi-layered angle-ply plates was discussed in Sect. 5.2 (Eq. 5.10 are applicable to all these plates). Accordingly, bending response of symmetric cross-ply and angle-ply plates is governed by the same equilibrium equations and boundary conditions as those for a specially laminated counterpart. As is shown below, the difference between these plates becomes apparent only at the phase of the stress analysis of individual layers where the constitutive relations are customized to reflect the actual layer orientation.

Consider a rectangular plate subject to an arbitrary distributed pressure $p(x, y)$ as shown in Fig. 2.1. The solution of the equation of equilibrium (5.11b) that is uncoupled from the in-plane equilibrium equations must satisfy the boundary conditions for bending:

$$\begin{aligned} x = 0, \quad x = a : w = 0, \quad & M_x = -\left(D_{11} w_{,xx} + D_{12} w_{,yy}\right) = 0 \;\rightarrow\; w_{,xx} = 0 \\ y = 0, \quad y = b : w = 0, \quad & M_y = -\left(D_{12} w_{,xx} + D_{22} w_{,yy}\right) = 0 \;\rightarrow\; w_{,yy} = 0 \end{aligned} \tag{5.21}$$

where the stress couples are obtained from (5.10) by substituting the expressions for the changes of curvatures and twist (1.29).

Limiting the analysis to the geometrically linear case, the solution is sought in double Fourier series (2.5) that satisfy the boundary conditions as is easily verified by substitution. The substitution of deflections given by (2.5) and pressure given by (2.3) into (5.11b) yields uncoupled equations for each harmonic. The amplitude of the mn-th harmonic is

$$W_{mn} = \frac{p_{mn}}{D_{11}\alpha_m^4 + 2(D_{12} + 2D_{66})\alpha_m^2\beta_n^2 + D_{22}\beta_n^4} \tag{5.22}$$

The stresses can now be calculated in each layer. Limiting the analysis to the plane stress state we obtain the strains in the k-th layer by substituting (2.5) into the strain-displacement expressions (1.31) where middle plane strains are equal to zero and the changes of curvature and twist are given by (1.29). These strains are subsequently substituted into (5.7) yielding

$$\begin{Bmatrix} \sigma_x \\ \sigma_y \\ \tau_{xy} \end{Bmatrix}_k = z\sum_{m=1}^{M}\sum_{n=1}^{N}\left\{ \begin{bmatrix} \bar{Q}_{11} & \bar{Q}_{12} & \bar{Q}_{16} \\ \bar{Q}_{12} & \bar{Q}_{22} & \bar{Q}_{26} \\ \bar{Q}_{16} & \bar{Q}_{26} & \bar{Q}_{66} \end{bmatrix}_k \begin{Bmatrix} \left(\frac{m}{a}\right)^2 \sin\alpha_m x \sin\beta_n y \\ \left(\frac{n}{b}\right)^2 \sin\alpha_m x \sin\beta_n y \\ -2\frac{mn}{ab} \cos\alpha_m x \cos\beta_n y \end{Bmatrix} \pi^2 W_{mn} \right\} \tag{5.23}$$

Although the stiffness terms A_{16}, A_{26}, D_{16}, D_{26} are not present in the equilibrium analysis, the transformed reduced stiffness $\bar{Q}_{16}$ and $\bar{Q}_{26}$ should be accounted for in the stress analysis of individual angle-ply layers. The stress transformation (5.6) can now be applied to evaluate the principal stresses in every layer that are subsequently employed in a strength criterion to check its strength.

5.4.2 *Bending of a Specially Orthotropic Plate Subject to Transverse Pressure p (x, y) Where Only Two Opposite Edges Are Simply Supported (Levy's Solution)*

Consider bending of a plate where the edges x = 0 and x = a are simply supported while the boundary conditions along the edges $y = -\frac{b}{2}$ and $y = \frac{b}{2}$ are arbitrary (Fig. 2.8). A similar problem was considered for isotropic plates in Sect. 2.3.

The solution of the equation of equilibrium (5.11b) is sought in the form similar to (2.19) that is reproduced here for convenience

$$w = \sum_{m=1}^{M} (f_m(y) + g_m(y)) \sin\alpha_m x \tag{5.24}$$

where $f_m(y)$ and $g_m(y)$ are unknown functions similar to those defined in Sect. 2.3. The boundary conditions along the simply supported edges are satisfied by (5.24).

The pressure acting on the plate is represented in single Fourier series (2.22). Subsequently, the substitution of (2.22) and (5.24) into (5.11b) yields the following nonhomogeneous equation:

$$\sum_{m=1}^{M}\left[D_{22}\frac{d^4 f_m}{dy^4}-2\left(D_{12}+2D_{66}\right)\alpha_m^2\frac{d^2 f_m}{dy^2}+D_{11}\alpha_m^4 f_m\right]\sin\alpha_m x$$
$$=\sum_{m=1}^{M} p_m(y)\sin\alpha_m x \tag{5.25}$$

This equation must be satisfied at arbitrary values of x. Accordingly, trigonometric functions cannot be assumed equal to zero. As a result, we obtain uncoupled equations for unknown functions $f_m(y)$:

$$D_{22}\frac{d^4 f_m}{d\,y^4}-2\left(D_{12}+2D_{66}\right)\alpha_m^2\frac{d^2 f_m}{dy^2}+D_{11}\alpha_m^4 f_m=p_m(y) \tag{5.26}$$

Realizing that any particular integral of (5.26) will satisfy the solution, finding such an integral usually does not present a problem. For example, if the load is independent of the y-coordinate, i.e. $p_m(y) = p_m = const.$, we can take $f_m = \frac{p_m}{D_{11}\alpha_m^4}$.

The corresponding homogeneous equation is

$$D_{22}\frac{d^4 g_m}{dy^4}-2\left(D_{12}+2D_{66}\right)\alpha_m^2\frac{d^2 g_m}{dy^2}+D_{11}\alpha_m^4 g_m=0 \tag{5.27}$$

The solution of the homogeneous equation (5.27) can be represented in the form

$$g_m=\sum_{j=1}^{4} C_j F_j(y) \tag{5.28}$$

where C_j are constants of integration and $F_j(y)$ are functions that depend on the roots of the characteristic equation

$$\lambda_{1,2}^2=\alpha_m^2\frac{D_{12}+2D_{66}\pm\left[\left(D_{12}+2D_{66}\right)^2-D_{11}D_{22}\right]^{\frac{1}{2}}}{D_{22}} \tag{5.29}$$

The following cases are distinguished dependent on the roots given by (5.29):

1. Real unequal roots occur if $(D_{12}+2D_{66})^2 > D_{11}D_{22}$. Then

$$F_1=\cosh\lambda_1 y,\quad F_2=\sinh\lambda_1 y,\quad F_3=\cosh\lambda_2 y,\quad F_4=\sinh\lambda_2 y \tag{5.30}$$

2. Real and equal roots $\lambda_1=\lambda_2\equiv\lambda$ occur if $(D_{12}+2D_{66})^2 = D_{11}D_{22}$. Then

$$F_1 = \cosh \lambda y, \quad F_2 = \sinh \lambda y, \quad F_3 = y \cosh \lambda y, \quad F_4 = y \sinh \lambda y \quad (5.31)$$

3. Complex conjugate roots occur if $(D_{12} + 2D_{66})^2 < D_{11}D_{22}$. They can be represented as in the form $\pm(\lambda_a \pm i\lambda_b)$ where the real and imaginary parts are easily available. Accordingly, in this case,

$$F_1 = (\cosh \lambda_a y)(\cos \lambda_b y), \quad F_1 = (\cosh \lambda_a y)(\sin \lambda_b y),$$
$$F_3 = (\sinh \lambda_a y)(\cos \lambda_b y), \quad F_4 = (\sinh \lambda_a y)(\sin \lambda_b y) \quad (5.32)$$

Constants of integration can be found from four boundary conditions along the edges $y = -\frac{b}{2}$ and $y = \frac{b}{2}$. Examples of such solutions can be found in a number of references, (e.g. Reddy 2004, and Whitney 1987).

5.4.3 Bending of Clamped Elliptical and Circular Anisotropic Plates

This is an important problem that has an analytical solution (Lekhnitskii 1968). The boundary conditions of elliptical and circular plates used in applications can sometimes be modeled as clamping. If such plates are manufactured from a laminated composite material, a cross-ply design is more practical than trying to manufacture a cylindrically orthotropic structure.

Consider a clamped symmetrically laminated cross-ply elliptical plate subject to a uniform pressure p_0 where the layers are oriented in the directions of the principal elliptical axes (Fig. 3.12). The equation of equilibrium of the plate is given by (5.11b) where $p(x, y) = p_0$. The kinematic boundary conditions are identical to those used in the analysis of isotropic elliptical plates (Sect. 3.6). Following the procedure similar to that used in the case of isotropic plates it is easy to verify that the solution satisfying both the equation of equilibrium as well as the boundary conditions is

$$w(x, y) = \frac{p_0 a^4}{8\left[3D_{11} + 2(D_{12} + 2D_{66})\left(\frac{a}{b}\right)^2 + 3D_{22}\left(\frac{a}{b}\right)^4\right]}\left(1 - \frac{x^2}{a^2} - \frac{y^2}{b^2}\right)^2 \quad (5.33)$$

The stress couples and stresses can easily be evaluated as functions of the deflection using the last three equations (5.10) and (5.7) where $\bar{Q}_{16} = \bar{Q}_{26} = 0$. In particular, the maximum bending stress couple found at the ends of the large semi-axis $x = \pm a,\ y = 0$ if $D_{11} > D_{22}\left(\frac{a}{b}\right)^2$ is

$$|M_x| = \frac{p_0 a^2 D_{11}}{3D_{11} + 2\left(D_{12} + 2D_{66}\right)\left(\frac{a}{b}\right)^2 + 3D_{22}\left(\frac{a}{b}\right)^4} \tag{5.34a}$$

If $D_{11} < D_{22}\left(\frac{a}{b}\right)^2$, the maximum bending stress couple occurs at the ends of the smaller semi-axis $x = 0,\ y = \pm b$ being equal to

$$\left|M_y\right| = \frac{p_0 a^2 D_{22}}{3D_{11} + 2\left(D_{12} + 2D_{66}\right)\left(\frac{a}{b}\right)^2 + 3D_{22}\left(\frac{a}{b}\right)^4}\left(\frac{a}{b}\right)^2 \tag{5.34b}$$

In the particular case of an isotropic plate the solution shown above converges to the result illustrated in Sect. 3.5.

If the cross-ply laminated plate is circular, the present solution is applicable using $a = b,\ r^2 = x^2 + y^2$:

$$w\left(x, y\right) = \frac{p_0 a^4}{8\left[3D_{11} + 2\left(D_{12} + 2D_{66}\right) + 3D_{22}\right]}\left(1 - \frac{r^2}{a^2}\right)^2 \tag{5.35}$$

5.5 Buckling Problems for Thin Composite Plates

5.5.1 *Linear Buckling Problems*

The choice of the buckling problem for thin composite plates considered here is based on the previous discussion where it was shown that the majority of laminated structures in applications can be analyzed using solutions for specially orthotropic plates with the bending stiffness terms reflecting the actual angle-ply or cross-ply lamination. The boundaries are often accurately modeled as simply supported. Accordingly, we consider buckling of a simply supported specially orthotropic plate subject to biaxial compression by the stress resultants $\bar{N}_x$ and $\bar{N}_y$ (see Fig. 2.11) where $p = \bar{N}_{xy} = 0$.

The equation of equilibrium follows from (5.11b) where transverse pressure is replaced with the projections of the applied in-plane stress resultants on the z-direction, similar to the buckling analysis in Chap. 2. Compressive stress resultants that result in buckling are assumed negative. Accordingly, the equation of equilibrium is

$$D_{11}w_{,xxxx} + 2\left(D_{12} + 2D_{66}\right)w_{,xxyy} + D_{22}w_{,yyyy} - \bar{N}_x w_{,xx} - \bar{N}_y w_{,yy} = 0 \tag{5.36}$$

In-plane stress resultants do not affect bending boundary conditions (5.21). In-plane equations of equilibrium and in-plane boundary conditions become relevant for symmetrically laminated plates only in the postbuckling geometrically nonlinear analysis.

The nontrivial solution that satisfies both the equation of equilibrium (5.36) as well as boundary conditions (5.21) is given by the same expression as in case of isotropic plates (Eq. h in Example 2.7), i.e. $w = W_{mn} \sin \alpha_m x \sin \beta_n y$.

The substitution of the buckling deflection into (5.36) and the requirement $W_{mn} \neq 0$ yields the buckling load combinations defined from

$$\bar{N}_{x,cr}\alpha_m^2 + \bar{N}_{y,cr}\beta_n^2 + D_{11}\alpha_m^4 + 2\left(D_{12} + 2D_{66}\right)\alpha_m^2\beta_n^2 + D_{22}\beta_n^4 = 0 \qquad (5.37)$$

Example 5.1: Uniaxial Compression of a Simply Supported Plate The solution is available from (5.37). Obviously, only the smallest eigenvalue, i.e. the smallest critical value of the applied stress resultant is of interest. Accordingly, it is evident that the plate buckles similarly to its isotropic counterpart forming one half-wave in the y-direction perpendicular to the applied load ($n = 1$). The solution becomes

$$\bar{N}_{x,cr} = -\left(\frac{\pi}{ma}\right)^2 \left[D_{11}m^4 + 2\left(D_{12} + 2D_{66}\right)m^2\lambda^2 + D_{22}\lambda^4\right], \quad \lambda = \frac{a}{b} \qquad \text{(a)}$$

The mode shape of buckling, i.e. the value of m corresponding to the buckling load, is now specified dependent on the plate aspect ratio (Berthelot 1999):

$$\begin{aligned}
m &= 1 \quad \text{if} \quad \lambda \leq \sqrt{2}\left(\frac{D_{11}}{D_{22}}\right)^{\frac{1}{4}} \\
m &= 2 \quad \text{if} \quad \sqrt{2}\left(\frac{D_{11}}{D_{22}}\right)^{\frac{1}{4}} \leq \lambda \leq \sqrt{6}\left(\frac{D_{11}}{D_{22}}\right)^{\frac{1}{4}} \\
m &= 3 \quad \text{if} \quad \sqrt{6}\left(\frac{D_{11}}{D_{22}}\right)^{\frac{1}{4}} \leq \lambda \leq \sqrt{12}\left(\frac{D_{11}}{D_{22}}\right)^{\frac{1}{4}}
\end{aligned} \qquad \text{(b)}$$

5.5.2 Geometric Nonlinearity and Initial Imperfections

The geometrically nonlinear problem can be obtained using the nonlinear stress function formulation presented in (5.14) with appropriate boundary conditions. Alternatively, it is possible to formulate a nonlinear counterpart of three equilibrium equations (5.11) in terms of displacements (Chia 1980). Examples of studies accounting for geometrically nonlinear deformations, usually in conjunction with other effects, such as shear deformability of the plate, thermal loading, presence of stiffeners, etc. are found in the papers of Reddy and Chao (1981), Starnes et al. (1985), Huang and Tauchert (1988), Sheinman and Frostig (1988), Savithri and Varadan (1993), Argyris and Tenek (1994) and Shen (2000).

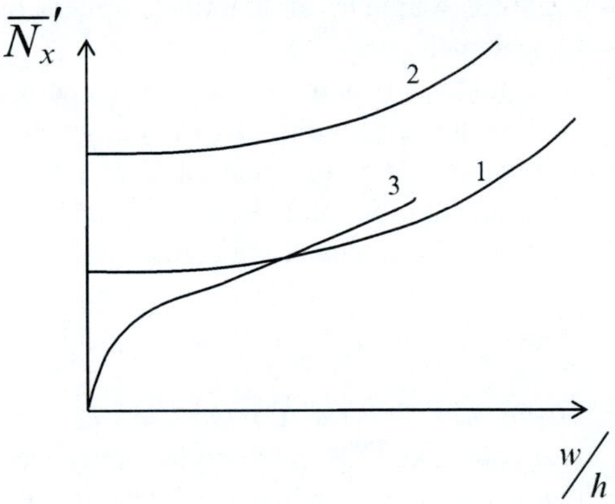

Fig. 5.6 Nondimensional buckling load $(\bar{N}_x')$ versus nondimensional deflection of axially compressed symmetrically laminated plates. Case *1*: Biaxial compression (the ratio $\frac{\bar{N}_y}{\bar{N}_x}$ is prescribed). Case *2*: Uniaxial compression. Case *3*: In-plane compression applied simultaneously with transverse pressure

Postbuckling behavior of composite plates that is often analyzed using finite element or finite difference methods is characterized by a stable load-deformation curve. The final failure occurs as a result of the loss of strength. Examples of relevant studies can be found in the papers of Shin et al (1993) and Han et al (2006).

The postbuckling behavior of symmetrically laminated plates schematically shown in Fig. 5.6 clearly reflects the prebuckling state where transverse deflections are absent and the postbuckling response characterized by a monotonous increase of deflections. In the presence of transverse pressure deflections develop even at small loads and buckling is replaced with bending.

The effect of initial imperfections on the response of composite plates is similar to that in their isotropic counterparts. Deflections develop even at small in-plane compressive and shear loads and they gradually increase as the loads become larger. Within the linear range, the equation of equilibrium of specially orthotropic, symmetric cross-ply and symmetric multi-layered angle-ply plates subject to biaxial compression are described by a generalization of Eq. 5.36:

$$D_{11}w_{,xxxx} + 2\left(D_{12} + 2D_{66}\right)w_{,xxyy} + D_{22}w_{,yyyy} - \bar{N}_x(w_{,xx} + w_{0,xx}) - \bar{N}_y(w_{,yy} + w_{0,yy}) = 0 \tag{5.38}$$

where as in case of isotropic plates, w is a deflection from the imperfect surface (w_0).

Imperfections do not introduce coupling between in-plane displacements and transverse deflections in the linear equations of equilibrium. Similar to isotropic plates, imperfections do not affect boundary conditions. Accordingly, the solution of the linear problem for a simply supported plate can be sought representing both initial imperfections and the deflections from the imperfect surface in double Fourier series as shown in Sect. 2.8 (the series for deflections satisfy boundary conditions (5.21)). The substitution of these series into (5.38) yields the amplitude of the mn-th harmonics:

$$W_{mn} = -\frac{\left(\bar{N}_x \alpha_m^2 + \bar{N}_y \beta_n^2\right) \overset{0}{W}_{mn}}{D_{11}\alpha_m^4 + 2\left(D_{12} + 2D_{66}\right)\alpha_m^2\beta_n^2 + D_{22}\beta_n^4 - \bar{N}_x\alpha_m^2 - \bar{N}_y\beta_n^2} \tag{5.39}$$

If the plate is isotropic, the result (5.39) converges to the corresponding solution in Sect. 2.8.

Example 5.2: Stress Analysis of the Plate Prior and After Buckling This discussion is related to the mode of failure, i.e. buckling vs. loss of strength. While the buckling loads of biaxially compressed plates are given by (5.37), a sufficiently stiff structure can fail if the in-plane stresses cause the loss of strength at load combinations that are smaller than those given by (5.37). The strength criteria discussed in Sect. 5.3 can be applied to check each layer at prebuckling conditions. There are no bending stresses in the plate prior to buckling (we assume that initial imperfections are absent). However, the analysis of layer-wise stresses is less trivial than in case of isotropic plates where the stress is obtained by dividing the in-plane load by the thickness. The chain of necessary calculations is shown below for the general case where the plate is subject to both biaxial compression and shear.

(a) In-plane strains uniformly distributed through the thickness for a specially orthotropic plate are (same formulae are applicable to symmetric cross-ply and symmetric multi-layered angle-ply configurations):

$$\begin{Bmatrix} \varepsilon_x^0 \\ \varepsilon_y^0 \\ \gamma_{xy}^0 \end{Bmatrix} = \begin{bmatrix} A_{11} & A_{12} & 0 \\ A_{12} & A_{22} & 0 \\ 0 & 0 & A_{66} \end{bmatrix} \begin{Bmatrix} \bar{N}_x \\ \bar{N}_y \\ \bar{N}_{xy} \end{Bmatrix} \tag{5.40}$$

(b) Layer-wise stresses obtained from (5.7) are (k-th layer):

$$\begin{Bmatrix} \sigma_x \\ \sigma_y \\ \tau_{xy} \end{Bmatrix}_k = \begin{bmatrix} \bar{Q}_{11} & \bar{Q}_{12} & \bar{Q}_{16} \\ \bar{Q}_{12} & \bar{Q}_{22} & \bar{Q}_{26} \\ \bar{Q}_{16} & \bar{Q}_{26} & \bar{Q}_{66} \end{bmatrix}_k \begin{Bmatrix} \varepsilon_x^0 \\ \varepsilon_y^0 \\ \gamma_{xy}^0 \end{Bmatrix} \tag{5.41}$$

(c) Stresses in the principal layer directions (along and perpendicular to the fibers) can now be obtained from the stress transformation equations (5.6). Note that in angle-ply plates the stiffness terms $\bar{Q}_{16}$, $\bar{Q}_{26}$are present, except for the layers oriented in the x or y-directions, while in cross-ply plates these terms are equal to zero. As is evident from the above equations, although the strain remains constant through the thickness prior to buckling, the stresses vary from layer to layer.

The stress analysis of the plate after buckling may be necessary since plates exhibit stable postbuckling behavior. The analysis involves the same steps as those shown for the prebuckling state. However, the plate acquires curvature in the postbuckling phase, so that Eq. 5.40 are replaced with the inverse of (5.10):

$$\begin{Bmatrix} \varepsilon_x^0 \\ \varepsilon_y^0 \\ \gamma_{xy}^0 \\ \kappa_x \\ \kappa_y \\ \kappa_{xy} \end{Bmatrix} = \begin{bmatrix} A_{11} & A_{12} & 0 & 0 & 0 & 0 \\ A_{12} & A_{22} & 0 & 0 & 0 & 0 \\ 0 & 0 & A_{66} & 0 & 0 & 0 \\ 0 & 0 & 0 & D_{11} & D_{12} & 0 \\ 0 & 0 & 0 & D_{12} & D_{22} & 0 \\ 0 & 0 & 0 & 0 & 0 & D_{66} \end{bmatrix}^{-1} \begin{Bmatrix} N_x \\ N_y \\ N_{xy} \\ M_x \\ M_y \\ M_{xy} \end{Bmatrix} \tag{5.42}$$

It is emphasized that in the postbuckling regime the in-plane stress resultants differ from the applied in-plane loads due to coupling between in-plane displacements and transverse deflections. Once the strains are specified from (5.42), the layer-wise stresses can be determined from (5.7).

5.6 Statics and Dynamics of Stringer-Reinforced Composite Plates

Reinforcing composite plates with stringers provides the same advantages as in isotropic metallic plates. The methodology of the analysis is similar to that for reinforced isotropic plates (Chap. 2). In this paragraph we illustrate examples of analytical solutions for an orthotropic plate with discrete stringers as well as the application of the smeared stiffeners technique.

A detailed study of composite plates reinforced by stringers was published by Lekhnitskii (1968); his original paper on orthotropic plates reinforced by stringers appeared in 1948 (Lekhnitskii 1948). More recent studies of stiffened rectangular composite plates were published by Starnes et al. (1985), Sheinman and Frostig (1988), Birman (1993, 1994), Wang et al. (2004, 2005) and Birman and Byrd (2008). Most published studies were conducted assuming that the plate is symmetrically laminated about its middle surface.

5.6.1 Bending of an Orthotropic Plate Reinforced by a System of Parallel Stringers

Consider a rectangular plate reinforced by parallel equally-spaced stringers as shown in Fig. 5.7 that is subject to pressure $p(x, y)$. The spacing between the stringers is large so that using the smeared stiffeners technique may result in a significant error. The stringers are bonded to the plate along the entire length and both the ends of the stringers as well as the edges $x = 0$ and $x = a$ are simply supported. The following solution was published by Lekhnitskii (1948).

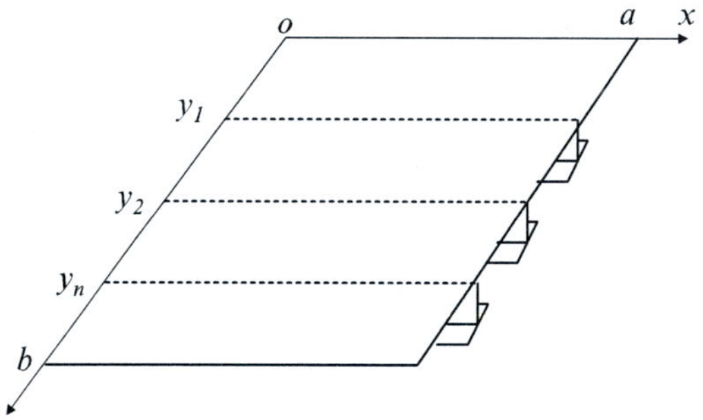

Fig. 5.7 Rectangular plate reinforced by equally-spaced parallel stringers

The plate is assumed specially orthotropic. As was previously emphasized, such model can be applied to symmetrically laminated cross-ply plates as well as symmetrically laminated angle-ply plates with a sufficiently large number of layers. The stringers can be either isotropic or laminated. It is noted that in aerospace composite structures stringers often have a hat cross section. Although these cross sections differ from T-shape stringers in Fig. 5.7, the analysis can be conducted by the procedure shown below.

The bending moment and the transverse force in the n-th stringer experiencing deformations jointly with the plate are

$$\bar{M}_n = -EI_n\bar{w}_{n,xx}, \quad \bar{Q}_n = -EI_n\bar{w}_{n,xxx} \tag{5.43}$$

where EI_n is the bending stiffness of the stringer and $\bar{w}_n = w(x, y = y_n)$ is a deflection of the plate under the n-th stringer that is equal to the deflection of the perfectly bonded stringer. The bending stiffness of the stringer is determined relative to its centroidal axis perpendicular to the xz-plane. If the stringer is isotropic, E is its modulus of elasticity. For a laminated stringer, E represents a flexural modulus.

Bending of each section of the plate between parallel n-th and (n + 1)-th stringers is governed by the equilibrium equation (5.11b) that becomes:

$$D_{11}w_{n,xxxx} + 2(D_{12} + 2D_{66})w_{n,xxyy} + D_{22}w_{n,yyyy} = p_n(x, y) \tag{5.44}$$

In (5.44) w_n is a plate (skin) deflection between the parallel stringers and the edges x = 0, x = a, and p_n is the pressure within the same plate section.

The solution must satisfy the boundary conditions along x = 0 and x = a, i.e. $w_n(x = 0) = w_n(x = a) = M_{xn}(x = 0) = M_{xn}(x = a) = 0$ where we refer to the stress couples in the n-th section of the plate. Similar conditions at the ends of the stringers that are bonded to the plate imply that $\bar{w}_n(x = 0) = \bar{w}_n(x = a) = \bar{M}_n(x = 0) = \bar{M}_n(x = a) = 0$ where the overbar identifies the moment in the stringer.

The deformation and force continuity conditions along the n-th stringer axis relate the deflections, slopes, bending stress couples and transverse shear stress resultants in adjacent sections of the plate to the corresponding functions for the stringer at $y = y_n$:

$$
\begin{aligned}
& w_n\,(x, y_n) = w_{n+1}\,(x, y_n) = \bar{w}_n(x) \\
& w_{n,y}\,(x, y_n) = w_{n+1,y}\,(x, y_n) \\
& M_{y(n+1)}\,(x, y_n) - M_{yn}\,(x, y_n) = C_n \bar{w}_{n,xxy} \\
& Q_{y(n+1)} + M_{xy(n+1),x} - Q_{yn} - M_{n,x} = E I_n \bar{w}_{n,xxxx}
\end{aligned} \tag{5.45}
$$

where C_n is a torsional stiffness of the stringer.

The set of solutions $w_n\,(x, y)$ for $1 \le n \le N+1$ where $N+1$ is the number of plate sections and N is the number of stringers must satisfy the equilibrium equations (5.44) for each plate section, the continuity conditions (5.45), the boundary conditions along x = 0 and x = a that are specified above and the conditions along the edges parallel to the stringers, i.e. along y = 0 and y = b. It can be observed that due to simply supported boundary conditions along the edges x = 0 and x = a the following solution resembles the Levy approach.

The pressure and the deflection within each section are represented in the Fourier series:

$$
\begin{aligned}
p_n\,(x, y) &= \sum_{m=1}^{M} p_{nm}(y) \sin \alpha_m x \\
w_n\,(x, y) &= \sum_{m=1}^{M} f_{nm} \sin \alpha_m x
\end{aligned} \tag{5.46}
$$

The expression for the deflection in (5.46) satisfies the boundary conditions along the edges x = 0, x = a. The substitution of (5.46) into the equilibrium equation (5.44) and the continuity conditions (5.45) yields the ordinary differential equation for each span of the plate as well as the continuity conditions along the edges $y = y_n$:

$$
D_{22} f_{nm,yyyy} - 2\,(D_{12} + 2D_{66})\,\alpha_m^2 f_{nm,yy} + D_{11}\alpha_m^4 f_{nm} = p_{nm}(y) \tag{5.47}
$$

and

$$
\begin{aligned}
& y = y_n: \qquad f_{nm} = f_{(n+1)m}, \qquad f_{nm,y} = f_{(n+1)m,y} \\
& f_{(n+1)m,yy} - f_{nm,yy} = \frac{C_n \alpha_m^2}{D_{22}} f_{nm,y}
\end{aligned}
$$

$$f_{(n+1)m,yyy} - f_{nm,yyy} = -\frac{EI_n}{D_{22}}\alpha_m^4 f_{nm} \tag{5.48}$$

In the case of open-profile stringers, their torsional stiffness can often be neglected, i.e. $C_n \approx 0$.

The solution of Eq. 5.47 involves four constants of integration for each span, i.e.

$$f_{nm} = \sum_{i=1}^{4} C_{inm} F_{inm}(y) \tag{5.49}$$

where $F_{inm}(y)$ are trigonometric or hyperbolic functions that can be specified dependent on the plate properties and geometry.

For a plate with N sections between the stringers the number of unknown constants of integration is 4N. These constants can be determined from 4(N-1) continuity conditions (5.48) along (N-1) span edges and 4 boundary conditions at the outer edges of the plate y = 0 and y = b. The numbers of boundary and continuity conditions being equal to the number of constants of integration, these constants are found from the system of 4N linear algebraic equations. Therefore, the solution of the problem is exact, with the accuracy limited only by the number of terms retained in series (5.46). If the deflections of the plate become too large, exceeding half-thickness, the geometrically linear solution shown here becomes too conservative and it is preferable to use a numerical solution accounting for geometric nonlinearity.

The solution shown above could be used in design of a stringer-reinforced plate as follows.

1. The check of the strength and deflections of sections of the plate between the stringers can be accomplished using the solution for deflections (5.49) to evaluate the strains and layer-wise stresses. Subsequently, one of the strength criteria discussed in Sect. 5.3 should be applied to check the strength of each layer.
2. In addition, it is necessary to check the strength of stringers. Although this is seldom a problem, i.e. the failure occurs in the sections of the plate, light stringers placed at a significant spacing may be unsafe. The corresponding check is easily conducted employing deflections given by Eq. 5.46 to find the bending moment and transverse force in the plane of the stringer according to (5.43). Subsequently, the analysis of the stringer is straightforward. For example, if the n-th stringer is composite, the bending stress in the j-th layer can be found as

$$\sigma_x^{(j)} = -z_j \frac{\bar{M}_n}{EI_n} E_{nx}^{(j)} \tag{5.50}$$

where $E_{nx}^{(j)}$ and z_j are the modulus of elasticity of the layer in the axial stringer direction and the distance from the layer to the centroid of the stringer.

5.6.2 Buckling and Free Vibrations of Stringer-Reinforced Cross-Ply and Functionally Graded Plates Analyzed by the Smeared Stiffeners Technique

In this section we consider the case where the plate shown in Fig. 5.7 is asymmetric about the middle plane and reinforced by identical closely spaced stringers. Although it is usually recommended to use symmetric laminates, possible examples of composites asymmetric about the middle plane are functionally graded structures referred to below in this paragraph and in Chap. 7. Composite plates symmetric about the middle plane and reinforced by stringers can be analyzed as a particular case of the present solution. The case where the plate is reinforced by stringers oriented in both x- and y-directions can be considered using a straightforward extension of this solution (Birman and Byrd 2008).

Geometrically linear equations of motion of a thin elastic plate including in-plane compressive static loads have already been developed in Chap. 4 (equations (4.37)). The mass per unit surface area in these equations is determined from

$$\hat{m} = m_p + \sum_n \delta(y - y_n) \int_z \rho_n(z) b_n(z) dz \tag{5.51}$$

where m_p is the mass of the plate without the stringers per unit surface area, ρ_n is the mass density of the stringer material and $b_n(z)$ is the width of the stringer. The z-coordinate is counted from the middle plane of the panel. It is also necessary to update the stiffness terms in equations (4.37) since the plate is composite, rather than isotropic as in Chap. 4.

The stress resultants and stress couples in the plate without stringers are given by relationships (5.10). The nonzero elements of the stiffness matrices contributed by the cross-ply panel without stringers are

$$A' = \begin{bmatrix} A'_{11} & A'_{12} & 0 \\ A'_{12} & A'_{22} & 0 \\ 0 & 0 & A'_{66} \end{bmatrix} \qquad B' = \begin{bmatrix} B'_{11} & B'_{12} & 0 \\ B'_{12} & B'_{22} & 0 \\ 0 & 0 & B'_{66} \end{bmatrix}$$

$$D' = \begin{bmatrix} D'_{11} & D'_{12} & 0 \\ D'_{12} & D'_{22} & 0 \\ 0 & 0 & D'_{66} \end{bmatrix} \tag{5.52}$$

where

$$\left\{A'_{ij}, B'_{ij}, D'_{ij}\right\} = \int_{-\frac{h}{2}}^{\frac{h}{2}} Q_{ij} \left\{1, z, z^2\right\} dz \tag{5.53}$$

Similar to Chap. 4, we assume that the contribution of the stringers affects only the stiffness in the direction of their axes. Furthermore, the torsional stiffness of the stringers is neglected. Therefore, the corresponding stiffness terms affected by the presence of composite stringers are

$$\left\{ A''_{11}, B''_{11}, D''_{11} \right\} = \sum_{n} \delta\left(y - y_n\right) \int_{z} b_n(z) E_n(z) \left\{ 1, z, z^2 \right\} dz \qquad (5.54)$$

where $E_n(z)$ is the modulus of elasticity of the stringer material and the coordinate is counted from the middle plane of the panel. The contributions of the stringers to other elements of the stiffness matrices are assumed negligible, i.e. $A''_{ij} = B''_{ij} = D''_{ij} = 0\ (ij \neq 11)$. Note that Eq. 5.54 differ from (2.62) or (2.63) since the latter equations were written for isotropic stringers, while (5.54) enable us to incorporate laminated stringers in the analysis. The elements of the overall stiffness matrices of the panel in Eq. 5.10 are obtained by summation the contributions of the skin and stringers according to (2.56).

Similar to paragraphs 2.9 and 4.5, all edges of the panel are assumed simply supported by frames or beams that have negligible torsional stiffness and negligible stiffness in the direction perpendicular to their axes, while their axial stiffness is very high. The corresponding boundary conditions are $w = M_x = N_x = v = 0$ along $x = 0,\ x = a$ and $w = M_y = N_y = u = 0$ along $y = 0,\ y = b$.

In this paragraph we limit ourselves to a particular case where one can apply the smeared stiffeners technique that has already been discussed in reference to isotropic stringer-reinforced plates. Consider the situation where the number of equally-spaced stringers is sufficiently large so that the spacing is small. Furthermore, the cross sections of the stringers are identical, as is often the case in applications. Then the contribution of stringers to the stiffness can be smeared over the surface of the panel. As was shown by Birman (1993), even composite plates with only three equally-spaced stringers can be adequately analyzed by the smeared stiffeners technique (of course, this technique should not be applied in problems where local bending between the stringers is essential).

The mode shapes of buckling and free vibrations can be represented by Eq. 4.38 where dynamic terms $\sin \omega t$ are excluded in the former problem. The substitution of (4.38) into (4.37) yields equations (4.39) that can be used to find the buckling load combinations or the natural frequencies.

Example 5.3: Buckling Loads of Functionally Graded Stringer-Reinforced Plates
Consider the effect of stringers on the buckling loads of a functionally graded plate discussed in Example 4.1. In all examples concerned with buckling the load was applied in the x-direction. The aspect ratio of plates considered in examples varied but the mode shape of buckling in all cases was dominated by a single half-wave in both x and y directions (this is also the mode corresponding to the fundamental frequency of the plate).

The blade stringer height necessary to achieve the required values of the buckling load evaluated using the analysis described above is shown in Fig. 5.8. The

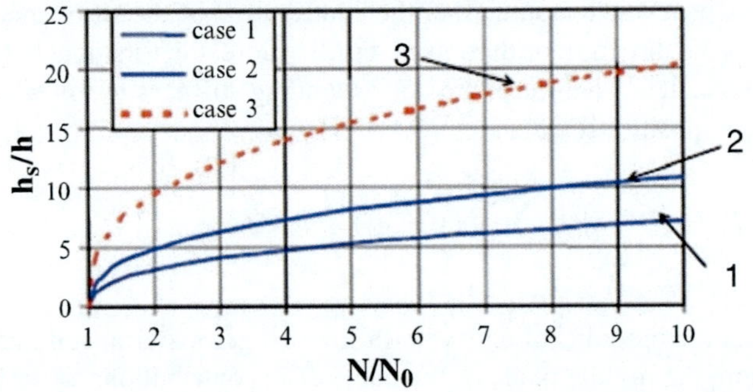

Fig. 5.8 Blade stringer height as a function of desired buckling load. The thickness of the plate is $h = 1.778mm$. The spacing of the stringers is equal to 0.1667 m. Case *1*: $a = 0.5m, b = 1.5m, N_0 = 4089\frac{N}{m}$. Case 2: $a = b = 1.0m, N_0 = 3312\frac{N}{m}$. Case *3*: $a = 1.0m, b = 0.5m, N_0 = 20700\frac{N}{m}$ (From Birman and Byrd 2008)

corresponding result relevant to the natural frequency was illustrated in Fig. 4.7. The buckling load referred to in Fig. 5.8 is normalized with respect to its value for the unstiffened plate. The vertical axis represents a ratio of the height of the blade stringer to the thickness of the plate. Both Figs. 5.8 and 4.7 clearly demonstrate the advantages of using stringer-reinforced plates, i.e. the possibility of achieving a much higher buckling load as well as an increased fundamental frequency at the expense of a relatively small increase in weight.

5.7 Shear Deformable Composite Plates

In the previous chapters we assumed that the plate is "thin," except for the discussion of shear deformable plates in Chap. 1. Accordingly, the previous solutions were developed using the classical (technical) theory of plates. The foundation for such approach is easily understood if one considers bending of a thin plate. If the plate is subject to lateral pressure p, the transverse normal stress at the loaded surface is equal and opposite to this pressure, i.e. $\sigma_z = -p$. At the opposite surface the stress $\sigma_z = 0$. Between two surfaces, the stress σ_z gradually changes from zero to$-p$.

The relative magnitude of stresses can easily be estimated on the example of a simply supported isotropic plate subject to a uniform pressure p_0. According to Timoshenko and Woinowsky-Krieger (1959), the maximum bending stress couples at the center of a square plate are $M_x = M_y = 0.0479p_0a^2$ where a is the in-plane dimension. The ratio between the maximum transverse normal and in-plane stresses can be estimated as $\left|\frac{\sigma_z^{max}}{\sigma_x^{max}}\right| = \left|\frac{\sigma_z^{max}}{\sigma_y^{max}}\right| = 3.48\left(\frac{h}{a}\right)^2$. Therefore, the maximum transverse normal stress can reach 5% of the maximum in-plane stresses only if the thickness to size ratio is smaller than 0.12. Such thick plates are seldom found in applications. Although transverse shear stresses can be larger than the transverse normal stress, isotropic structures usually do not require a three-dimensional solution. An exception may be found in the vicinity of discontinuities, such as rivet holes where three-dimensional effects can be present.

Contrary to isotropic materials, composite laminates often possess relatively low shear stiffness. This results in higher transverse shear strains that may invalidate the classical plate theory based on the assumption that transverse shear is negligible.

The theory of shear deformable plates can be traced back to pioneering work of Reissner (1945) and Mindlin (1951). Predictably, these early studies were concerned with isotropic plates and limited to the first-order theory. The extension of the analysis to anisotropic (composite) plates was undertaken by Reissner and Stavsky (1961), Stavsky (1961) and Yang et al. (1966). The subject of shear deformable plates and shells, including first-order as well as higher-order theories, was comprehensively reviewed in the monograph of Reddy (2004).

According to higher-order shear deformable theories, the displacements of the plate are represented by

$$\begin{aligned} u(x,y) &= u_0(x,y) + z\psi_x(x,y) + \sum_{i=2} z^i \eta_i(x,y) \\ v(x,y) &= v_0(x,y) + z\psi_y(x,y) + \sum_{i=2} z^i \mu_i(x,y) \\ w &= w_0(x,y) \end{aligned} \tag{5.55}$$

Transverse deflections remain constant throughout the thickness as was the case in the classical plate theory. In-plane displacements include the terms proportional to the first power of z, reflecting rotations about the middle plane. The higher-order terms proportional to higher power of z account for a possible warping of the cross section (these terms are associated with the higher-order theories). The number of functions η_i and μ_i in higher-order theories can be reduced by requiring that transverse shear stresses and transverse normal stresses on the surfaces of the plate must be equal to zero. In particular, such requirement to shear stresses is employed in the third-order theory where the highest order of the power in (5.55) is equal to $i = 3$. In the contrary, the assumption that transverse shear strains remain constant throughout the thickness of the plate adopted in the first order shear deformation theory results in the violation of the boundary condition for transverse shear and normal stresses on the surfaces.

This book does not include a review of higher-order theories and the discussion is limited to the first-order shear deformation theory. This theory is usually adequate for the study of typical composite and sandwich plates, except for plates that are subject to local concentrated loads or have discontinuities resulting in a local 3-D state of stress. Also, the first order theory cannot characterize sandwich plates with a "soft" core (Frostig et al. 2005, and Li and Kardomateas 2008).

The following analysis is conducted by assumption that deformations of the plate remain small, justifying the use of the geometrically linear theory. Shear deformable plates are usually relatively thick and as a result, they are quite stiff. Such plates often fail prior to achieving large deformations that would justify the use of geometrically nonlinear strain-displacement relations.

According to the first order theory, the in-plane strains in the plate can be written utilizing Eqs. 1.46 and 1.48 in the form:

$$\begin{aligned} \varepsilon_x^0 &= u_{0},_{x}, & \kappa_x &= \psi_{x},_{x} \\ \varepsilon_y^0 &= v_{0},_{y}, & \kappa_y &= \psi_{y},_{y} \\ \gamma_{xy}^0 &= u_{0},_{y} + v_{0},_{x}, & \kappa_{xy} &= \psi_{x},_{y} + \psi_{y},_{x} \end{aligned} \tag{5.56}$$

Transverse shear strains are presented by (1.45). These strains do not vary throughout the thickness of the plate reflecting the assumptions of the first-order theory. As a result, transverse shear stresses are present on the plate surfaces reflecting the approximate nature of the theory. Although the transverse normal strain is assumed equal to zero, $\varepsilon_z = w_{0},_{z} = 0$, it is possible to calculate the stress σ_z on the plate surfaces either integrating the corresponding elasticity equilibrium equation $\tau_{xz},_{x} + \tau_{yz},_{y} + \sigma_{z},_{z} = 0$ or the using three-dimensional stress-strain relationships. This stress should be equal to zero or to the applied pressure on the surface of the plate. Therefore, this boundary condition is also violated in the first-order theory. The inaccuracy introduced in the solution through the violation of the boundary conditions on the surface is partially compensated by an appropriate choice of the shear correction factor as explained below.

The strains are now substituted into the constitutive relations (5.7) and (5.8) that are not affected by the kinematics of the plate resulting in in-plane stresses and the corresponding in-plane stress resultants and bending and twisting stress couples. Additionally, transverse shear stresses in the k-th layer are:

$$\begin{Bmatrix} \tau_{yz} \\ \tau_{xz} \end{Bmatrix}_k = \begin{bmatrix} \bar{Q}_{44} & \bar{Q}_{45} \\ \bar{Q}_{45} & \bar{Q}_{55} \end{bmatrix}_k \begin{Bmatrix} \gamma_{yz} \\ \gamma_{xz} \end{Bmatrix} \tag{5.57}$$

where the corresponding transformed reduced stiffnesses are evaluated as functions of the reduced stiffnesses and the lamination angle of the layer:

$$\begin{aligned} \bar{Q}_{44} &= Q_{44}\cos^2\theta + Q_{55}\sin^2\theta \\ \bar{Q}_{45} &= (Q_{55} - Q_{44})\cos\theta\sin\theta \\ \bar{Q}_{55} &= Q_{55}\cos^2\theta + Q_{44}\sin^2\theta \end{aligned} \tag{5.58}$$

Each term in (5.58) refers to a particular layer (the layer index "k" is omitted). The reduced stiffnesses in (5.58) are the following functions of the properties of the corresponding layer:

$$Q_{44} = G_{23}, \quad Q_{55} = G_{13}, \tag{5.59}$$

where G_{23} is the shear modulus in the plane perpendicular to the fibers and G_{13} is the shear modulus in the plane oriented along the fibers and perpendicular to the plane of the layer. Typically, $G_{13} = G_{12}$, unless special measures are taken to ensure a different fiber density in the corresponding planes.

The transverse shear stress resultants are obtained by integrating the corresponding stresses throughout the thickness of the plate (see Eq. 1.52):

$$Q_y = k \int_{-\frac{h}{2}}^{\frac{h}{2}} \tau_{yz} dz, \quad Q_x = k \int_{-\frac{h}{2}}^{\frac{h}{2}} \tau_{xz} dz, \tag{5.60}$$

where k is the so-called shear correction factor introduced to account for the inaccuracy due to the assumptions employed in the first order theory that is discussed below.

The substitution of (5.57) into (5.60) yields the constitutive relations for the transverse shear stress resultants. In most applications, such as cross-ply and symmetrically laminated angle-ply composites, coupling between transverse shear strains is absent, so that

$$\begin{aligned} Q_y &= kA_{44}\left(w_{,y} + \psi_y\right) \\ Q_x &= kA_{55}\left(w_{,x} + \psi_x\right) \end{aligned} \tag{5.61}$$

where $\{A_{44}, \ A_{55}\} = \int_{-\frac{h}{2}}^{\frac{h}{2}} \{\bar{Q}_{44}, \ \bar{Q}_{55}\}\, dz.$

Equations of motion for shear deformable plates were derived in Sect. 1.6 (Eq. 1.81). Dynamic equations are affected by inertial terms that take into account the in-plane, translational and rotational inertias. Accounting for these contributions in composite laminates, Eq. 1.81 are modified:

$$\begin{aligned} &N_{x,x} + N_{xy,y} = \hat{m}u_{0,tt} + I_1\psi_{x,tt} \\ &N_{xy,x} + N_{y,y} = \hat{m}v_{0,tt} + I_1\psi_{y,tt} \\ &M_{x,x} + M_{xy,y} - Q_x = I_1u_{0,tt} + I_2\psi_{x,tt} \\ &M_{xy,x} + M_{y,y} - Q_y = I_1v_{0,tt} + I_2\psi_{y,tt} \\ &Q_{x,x} + Q_{y,y} + \underline{N_x w_{,xx} + 2N_{xy}w_{,xy} + N_y w_{,yy}} \\ &\quad + \underline{\underline{\left(N_{x,x} + N_{xy,y}\right)w_{,x} + \left(N_{xy,x} + N_{y,y}\right)w_{,y}}} = -p + \hat{m}w_{,tt} \end{aligned} \tag{5.62}$$

where

$$\{\hat{m}, I_1, I_2\} = \int_{-\frac{h}{2}}^{\frac{h}{2}} \rho(z) \{1, z, z^2\} \, dz \tag{5.63}$$

$\rho(z)$ being a mass density that can vary from layer to layer, contrary to isotropic materials where it is constant. If the mass density is constant throughout the plate, Eq. 5.62 are reduced to (1.81) where $I = I_2$.

Although all terms in (5.62), except for those proportional to time derivatives of rotations, are present in the classical plate theory, the stress resultants and stress couples are different. Considering a relatively small effect of in-plane and rotational inertias on the in-plane equations of motion, the doubly-underlined terms in the last equation (5.62) are often disregarded. Furthermore, in the absence of nonlinear effects, the singly-underlined terms are present only if in-plane loads $\bar{N}_x, \bar{N}_y, \bar{N}_{xy}$ are applied to the plate so that $N_x = \bar{N}_x, N_y = \bar{N}_y, N_{xy} = \bar{N}_{xy}$.

Limiting the analysis to the static case and substituting Eqs. 5.8 and 5.60 into (5.62) we obtain the equations of equilibrium. In the general case, these equations are quite long and they cannot be analytically integrated. Accordingly, we show here the solution for a specially orthotropic plate that is also applicable to cross-ply symmetrically laminated plates and to multilayer, angle-ply symmetrically laminated plates as discussed above. In such case, $A_{16} = A_{26} = D_{16} = D_{26} = 0$ and $[B] = 0$ so that equations of equilibrium become uncoupled, i.e. two equations for in-plane displacements are independent of three equations for the deflection and rotations:

$$\begin{aligned} &A_{11}u_{0,xx} + A_{66}u_{0,yy} + (A_{12} + A_{66})\, v_{0,xy} = 0 \\ &(A_{12} + A_{66})\, u_{0,xy} + A_{66}v_{0,xx} + A_{22}v_{0,yy} = 0 \end{aligned} \tag{5.64}$$

and

$$\begin{aligned} &-kA_{55}w_{,x} + D_{11}\psi_{x,xx} + D_{66}\psi_{x,yy} - kA_{55}\psi_x + (D_{12} + D_{66})\, \psi_{y,xy} = 0 \\ &-A_{44}w_{,y} + (D_{12} + D_{66})\, \psi_{x,xy} + D_{66}\psi_{y,xx} + D_{22}\psi_{y,yy} - kA_{44}\psi_y = 0 \\ &k\left(A_{55}w_{,xx} + A_{44}w_{,yy}\right) + \bar{N}_x w_{,xx} + 2\bar{N}_{xy}w_{,xy} + \bar{N}_y w_{,yy} + kA_{55}\psi_{x,x} \\ &\quad + kA_{44}\psi_{y,y} = -p \end{aligned} \tag{5.65}$$

In addition, boundary conditions must be specified along the edges. Here we refer to several possible sets of boundary conditions for the edges $x = 0$ and $x = a$. The conditions for the edges y = constant can be formulated by analogy. Note that there are five conditions along each edge reflecting the presence of five unknowns, i.e. three displacements u_0, v_0, w and two rotations ψ_x, ψ_y in the equations of equilibrium.

1. Simply supported edges (the support structure has infinite stiffness in the plane of the edge and negligible out-of-plane stiffness):

$$w = \psi_y = v_0 = N_x = M_x = 0 \tag{5.66}$$

Note that a variation of this condition obtained by replacing $v_0 = 0$ with $N_{xy} = 0$ implies that the support structure has a negligible stiffness along the edge; such situations are not encountered in practice.

2. Simply supported edges (the support structure has infinite stiffness in the plane of the edge and its out-of-plane displacements are prevented):

$$w = \psi_y = v_0 = u_0 = M_x = 0 \tag{5.67}$$

Such condition can occur as a result of adjacent plates supported at the same edge whose in-plane stiffness restrains displacements of the boundaries.

3. Clamped edge (such effect can be achieved due to symmetry of both the adjacent plates as well as the load about the edge, as explained in Sect. 2.2):

$$w = u_0 = v_0 = \psi_x = \psi_y = 0 \tag{5.68}$$

4. Free edge (such edge should be avoided due to a potential for delamination of the layers as a result of local concentration of transverse shear and normal stresses):

$$Q_x = N_x = M_x = N_{xy} = M_{xy} = 0 \tag{5.69}$$

It can be observed that for symmetrically laminated plates considered here, in the linear formulation, in-plane boundary conditions in terms of in-plane displacements and stress resultants are uncoupled from the conditions referring to transverse deflections, rotations, stress couples and transverse shear stress resultants. If the plate with $A_{16} = A_{26} = D_{16} = D_{26} = [B] = 0$ is subject to transverse loads, such as lateral pressure or concentrated forces, in-plane displacements $u_0 = v_0 = 0$.

The analytical solution is available if the out-of-plane boundary conditions correspond to those in (5.66), i.e. $w = \psi_y = M_x = 0$ along x = 0, x = a and $w = \psi_x = M_y = 0$ along y = 0, y = b. The plate considered below is subject to an arbitrary distributed pressure and to in-plane stress resultants $\bar{N}_x$ and $\bar{N}_y$.

It is evident that the solution in double Fourier series

$$\begin{aligned}
w &= \sum_{m=1}^{M} \sum_{n=1}^{N} W_{mn} \sin \alpha_m x \sin \beta_n y \\
\psi_x &= \sum_{m=1}^{M} \sum_{n=1}^{N} F_{mn} \cos \alpha_m x \sin \beta_n y \\
\psi_y &= \sum_{m=1}^{M} \sum_{n=1}^{N} P_{mn} \sin \alpha_m x \cos \beta_n y
\end{aligned} \tag{5.70}$$

satisfies the boundary conditions. The substitution of (5.70) into (5.65) yields a system of three algebraic equations with respect to the amplitudes of the corresponding harmonics for each couple $(m,\ n)$:

$$[L_{ij}]\{W_{mn}\ F_{mn}\ P_{mn}\}^T = \{0\ 0\ p_{mn}\}^T \tag{5.71}$$

where $L_{ij} = L_{ji}$, the superscript "T" identifies a transpose of the matrix or vector, and p_{mn} is the amplitude in the double Fourier series (2.3) representing an arbitrary applied pressure.

The elements of the matrix of coefficients in (5.71) are

$$\begin{aligned}
L_{11} &= D_{11}\alpha_m^2 + D_{66}\beta_n^2 + kA_{55}\\
L_{12} &= (D_{12} + D_{66})\,\alpha_m\beta_n\\
L_{13} &= kA_{55}\alpha_m\\
L_{22} &= D_{66}\alpha_m^2 + D_{22}\beta_n^2 + kA_{44}\\
L_{23} &= kA_{44}\beta_n\\
L_{33} &= \left(kA_{55} + \bar{N}_x\right)\alpha_m^2 + \left(kA_{44} + \bar{N}_y\right)\beta_n^2
\end{aligned} \tag{5.72}$$

Note that the series solution shown here is not possible in case where in-plane shear stress resultant $\bar{N}_{xy}$ is applied to the plate. Both bending as well as buckling problems can be solved using (5.72). In case where transverse pressure is applied to the plate this system of three equations is solved with respect to the amplitudes W_{mn}, F_{mn}, P_{mn}. The presence of in-plane tensile or compressive loads does not change the approach to the solution of the bending problem since they are included in the coefficient L_{33}. If the load is represented by in-plane compressive stress resultants and transverse pressure is absent, critical combinations of $\bar{N}_x$ and $\bar{N}_y$ corresponding to the mode shape of buckling with m and n half-waves in the x and y directions respectively, are determined from the nonzero requirement to the vector of amplitudes. This implies that the determinant formed by the coefficients L_{ij} in (5.71) must be equal to zero.

The following observations can be made:

1. In dynamic problems, the counterparts of Eqs. 5.64 and 5.65 are uncoupled as long as the plate is symmetric about the middle plane. This is because the rotational inertia coefficient $I_1 = 0$ due to symmetry.
2. Besides the solution for a specially orthotropic plate shown above, exact solutions of geometrically and physically linear problems in double Fourier series are available for the following cases:

   – Antisymmetric cross-ply plate with boundary conditions (5.66). See Bert et al. (1981).

– Antisymmetric angle-ply plates with boundary conditions $w = \psi_y = u = N_{xy} = M_x = 0$ along the edges $x = 0$ and $x = a$ and similar conditions along the perpendicular edges $y = 0$ and $y = b$. See Bert and Chen (1978).

Additionally, Levy's approach to the analysis of antisymmetric angle-ply and cross-ply plates with one pair of simply supported opposite edges has been published (Reddy 2004).

It remains to specify the shear correction factor k that appears in Eq. 5.60. This factor originally introduced for isotropic structures by Timoshenko in 1921 has been a subject of numerous studies (see for example, a discussion in the papers of Bert and Gordaninejad (1983) and Birman and Bert (2002)). The most popular methods of the evaluation of the shear correction factor include:

1. Comparison of the transverse shear stress according to the theory of elasticity solution for a homogeneous orthotropic plate to the stress given by the first-order shear deformation theory (essentially, an extension of the Timoshenko approach for isotropic beams);
2. Comparison of the strain energies for a homogeneous orthotropic plate derived in terms of transverse shear stress resultants according to the elasticity formulation and according to the first-order theory (extension of the Reissner approach for isotropic beams).

The former method yields $k = \frac{2}{3}$, while the latter method results in $k = \frac{5}{6}$ (Whitney 1987).

Dynamic analyses of shear deformable structures with the goal of specifying the shear correction factor have also been undertaken (see Birman and Bert (2002) for relevant references). It was shown that the shear correction factor may be a function of the plate material constants and aspect ratio. In general, the shear correction factor calculated by presently available methods lies in the range between 2/3 and 1.

An example of the comparison between exact solutions obtained by the theory of elasticity and solutions generated by the classical and first-order shear deformation theories was presented for cross-ply rectangular laminates experiencing bending as a result of a sinusoidal pressure in the book of Whitney (1987). It was shown that the classical theory of thin plates becomes inaccurate at the size-to-thickness ratio smaller than approximately 25. In the contrary, the first-order theory yielded results that are in close agreement with the exact elasticity solution. The results obtained by the first-order theory were relatively insensitive to the value of the shear correction factor. The first-order shear deformation theory was also found accurate for the analysis of a sandwich plate discussed in the next paragraph (for a detailed analysis of the applicability of the first-order shear deformation theory to the analysis of sandwich plates, see the recent paper of Nguyen et al (2010)).

A detailed comparison of the accuracy of the higher-order (third-order), first-order and classical plate theories with available three-dimensional (3-D) elasticity solutions was presented in the monograph of Reddy (2004). Similar to the observations that follow from Whitney (1973), a discrepancy between deflections and in-plane bending stresses obtained by various theories can become essential if the side-to-thickness ratio is smaller than approximately 25. The deflections and stresses obtained by higher-order, first-order and classical plate theories are unconservative, the dangerous error increasing for the first-order and particularly, classical theories. Both the classical and the first-order theories overpredict the eigenvalues (buckling load and natural frequencies) of shear deformable plates.

In spite of the observations made above, the first-order shear deformation theory can be sufficiently accurate for the strength, buckling and vibration analyses of numerous composite structures found in engineering applications, with the exception of cases of sandwich plates with a "soft" core (e.g., Li and Kardomateas 2008) and the situations where the structure is subject to concentrated loads. This theory cannot be relied on for the stress analysis in the vicinity of discontinuities with a local concentration of 3-D stresses. Even if a composite structure is "thin," material properties may superimpose a requirement to use first-order or higher-order theories, rather than the classical theory of thin plates. For example, Greenberg and Stavsky (1980) showed that relatively long cylindrical shells should be analyzed by the first-order theory if they have a low transverse shear modulus.

5.8 Sandwich Plates

The concept of sandwich construction can be traced to I-beams or truss structures. A beam of a rectangular cross section working in bending is usually a bad design. It is easy to observe that the bending stress in such beam varies from a maximum tensile stress on one surface to a maximum compressive stress on the opposite surface leaving the central part of the cross section underloaded. Although the maximum transverse shear stress occurs at the neutral axis of such beam, i.e. half-way from each of the outer surfaces, this stress does not govern the design, except for very short and deep configurations.

The necessity to increase the moment of inertia of a beam subject to bending, simultaneously increasing the buckling load and the fundamental frequency, resulted in an I-beam, i.e. the structure where a light and slender web carries a small fraction of the load that is mostly carried by the flanges. A version of the concept of an I-beam is found in a truss where the "web" consists of a system of diagonal members.

A sandwich structure typically consists of two facings joined by a light core of different material (Fig. 5.9). The facings are designed to mostly carry in-plane and/or bending loads, while the core joins the facings into a cohesive structure and carries a lion share of the transverse shear load. Accordingly, the facings are usually manufactured from heavier materials with high strength and stiffness, while a lighter core has lower strength and stiffness. The effectiveness of a sandwich structure can qualitatively be illustrated using the approach of Vinson (1999) who compared the bending stiffness and the maximum bending stress in a monocoque (single layer) isotropic beam to those in the beam where the cross section is divided into two half-thickness facings that are separated and located at a distance from each other (Fig. 5.10). The core between two facings does not carry the load and its weight is

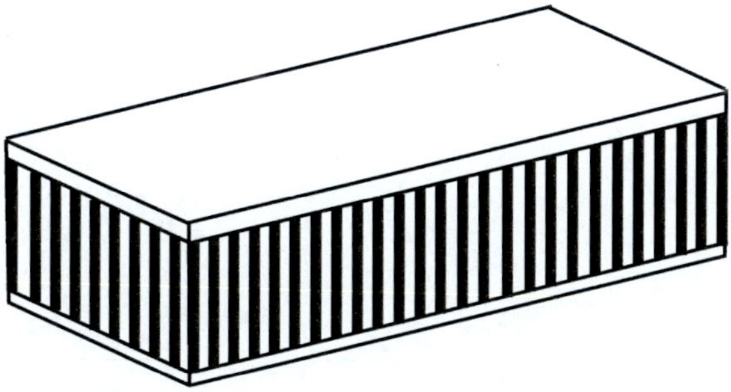

Fig. 5.9 Sandwich plate showing two facings and a honeycomb core

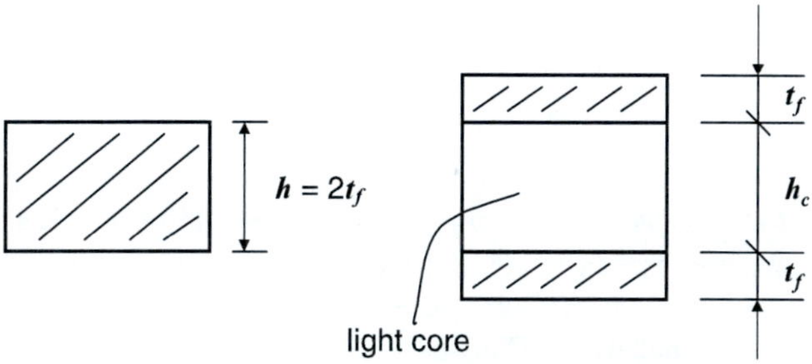

Fig. 5.10 Illustration of the sandwich effect. *Left*: solid (monocoque) plate. Right: The solid cross section is replaced with a sandwich consisting of two half-thickness facings separated by a light core of a negligible stiffness

assumed small compared to that of the facings. Therefore, we compare two designs of an approximately equal weight. As was shown by Vinson (1999), the ratio of the bending stiffness of the sandwich and monocoque designs is equal to

$$\frac{D_{sand}}{D_{mon}} = \frac{3}{4}\left(\frac{h_c}{t_f}\right)^2 \tag{5.73}$$

where h_c and t_f are the thicknesses of the core and one of the facings, respectively.

Furthermore, the ratio of the bending stress is

$$\frac{\sigma_{sand}}{\sigma_{mon}} = \frac{2t_f}{3h_c} \tag{5.74}$$

Obviously, the stiffness of the sandwich beam is three times higher than that of the monocoque beam if the core is twice thicker than each facing. In reality, the core-to-facing thickness ratio is usually larger. If it is equal to 5 (e.g. the sandwich consisting of two 2-mm facings separated by a 10-mm thick core), the stiffness increases by a factor of 18.75. For $\frac{h_c}{t_f} = 10$, the stiffness ratio is 75(!). The increase in the strength

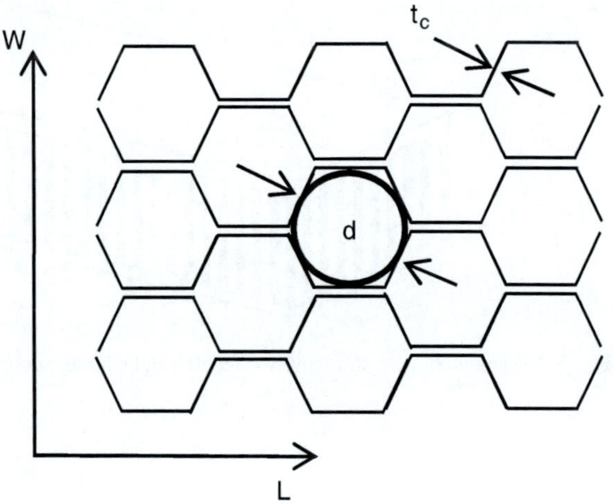

Fig. 5.11 Plainview of a hexagonal honeycomb core

is not as high as that in the stiffness. However, as is evident from (5.74), for $\frac{h_c}{t_f} = 10$, the strength increases by a factor of 15, i.e. the gains are very impressive.

Typical materials used for facings are metals (steel, aluminum and titanium alloys), composites and wood. Composite facings may be laminated or woven; examples of facing materials are carbon/epoxy or graphite/epoxy (aerospace applications) and less expensive glass/epoxy and glass/vinylester (naval and civil applications). The properties of typical facing materials can be found in the monographs of Vinson (1999) or Zenkert (1997).

The core of a sandwich plate should satisfy a number of requirements, including low weight, the capacity to carry transverse shear and transverse normal loads and preventing wrinkling or kinking of the facings. Typical designs of the core include foam, honeycomb, web and truss constructions (Vinson 1999). In the following, the discussion is limited to honeycomb and foam core sandwich plates, excluding web and truss core sandwiches, although such designs have been entertained by civil engineers. Honeycomb core are widely used in aerospace applications, while foam core are found in civil engineering. Sandwich structures in naval applications often employ balsa cores. The honeycomb core can have hexagonal, square and other shapes. The geometry of the hexagonal core is depicted in Fig. 5.11. It is noted that the stiffness and strength in the in-plane transverse (W) and longitudinal (L) directions are different, enabling a designer to better tailor the response of the structure. While the properties of foam cores are predictable, the effective properties of a honeycomb core depend on the material of the core as well as on the size and shape of the cells. Table 5.3 presents selective properties of typical sandwich cores.

The assumptions employed in the analysis of sandwich structures vary dependent on the employed theory. It is usually accepted that the solution should account for shear deformability of the structure since transverse shear moduli of the core are low. Reviews of the history and methods of the analysis of sandwich structures can be

Table 5.3 Properties of balsa, honeycomb and foam core materials (Zenkert 1997; Divinycell Technical Data Sheet, www.diabgroup.com)

Core	Density (kg/m^3)	Transverse shear modulus (MPa) L-direction	Transverse shear modulus (MPa) W-direction	Shear strength (MPa) L-direction	Shear strength (MPa) W-direction
Balsa	96	108	108	1.85	1.85
Balsa	130	134	134	2.49	2.49
Balsa	180	188	188	3.46	3.46
Al. alloy honeycomb	32	180	98	0.83	0.48
Al. alloy honeycomb	70	460	200	2.2	1.5
Al. alloy honeycomb	130	930	370	5.0	3.1
Foam: Divinycell H 60	60	20	20	0.76	0.76
Foam: Divinycell H100	100	35	35	1.6	1.6
Foam: Divinycell H200	200	85	85	3.5	3.5

Note: Balsa exists in different densities; denser balsa has better stiffness and strength. The properties of balsa and foam cores are independent of the in-plane direction)

found in the monographs by Plantema (1966), Allen (1969), Zenkert (1997), Vinson (1999) and Carlsson and Kardomateas (2011). The review by Noor et al (1996) and the ASME volume by Rajapakse et al (2000) contain additional information. There are also special International Conferences on Sandwich Construction that publish valuable proceeding. In this paragraph we do not attempt to review all available theories; instead, we discuss the approach based on the first-order shear deformation theory and comment on the modes of failure of sandwich plates.

The assumptions employed in the analysis of sandwich structures by the first-order shear deformation theory can be subdivided in two classes:

1. Kinematic assumptions of the first-order shear deformation theory are repeated here for convenience:
   - A cross section perpendicular to the undeformed middle plane of the sandwich plate rotates about this plane during deformations but remains flat (no warping). This implies that in-plane displacements are linear functions of the thickness coordinate;
   - The transverse deflection is uniform throughout the thickness, i.e. the transverse normal strain $\varepsilon_z = 0$.
2. Assumptions reflecting geometry and material properties:
   - In-plane stresses in the core are negligible;

– Transverse shear stresses in the facings can be neglected (note that it is possible to apply the first-order theory without employing this assumption).

The problem is assumed both geometrically and physically linear. Effects of local failure modes on the global response are neglected in the linear formulations (for example, it is assumed that buckling of the plate is uncoupled from wrinkling of the facings). The thickness of the facings and that of the core are usually assumed constant. Perfect bonding is assumed between the facings and the core so that they deform as a unit, without sliding with respect to each other.

Using these assumptions, we can apply the first-order shear deformation theory discussed in the previous paragraph to the analysis of sandwich plates. The second group of assumptions listed above affects the computation of the material stiffness terms. In particular, the extensional, coupling and bending stiffness terms that affect the stress resultants and stress couples produced by in-plane stresses are calculated by integrating the corresponding transformed reduced stiffnesses through the thickness of the facings only:

$$\{A_{ij},\ B_{ij},\ D_{ij}\} = \int_{-\left(\frac{h_c}{2}+t_f\right)}^{-\frac{h_c}{2}} \bar{Q}_{ij}\left\{1,\ z,\ z^2\right\} dz + \int_{\frac{h_c}{2}}^{\left(\frac{h_c}{2}+t_f\right)} \bar{Q}_{ij}\left\{1,\ z,\ z^2\right\} dz,$$

$$ij = 11,\ 12,\ 22,\ 16,\ 26,\ 66 \qquad (5.75)$$

where h_c and t_f are shown in Fig. 5.10 and the middle plane $z=0$ is located at the mid-height of the core. If the facings are either symmetric cross-ply laminated or multilayered symmetric angle-ply laminated, $A_{16} = A_{26} = B_{ij} = D_{16} = D_{26} = [B] = 0$ in Eq. 5.75.

Transverse shear stress resultants produced by stresses τ_{xz} and τ_{yz} are calculated by integrating these stresses through the thickness of the core. The corresponding extensional stiffness terms are

$$\{A_{44},\ A_{55}\} = \int_{-\frac{h_c}{2}}^{\frac{h_c}{2}} \left\{\bar{Q}_{44},\ \bar{Q}_{55}\right\} dz \qquad (5.76)$$

The choice of the shear correction factor in sandwich structures is sometimes dictated by the experience with ordinary shear deformable plates where it is typically taken equal to 5/6 or 2/3. Birman and Bert (2002) compared the shear correction factors in sandwich beams derived by six different methods:

1. Modeling the structure as a discrete mass system;
2. Comparison of the shear strain energies;
3. Comparison of the average shear strains;
4. Comparison of natural frequencies;

5. Comparison of average transverse shear stresses;
6. Minimization of the quadratic error of the shear stresses.

In each case, the results obtained by the first-order shear deformation theory were compared to the available exact solution. The results of these comparisons illustrated that the first three methods lack universality, i.e. they were applicable only within certain ranges of geometry and material properties. The latter three methods yielded the shear correction factor equal to 1.0 for arbitrary geometry and materials. Based on these comparisons, the authors recommended using the factor equal to 1.0 in the analysis of sandwich structures.

Using the values of stiffnesses given in (5.75) and (5.76) the analysis of displacements and rotations in sandwich structures can be conducted as shown in Sect. 5.7. Numerical solutions become necessary if boundary conditions prevent the analysis in double Fourier series. Sandwich structures with a soft core, local concentrated loads and discontinuities can be analyzed by higher-order theories or three-dimensional theory of elasticity.

The features of the results obtained by the first order shear deformation analysis discussed above are:

1. In-plane strains $\varepsilon_x,\ \varepsilon_y,\ \gamma_{xy}$ are linear functions of the thickness coordinate;
2. In-plane stresses $\sigma_x,\ \sigma_y,\ \tau_{xy}$ are equal to zero within the core. In the facings, in-plane stresses change abruptly from layer to layer reflecting the changes in the transformed reduced stiffness. Accordingly, the maximum stresses do not necessarily occur at the outer surfaces of the plate as would be the case in metallic structures.
3. Transverse shear stresses are neglected in the facings. Transverse shear strains evaluated by (1.45), according to the assumptions of the first order theory remain constant throughout the thickness of the core. Accordingly, the stresses $\tau_{xz},\ \tau_{yz}$ do not depend on the z-coordinate. In reality, transverse shear stresses do not remain independent of the thickness coordinate and this simplification of the first order theory may become the source of an error. Higher-order theories account for a nonuniform through the thickness transverse shear stress distribution.

The previous discussion does not refer to possible local modes of failure that can occur both in bending and in buckling problems. Therefore, it is important to discuss these non-global modes of failure that are unique to sandwich structures. The mathematical treatment of these modes is excluded; see for example, the monograph of Vinson (1999) for details regarding the first three modes.

1. Wrinkling occurs in the facings subject to compression or in-plane shear. Wrinkling is characterized by narrow waves in the facings. If the sandwich plate is undergoing uniaxial compression, wrinkles are perpendicular to the direction of the compressive stress (Fig. 5.12). In the case of biaxial compression, wrinkling was analyzed by Birman and Bert (2004). Wrinkling may occur even if the plate is subject to bending since one of the facings experiences compression.
2. Core shear instability or crimping is schematically shown in Fig. 5.13. This mode of failure occurs when both facings are subject to compression.

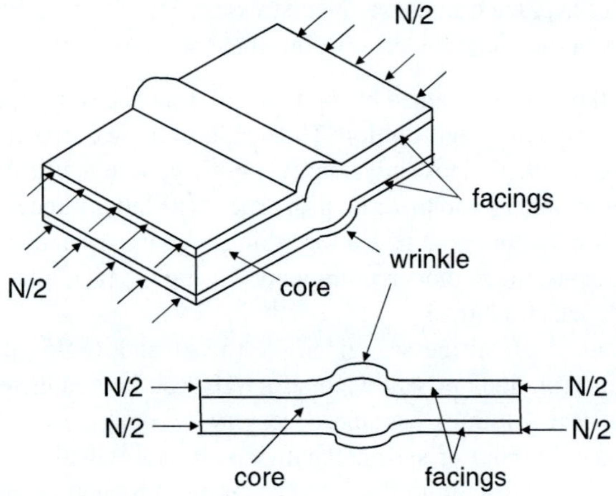

Fig. 5.12 Facing wrinkling under compression. Note that compressive forces cannot be applied to the core, so that joints between sandwich and adjacent structures have to be designed accordingly to distribute these forces between the facings

Fig. 5.13 Core shear instability

3. Face dimpling could occur in honeycomb-core sandwich plates. It is manifested by a local instability of the facing over one cell of the honeycomb core. Such mode of failure is unlikely in sandwich plates manufactured using conventional geometries and materials.
4. Core-face debond (e.g., Camps et al. 2000). This mode of failure results in the loss of shear transfer between the core and the adjacent facing and a violation of the overall integrity of the sandwich structure. Debond in foam-core plates usually involves a crack propagating in the foam core near the interface with the facing. Therefore, this is a fracture problem that requires us to specify the debond toughness. In honeycomb sandwich structures, debond involves a separation of the honeycomb core from the facing.
5. Delamination in the facing can occur as a result of high local transverse stresses, such as those experienced during impact. This mode of failure is sometimes accompanied by other modes, such as the core-facing debond. Delamination can sometimes be avoided in impact-prune structures by using woven (textile) facings that do not include multiple layers.

5.9 Design Philosophy and Recommendations

The subject of composite material structures and in particular, composite plates involves numerous multidisciplinary and multiscale aspects. Each of the relevant areas, including the choice of the material, design, joints, the checks that are necessary to perform dependent on the loading conditions, manufacturing issues, maintenance requirements, environmental effects, etc. often present a much broader spectrum of problems than those encountered in isotropic plates. In spite of these complications, composites provide a potential for lighter structures capable of withstanding higher loads and offering longer life-spans. While it is impossible to address all issues listed above in this paragraph, we outline here some of the fundamental concerns that should be considered by designers of composite plate structures. Most of the subjects reviewed below are applicable to composite structures of arbitrary geometries, i.e. shells, stringers, frames, etc., besides composite plates.

Composite materials possess a higher specific strength and specific stiffness than typical metals and alloys. While assessing the comparison between relative strength and stiffness of composite and metallic materials in Fig. 5.1, it is also important to remember that composite properties can be tailored in the manner that is impossible in metals and alloys. It is possible to optimize the response of a composite plate by varying the orientation and volume fraction of fibers in the layers or by using functionally graded materials. A recently suggested method of optimizing the response is based on the introduction of stiff particles in the fiber-reinforced matrix (Genin and Birman 2009). Another important advantage of composite materials is related to their high fatigue resistance. Polymer-matrix composites (PMC) are free from corrosion problems, while metal-matrix (MMC) and ceramic matrix (CMC) materials can withstand high temperature.

The advantages of composites should be compared to their shortcomings. Among those we can point to a high manufacturing cost, delamination tendencies, and difficulties involved in periodic inspections necessary to detect delaminations and other local damage. For example, a review of costs of composite materials compared to aluminum alloys in aerospace structures was presented by Resetar et al. (undated) in mid-1990s. Although this data is partially outdated, some of the conclusions remain valid. Two types of costs were compared, namely nonrecurring and recurring. The former costs refer to expenses involved in the study, analysis, design, tooling and evaluation, while the latter costs are consumer support, repair, etc. Both nonrecurring engineering as well as nonrecurring tooling for composites was (and remains) more expensive than those for metals. At the same time, the rate of reduction of the cost of composite materials with time is higher than that of metals (in other words, the cost of composites becomes less burdensome every year). A more expensive composite structure may still be advantageous considering the benefits due to its lower weight that affects the lifetime cost. An example is Boeing 787 Dreamliner with the composite airframe that is 20% lighter than a similar aluminum airframe as a result of relying on carbon reinforced plastics and other composites (Hale 2006). Accordingly, the fuel cost for such composite airplane is much lower than the cost for a counterpart built using aluminum alloys.

Delamination is a serious consideration for engineers adopting composites. This damage is constituted by inter-layer cracks that "unzip" the structure. The source of delamination is usually an impact or a local concentration of transverse stresses, such as that encountered at the free edge of a laminate or in the vicinity to discontinuities. Delamination cracks are usually impossible to detect by visual inspection. Accordingly, a non-destructive technique, such as ultrasound testing, is often required.

General design considerations, although by no means complete, include the following observations and recommendations that are essential for a successful design of composite plates:

1. The majority of composite structures are symmetrically laminated about the middle surface (middle plane in plate structures). The number of layers in laminates found in engineering applications is usually large. The coupling stiffness matrix as well as some extensional and bending stiffness terms in symmetric multilayered laminates are equal to zero, i.e. $[B] = A_{16} = A_{26} = D_{16} = D_{26} = 0$. The absence of coupling is beneficial for the structure compared to same-weight same-material asymmetric structures subject to identical load in a number of ways:

   (a) Smaller deformations;
   (b) Smaller stresses;
   (c) Higher buckling load;
   (d) Higher fundamental frequency (this is often desirable to avoid resonance);
   (e) Higher resistance against impact, etc.

   Although a symmetric laminate is usually preferable, the symmetry can be violated if the structure is subject to a thermal field resulting in a nonuniform through the thickness temperature distribution. Properties of all materials are affected by temperature (see Sect. 1.9). However, this effect is more pronounced in polymeric matrix composites (PMC) where the stiffness of the matrix can be significantly affected, as a result of even relatively small temperature variations. Therefore, a small temperature gradient in the thickness direction in PMC plates can result in a transformation of the symmetrically laminated structure into an asymmetric laminate and a significant reduction of the stiffness.
2. The danger of delamination damage dictates the necessity to avoid discontinuities or free unsupported edges that have a potential for a local concentration of transverse stresses acting in the thickness direction.
3. The stiffness and strength of a composite plate can be increased by the following methods:

   (a) Using a stiffer material with a higher strength. Although this may be an acceptable solution, cost considerations may interfere in such replacement of the material. Stringer-reinforced and sandwich designs are often more weight-efficient and they have a potential for a higher load-carrying capacity.

(b) Increasing the thickness of the plate to increase its strength and stiffness. While in isotropic structures bending stiffness is proportional to the third power of the thickness, the situation is different in composites where the stiffness of each layer is different. Optimizing the orientation of the layers, it may be possible to achieve an even better result than in isotropic structures. However, the use of stringers or a sandwich design support the plate are usually more efficient.

(c) Using functionally graded structures where the proportion of constituent phases varies through the thickness (Birman and Byrd 2007) or spatially tailored structures with in-plane variable stiffness (Birman et al. 2008).

(d) Stringer-reinforced structures can almost always be designed either lighter or with a higher load-carrying capacity than plates of the same total weight without stringers. In other words, it is beneficial to redistribute the material making the skin thinner and reinforcing it by stringers. The advantages of stringer-reinforced composite plates are similar to those in isotropic counterparts, including higher strength and buckling load and higher natural frequencies. However, these structures may exhibit unique modes of damage the designer should be aware of. In particular, a resin-rich region can form in co-cured stringer-reinforced plates at the junction of the web and the plate. Local cracks may originate from this resin-rich region and propagate throughout the plate. In addition, a three-dimensional local state of stress with large transverse stresses at the edge of the stringer flange sometimes results in a development of delaminations. An example of such damage is shown in Fig. 5.14 where the crack has already propagated throughout the entire width of the flange-to-plate skin surface. The methods of preventing such local delaminations and damage include using z-pins (e.g. Greenhalgh et al 2006).

While reinforced composite plates have high bending load-carrying capacity, the stress analysis should include the check of strength of the stringers, in addition to that for the skin. In a reinforced plate experiencing bending the strains in the stringer at the location that is most remote from the skin will be larger than those in the skin. Hence the stresses in the stringers may be larger than those in the skin. Whether such higher stresses materialize depends on the plate. This is because in composite laminates higher strains do not automatically imply higher stresses, the latter being dependent on the stiffness of a particular layer.

(e) Sandwich structures represent a valid alternative to stringer-reinforced plates. While these structures have a number of peculiar failure modes, they also offer a very significant weight reduction and strength and stiffness gains that may rival and exceed those available in stringer-reinforced counterparts. Designers have a broad choice of sandwich structural arrangements, including various core designs, relative thickness of facings and the core, dissimilar facings, etc. As a result, sandwich plates and shells have found numerous applications in design of aerospace, naval and civil engineering structures.

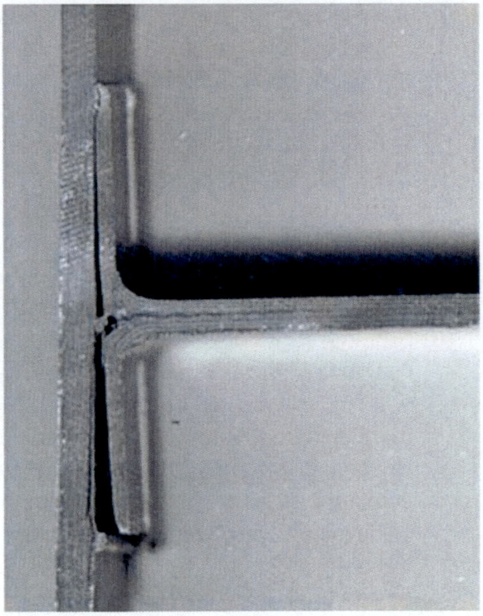

Fig. 5.14 Cross section of a co-cured stringer and adjacent plate with clearly shown web delamination. The material of the plate and stringer is graphite/epoxy Hexcel T800/M21 (From Greenhalgh et al. 2006)

Design of joints in sandwich structures represents a serious problem (this problem is not discussed in detail since it is outside the scope of the book). For example, if the structure is subject to tension, in-plane shear or compression, it is vital to avoid any in-plane loads directly applied to the core because of its low in-plane stiffness and strength. Some of the relevant issues are discussed in literature on sandwich structures referred to in this chapter.

Overall, in spite of a more complicated design process, manufacturing issues and lifetime inspection requirements, composite plates have numerous advantages that make them attractive in numerous applications. This is reflected in a steady increase in the use of composites in various areas of engineering.

References

Allen, H. G. (1969). *Analysis and design of structural sandwich panels*. Oxford: Pergamon Press.

Argyris, J., & Tenek, L. (1994). Linear and geometrically nonlinear bending of isotropic and multilayered composite plates by the natural mode method. *Computer Methods in Applied Mechanics and Engineering, 113*, 207–251.

Barbero, E. J. (1998). *Introduction to composite materials design*. Philadelphia: Taylor & Francis.

Bert, C. W., & Chen, T. L. C. (1978). Effect of shear deformation on vibration of antisymmetric angle-ply laminated rectangular plates. *International Journal of Solids and Structures, 14*, 465–473.

Bert, C. W., & Gordaninejad, F. (1983). Transverse shear effects in bimodular composite laminates. *Journal of Composite Materials, 17*, 282–298.

Bert, C. W., Reddy, J. N., Chao, W. C., & Reddy, V. S. (1981). Vibration of thick rectangular plates of bimodulus composite material. *Journal of Applied Mechanics, 48*, 371–376.

Berthelot, J.-M. (1999). *Composite materials, mechanical behavior and structural analysis*. New York: Springer.

Birman, V. (1993). Active control of composite plates using piezoelectric stiffeners. *International Journal of Mechanical Sciences, 35*, 387–396.

Birman, V. (1994). Analytical models of sandwich plates with piezoelectric strip-stiffeners. *International Journal of Mechanical Sciences, 36*, 567–578.

Birman, V., & Bert, C. W. (2002). On the choice of shear correction factor in sandwich structures. *Journal of Sandwich Structures and Materials, 4*, 83–95.

Birman, V., & Bert, C. W. (2004). Wrinkling of composite-facing sandwich panels under biaxial loading. *Journal of Sandwich Structures and Materials, 6*, 217–237.

Birman, V., & Byrd, L. W. (2007). Modeling and analysis of functionally graded materials and structures. *Applied Mechanics Reviews, 60*, 195–216.

Birman, V., & Byrd, L. W. (2008). Stability and natural frequencies of functionally graded stringer-reinforced panels. *Composites. Part B: Engineering, 39*, 816–825.

Birman, V., Chona, R., Byrd, L. W., & Haney, M. A. (2008). Response of spatially tailored structures to thermal loading. *Journal of Engineering Mathematics, 61*, 201–217.

Camps, G. M. V., Carlsson, L. A., & Li, X. (2000). Influence of core on face/core debond toughness. In Y. D. S. Rajapakse., G. A. Kardomateas & V. Birman (Eds.), *Mechanics of sandwich structures* (pp. 79–87). New York: ASME Press, Vol. AD-62 and AMD-245.

Carlsson, L. A., & Kardomateas, G. A. (2011). *Structural and failure mechanics of sandwich composites*. Dordrecht, Heidelberg, London, New York: Springer.

Chia, C.-T. (1980). *Nonlinear analysis of plates*. New York: McGraw-Hill.

Christensen, R. M. (1997). Stress based yield/failure criteria for fiber composites. *International Journal of Solids and Structures, 34*, 529–543.

Christensen, R. M. (2005). *Mechanics of composite materials* (2nd ed.). Mineola: Dover.

Frostig, Y., Thomsen, O. T., & Sheinman, I. (2005). On the non-linear theory of unidirectional sandwich panels with a transversely flexible core. *International Journal of Solids and Structures, 42*, 1443–1463.

Genin, G. M., & Birman, V. (2009). Micromechanics and structural response of functionally graded, particulate-matrix, fiber-reinforced composites. *International Journal of Solids and Structures, 46*, 2136–2150.

Gibson, R. F. (2007). *Principles of composite materials mechanics* (2nd ed.). Boca Raton: CRC Press.

Greenberg, J. B., & Stavsky, Y. (1980). Vibrations of axially compressed laminated orthotropic cylindrical shells, including transverse shear deformations. *Acta Mechanica, 37*, 13–28.

Greenhalgh, E., Lewis, A., Bowen, R., & Grassi, M. (2006). Evaluation of toughening concepts at structural features in CFRP – Part I: Stiffener pull-off. *Composites: Part A, 37*, 1521–1535.

Hale, J. (2006). Boeing 787 from the ground up. *Aero Quarterly, Boeing.com Commercial Aeromagazine, 4*, 17–23.

Han, S.-C., Lee, S.-Y., & Rus, G. (2006). Postbuckling analysis of laminated composite plates subjected to the combination of in-plane shear, compression and lateral loading. *International Journal of Solids and Structures, 43*, 5713–5735.

Hashin, Z. (1980). Failure criteria for unidirectional fiber-reinforced composites. *Journal of Applied Mechanics, 47*, 329–334.

Hinton, M. J., Kaddour, A. S., & Soden, P. D. (2002). A comparison of the predictive capabilities of current failure theories for composite laminates, judged against experimental evidence. *Composites Science and Technology, 62*, 1725–1797.

Huang, N. N., & Tauchert, T. R. (1988). Large deformation of antisymmetric angle-ply laminates resulting from nonuniform temperature loading. *Journal of Thermal Stresses, 11*, 287–297.

Jones, R. M. (1998). *Mechanics of composite materials* (2nd ed.). Philadelphia: Taylor & Francis.

Jones, R. M. (1999). *Mechanics of composite materials*. Philadelphia: Taylor & Francis.

Kelly, A. (1987). Composites for the 1990s. *Philosophical Transactions of the Royal Society of London, Series A, Mathematical and Physical Sciences, 322*, 409–423.

Lekhnitskii, S. G. (1948). Bending of a rectangular orthotropic plate with parallel rigid ribs. *Applied Mathematics and Mechanics*, 12(3), (In Russian: Prikladnaya Mathematika i Mekhanika).

Lekhnitskii, S. G. (1968). *Anisotropic plates*. New York: Gordon and Breach.

Li, R., & Kardomateas, G. A. (2008). Nonlinear high-order core theory for sandwich plates with orthotropic phases. *AIAA Journal, 46*, 2926–2934.

Liew, K. M., Yang, J., & Kitipornchai, S. (2003). Postbuckling of piezoelectric FGM plates subject to thermo-electro-mechanical loading. *International Journal of Solids and Structures, 40*, 3869–3892.

Mindlin, R. D. (1951). Influence of rotatory inertia and shear on flexural motions of isotropic elastic plates. *Journal of Applied Mechanics, 18*, 31–38.

Narayanaswami, R., & Adelman, H. M. (1977). Evaluation of the tensor polynomial and Hoffman strength theories for composite materials. *Journal of Composite Materials, 11*, 366–377.

Nguyen, C. H., Chandrashekhara, K., & Birman, V. (2010). Enhanced static response of sandwich panels with honeycomb core through the use of Stepped Facings. *Journal of Sandwich Structures and Materials*. Published online, June 24, 2010 (JSM 369615).

Noor, A. K., Burton, W. S., & Bert, C. W. (1996). Computational models for sandwich panels and shells. *Applied Mechanics Reviews, 49*, 155–199.

Plantema, F. J. (1966). *Sandwich construction: The bending and buckling of sandwich beams, plates and shells*. New York: Wiley.

Puck, A., & Schurmann, H. (2002). Failure analysis of FRP laminates by means of physically based phenomenological models. *Composites Science and Technology, 62*, 1633–1662.

Rajapakse, Y. D. S., Kardomateas, G. A., & Birman, V. (Eds.). (2000). *Mechanics of sandwich materials* (Vol. AD-62/AMD-245). New York: ASME Press.

Reddy, J. N. (2004). *Mechanics of laminated composite plates and shells. Theory and analysis* (2nd ed.). Boca Raton: CRC Press.

Reddy, J. N., & Chao, W. C. (1981). Nonlinear bending of thick rectangular laminated composite plates. *International Journal of Non-Linear Mechanics, 16*, 291–301.

Reissner, E. (1945). The effect of transverse shear deformation on the bending of elastic plates. *Journal of Applied Mechanics, 12*, 69–77.

Reissner, E., & Stavsky, Y. (1961). Bending and stretching of certain types of aeolotropic elastic plates. *Journal of Applied Mechanics, 28*, 402–408.

Resetar, S. A., Rogers, J. C., & Hess, R. W. (Undated). *Advanced airframe structural materials. A primer and cost estimating methodology* (Rand report, R-4016-AF).

Savithri, S., & Varadan, T. K. (1993). Large deflection analysis of laminated composite plates. *International Journal of Non-Linear Mechanics, 28*, 1–12.

Sheinman, I., & Frostig, Y. (1988). Post-buckling analysis of stiffened laminated panel. *Journal of Applied Mechanics, 55*, 635–640.

Shen, H.-S. (2000). Nonlinear analysis of composite laminated thin plates subjected to lateral pressure and thermal loading and resting on elastic foundation. *Composite Structures, 49*, 115–128.

Shin, D. K., Griffin, O. H., Jr., & Gurdal, Z. (1993). Postbuckling response of laminated plates under uniaxial compression. *International Journal of Non-Linear Mechanics, 28*, 95–115.

Starnes, J. H., Jr., Knight, N. F., Jr., & Rouse, M. (1985). Post-buckling behavior of selected flat stiffened graphite-epoxy panels loaded in compression. *AIAA Journal, 23*, 1236–1246.

Stavsky, Y. (1961). Bending and stretching of laminated aeolotropic plates. *Journal of Engineering Mechanics, 87*, 31–56.

Timoshenko, S. P. (1921). On the correction for shear of the differential equation for transverse vibrations of prismatic bars. *Philosophical Magazine, Ser. 6, 4*, 742–746.

Timoshenko, S., & Woinowsky-Krieger, S. (1959). *Theory of plates and shells*. New York: McGraw-Hill.

Tong, L., & Steven, J. (1999). *Analysis and design of structural bonded joints*. Hingham: Kluwer.

Vinson, J. R. (1999). *The behavior of sandwich structures of isotropic and composite materials*. Lancaster: Technomic.

Wang, X., Hansen, J. S., & Oguamanam, D. C. D. (2004). Layout optimization of stiffeners in stiffened composite plates with thermal residual stresses. *Finite Elements in Analysis and Design, 40*, 1233–1257.

Wang, X., Oguamanam, D. C. D., & Hansen, J. S. (2005). Shape optimization of stiffeners in stiffened composite plates with thermal residual stresses. *Structural and Multidisciplinary Optimization, 30*, 38–42.

Whitney, J. M. (1987). *Structural analysis of laminated anisotropic plates*. Lancaster: Technomic.

Yang, P. C., Norris, C. H., & Stavsky, Y. (1966). Elastic wave propagation in heterogeneous plates. *International Journal of Solids and Structures, 2*, 665–684.

Yang, D.-S., Huang, Y., & Ren, X.-X. (2006). Analysis of symmetric laminated rectangular plates in plane stress. *Applied Mathematics and Mechanics, 27*, 1719–1726.

Zenkert, D. (1997). *Sandwich construction*. London: EMAS (Engineering Materials Advisory Services Publishing)/Chameleon Press, Ltd.

Chapter 6
Thermoelastic Problems in Isotropic and Composite Plates

Plates often operate in high-temperature environments. Thermal loading can be applied instantaneously (thermal shock), as a continuous function of time or statically. Besides being a function of time, temperature may vary through the thickness or over the planform of the plate. Relevant issues that have to be addressed by a designer include the following steps:

1. Solution of the problem of heat conduction throughout the plate in association with appropriate thermal boundary conditions;
2. Solution for thermally-induced displacements and stresses in conjunction with appropriate structural boundary conditions;
3. Application of the strength criterion.

The steps outlined above cannot be implemented without accounting for the effect of temperature on the material properties discussed in Sect. 1.9. In particular, an accurate distribution of temperature throughout the plate should reflect the influence of temperature on thermal conductivity. The effect of temperature on the stiffness requires adjustments to the tensor of stiffness, and the strength is also a function of temperature. The solution can be simplified if functional relationships between material properties and temperature are known, as in the case shown in Eq. 1.109. However, such information is not available for numerous materials used in engineering applications complicating the design process. Note that elastic deformations may be coupled with the heat transfer problem, but such coupling is usually neglected (e.g., Boley and Weiner 1960; Nowinski 1978). While the effect of temperature on the strength of the plate material is not discussed in this book, the designer should be aware that it may be considerable. For example, composite plates collapse if temperature approaches the glass transition value. In metallic plates, the material melting temperature can be identified with the collapse thermal loading, while the bound of the dangerous temperature range should be specified dependent on the load and material properties.

In this chapter we begin with the formulation of the heat transfer problem for a three-dimensional medium (such medium can represent a plate where temperature varies both over the surface as well as through the thickness). A representative

V. Birman, *Plate Structures*, Solid Mechanics and Its Applications 178,
DOI 10.1007/978-94-007-1715-2_6,

problem illustrates a one-dimensional heat transfer in a functionally graded plate (plate composed of two materials with variable through the thickness volume fractions) where temperature varies in the thickness direction. Subsequently, we concentrate on thermal problems of isotropic and composite plates, including bending and buckling. A representative problem that has to be addressed in numerous applications, i.e. the effect of fire on composite plates, is described in detail. At the end of the chapter we provide a number of practical recommendations for designers of plates operating at high temperature.

6.1 Heat Transfer Problem

Consider heat conduction in an anisotropic three-dimensional medium using a Cartesian coordinate system. If the internal heat generation is absent, the equation of heat transfer is (Ozisik 1993):

$$k_{11}\frac{\partial^2 T}{\partial x^2} + k_{22}\frac{\partial^2 T}{\partial y^2} + k_{33}\frac{\partial^2 T}{\partial z^2} + (k_{12} + k_{21})\frac{\partial^2 T}{\partial x \partial y} + (k_{13} + k_{31})\frac{\partial^2 T}{\partial x \partial z} + (k_{23} + k_{32})\frac{\partial^2 T}{\partial y \partial z} = \rho c_p \frac{\partial T}{\partial t} \tag{6.1}$$

where k_{ii} and $k_{ij} = k_{ji}$ are elements in the thermal conductivity tensor (thermal conductivity coefficients), ρ is the density and c_p is the specific heat. Equation 6.1 is written by assumption that the properties of the medium (thermal conductivities) are not affected by local temperature. If the effect of temperature on the principal conductivities is accounted for, Eq. 6.1 should be modified accordingly. For example, if the principal axes of the conductivity tensor coincide with the coordinate system, i.e. $k_{ij} = 0$, and $k_{ii}(T)$ are the so-called principal conductivities, the equation of heat transfer becomes

$$\frac{\partial}{\partial x}\left[k_{11}(T)\frac{\partial T}{\partial x}\right] + \frac{\partial}{\partial y}\left[k_{22}(T)\frac{\partial T}{\partial y}\right] + \frac{\partial}{\partial z}\left[k_{33}(T)\frac{\partial T}{\partial x}\right] = \rho c_p(T)\frac{\partial T}{\partial t} \tag{6.2}$$

Equation 6.2 is applicable to the analysis of layers in cross-ply plates, while in angle-ply layers $k_{ij} \neq 0$ and the corresponding equation is more complicated.

In a multilayered composite plate the orientation of the principal axes of the conductivity tensor varies from layer to layer. Accordingly, Eqs. 6.1 or 6.2 should be used for each layer. The solutions of these equations must satisfy thermal boundary conditions as well as the conditions of thermal continuity at the interfaces of adjacent layers.

Thermal boundary conditions should be specified both on the plate surfaces as well as along its edges. The latter are often disregarded in large plates assuming that heat exchange with edges structures does not affect the overall response. However, it is necessary to remember that heat flow to or from edge structures may affect the distribution of temperature.

Boundary conditions on the surfaces of the plate can include convection, radiation and external heat supply. In particular, in the case of convection to an ambient environment at temperature T_∞, the heat flux measured in units of energy per surface area (W/m^2) is

$$q_c = h\,(T - T_\infty) \tag{6.3}$$

where h is a heat transfer coefficient that is also called boundary conductance (it may be influenced by the surface temperature T).

The heat flux due to radiation between a convex surface and the surroundings is evaluated by

$$q_r = \varepsilon\sigma\left(T^4 - T_\infty^4\right) \tag{6.4}$$

where ε is the surface emissivity and $\sigma = 5.6697 \times 10^{-8}\ W/(m^2K^4)$ is the Stefan-Boltzmann constant. Note that temperatures in (6.4) are measured in degrees Kelvin (absolute temperature scale). In certain situations, Eq. 6.4 can be reduced to the form (6.3) where $h = h_r$ is called the radiation boundary conductance (e.g., Boley and Weiner 1960).

The heat flux conducted through the surface in the normal direction is a function of the conductivity in this direction (the conductivity normal to the plate surface, i.e. $k_n \equiv k_z \equiv k_{33}$):

$$q_n = -k_{33}(T)\frac{\partial T}{\partial z} \tag{6.5}$$

The balance of energy on the surface exposed to all three mechanisms of heat exchange, i.e. convection, radiation and heat conduction as well as heat flux supplied by an external source q_{ext} results in a single equation where the inflow is equal to outflow by convection and radiation (Ozisik 1993).

Particular cases of boundary conditions include

1. Prescribed surface temperature $T\,(x, y, t)$.
2. Prescribed heat flux from outside (e.g., heat from fire):

$$q_{ext}\,(x, y, t) = -k_{33}(T)\frac{\partial T\,(x, y, t)}{\partial z} \tag{6.6}$$

3. Perfectly insulated surface:

$$\frac{\partial T\,(x, y, t)}{\partial z} = 0 \tag{6.7}$$

4. Convection boundary condition that implies the balance between the conduction of heat in the direction perpendicular to the plate surface and the convection

into the ambient environment. For a plate cross section with the surfaces 1 and 2 perpendicular to the z-axis that is oriented from surface 1 to surface 2 this condition is as follows:

At surface 1

$$k_{33}(T)\frac{\partial T\,(x,y,t)}{\partial z} = h\,[T\,(x,y,t) - T_\infty(t)] \tag{6.8a}$$

At surface 2

$$-k_{33}(T)\frac{\partial T\,(x,y,t)}{\partial z} = h\,[T\,(x,y,t) - T_\infty(t)] \tag{6.8b}$$

Equation 6.8 can be extended to account for the effect of radiation. If the ratio of the difference between the surface and ambient temperatures to the latter temperature is much smaller than unity, the corresponding equations are

$$k_{33}(T)\frac{\partial T\,(x,y,t)}{\partial z} = h\,[T\,(x,y,t) - T_\infty(t)] + h_r\,[T\,(x,y,t) - T_\infty(t)] \tag{6.9a}$$

$$-k_{33}(T)\frac{\partial T\,(x,y,t)}{\partial z} = h\,[T\,(x,y,t) - T_\infty(t)] + h_r\,[T\,(x,y,t) - T_\infty(t)] \tag{6.9b}$$

where the heat transfer radiation coefficient is

$$h_r = 4\varepsilon\sigma T_\infty^3 \tag{6.10}$$

The interface thermal continuity conditions between adjacent layers of a composite plate require the continuity of both the temperature and the heat flux through the interface. Therefore, for the interface between the $(i-1)^{th}$ and i^{th} layers,

$$\begin{aligned} T_{i-1}\,(x,y,t) &= T_i\,(x,y,t) \\ k_{33}^{(i-1)}\,(T_{i-1})\,\frac{\partial T_{i-1}\,(x,y,t)}{\partial z} &= k_{33}^{(i)}\,(T_i)\,\frac{\partial T_i\,(x,y,t)}{\partial z} \end{aligned} \tag{6.11}$$

Equations presented in this paragraph can be employed to formulate the heat transfer problem for a multilayered plate. In the general case of a composite angle-ply plate the solution can be obtained analytically (e.g., Ozisik 1993) or more often, numerically. If the heat flow is limited to the thickness direction only, the solution becomes easier since the mathematical formulation is reduced to the ordinary differential equation available from (6.1) or (6.2) in the static case and to the partial differential equation involving the z-coordinate and time in the dynamic case. An example is illustrated in the next paragraph.

6.2 Representative Problem: Heat Transfer in a Functionally Graded Plate Subject to a Uniform over the Surface Thermal Loading

Functionally graded material (FGM) structures have a number of advantages, such as a reduced or eliminated delamination tendency and improved thermal properties (for a detailed analysis of applications of FGM, see the review by Birman and Byrd 2007). In typical FGM structures, these improved properties are achieved by a graded distribution of constituent phases through the thickness. Structures designed with in-plane varying properties are sometimes called spatially tailored. An extended discussion of FGM plates is presented in Chap. 7, while the present paragraph merely illustrates the static one-dimensional conduction in a FGM plate as an example of heat transfer problems that have to be addressed in the analysis of such structures.

In most high-temperature applications of FGM structures the phase with a low thermal conductivity coefficient dominates the region that is subject to high temperature. For example, in ceramic-metal FGM structures ceramic particles (ceramic phase) would be concentrated close to the exposed surface, while the volume fraction of the metal phase gradually increases towards the colder surface. In this paragraph we illustrate a one-dimensional (through the thickness) steady state heat transfer problem for a FGM plate. Such problem arises when the plate is subject to a uniform over the surface thermal loading and the heat exchange between the plate and the boundary structures can be neglected (accordingly, there is no in-plane heat flow). In such case, if volume fractions of the constituent phases vary only in the thickness direction, the assumption of a one-dimensional heat transfer is justified.

The thermal conductivity of a particulate material can be determined by the method proposed by Hatta and Taya (1985) that was later employed by Vel and Batra (2003) in the study of transient thermal stresses in FGM plates. According to this approach, if the particles and matrix are isotropic,

$$k = k_1 + \frac{(k_2 - k_1)\, V_2}{1 + (1 - V_2)\,(k_2 - k_1)\,(3k_1)^{-1}} \tag{6.12}$$

where the indices 1 and 2 refer to the matrix and particulate phases, respectively, V_2 is a volume fraction of particles and k_i are the thermal conductivities.

Alternatively, thermal conductivity of a material with spherical particles can be determined by the model of Yin et al. (2005):

$$k = k_1 \frac{V_2 \bar{\alpha}\left(1 + 0.25 V_2 \bar{\beta}^2\right) + 1 - V_2}{V_2\left(\frac{k_1}{k_2}\right)\bar{\alpha}\left(1 + 0.25 V_2 \bar{\beta}^2\right) + 1 - V_2} \tag{6.13}$$

where

$$\bar{\alpha} = \frac{3k_2}{2k_1 + k_2} \quad \bar{\beta} = \frac{k_2 - k_1}{2k_1 + k_2} \tag{6.14}$$

Numerical examples illustrated that the conductivities predicted by Eqs. 6.12 and 6.13 are almost identical.

Consider now one-dimensional quasi-static heat conduction in a FGM plate subject to prescribed temperatures on the opposite surfaces, so that $T(z_0) = T_0$, $T(z_N) = T_1$ where the coordinate is counted from the hotter surface ($z_0 = 0$). The coordinate of the colder surface is $z = z_N \equiv \tilde{h}$ ($\tilde{h}$ being the thickness of the plate, it differs from the notation employed for the thickness elsewhere in the book to distinguish it from the heat transfer coefficient in convection problems). Other thermal boundary conditions could also be considered using the approach illustrated below.

The heat transfer equation written by assumption of the absence of internal heat generation is obtained from (6.2):

$$\frac{d}{dz}\left[k(T, z)\frac{dT(z)}{dz}\right] = 0 \tag{6.15}$$

Substituting (6.12) or (6.13) into (6.15) and using (1.109) or an alternative analytical thermal conductivity-temperature relation to characterize temperature dependence of the conductivities of the individual phases yields the mathematical problem that can be solved numerically.

Consider for example the case where the thermal conductivity of the composite material can be written as a product of functions of the thickness coordinate z and the local temperature T. Then the substitution of $k(z, T) = f_1(z) \times f_2(T)$ into (6.15) yields $\Phi(T) = C_1\theta(z) + C_2$ where

$$\Phi(T) = \int f_2(T)dT; \quad \theta(z) = \int \frac{dz}{f_1(z)} \tag{6.16}$$

In the particular case where $f_1(z) = k_o + \frac{\Delta k}{\tilde{h}} z$, $f_2(T) = a_o + a_1 T$ the solution for the temperature distribution is (this solution was obtained by Dr. L.W. Byrd who also provided Figs. 6.1–6.4)

$$T(z) = \frac{-a_o \pm \sqrt{a_o^2 + 2a_1(C_1\theta(z) + C_2)}}{a_1} \tag{6.17}$$

where

$$C_1 = \frac{\Delta k}{\tilde{h}\ln\left(\frac{k_1}{k_o}\right)}\left[a_o(T_1 - T_0) + \frac{a_1}{2}\left(T_1^2 - T_0^2\right)\right]$$

$$C_2 = \left(a_o T_0 + \frac{a_1}{2}T_0^2\right) - \frac{C_1\tilde{h}}{\Delta k}\ln(k_0) \tag{6.18}$$

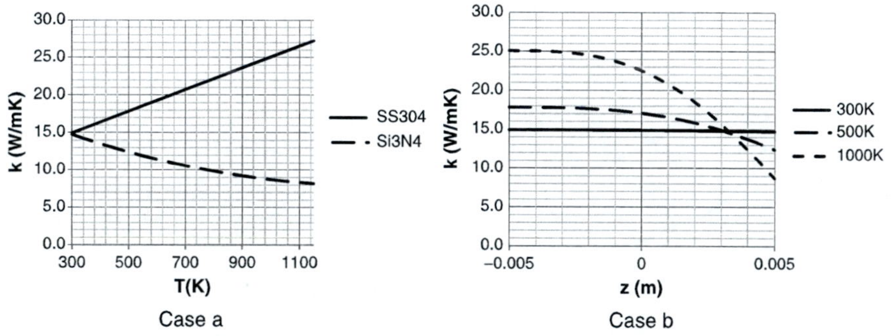

Fig. 6.1 Thermal conductivities of silicon nitride (Si_3N_4), metal (SS304) and FGM Si_3N_4/SS304 plates as functions of temperature. *Case a*: Conductivities of ceramic (Si_3N_4) and metal (SS304). *Case b*: Conductivities throughout the thickness of a FGM plate subject to various uniform temperatures

In this solution $0 \leq z \leq \tilde{h}$, while k_0, k_1 are the values of the thermal conductivity at $z = 0$ and $z = \tilde{h}$, respectively.

The 0.01 m-thick ceramic-metal plate considered in numerical examples contained silicon nitride (Si_3N_4) particles embedded within metal (SS304) matrix. The conductivities of ceramic and metal phases are shown as functions of temperature in Fig. 6.1a using data from Incropera and DeWitt (1996). Although the conductivities of two materials are almost identical at room temperature, they diverge at elevated temperatures. The volume fraction of the ceramic phase considered in this example was distributed through the thickness according to the power law $V_2(z) = \left(\frac{z}{h} + \frac{1}{2}\right)^n$, the z-coordinate being counted from the middle surface of the plate and the power $n = 3$. In Fig. 6.1b, the conductivity of the FGM plate determined by Eq. 6.12 is shown as a function of the thickness coordinate and a uniform through-the-thickness temperature.

The effect of the material distribution (power n in the law $V_2(z)$) on the conductivity of a 0.01 m thick FGM plate subject to a uniform temperature (1,000 K) is shown in Fig. 6.2. In this and subsequent figures, $z = -0.005$ m is the colder surface and $z = 0.005$ m is the hotter surface. A lower overall conductivity is preferred in applications where it is desirable to maintain a low temperature of the colder side of the plate for a longer time (this includes the case where the colder side is actively cooled). As follows from Fig. 6.2, a lower power n is desirable in such applications. This is not surprising since lower values of n correspond to a larger overall content of ceramic particles.

Temperature distribution in the plate subject to a heat flux on the ceramic-rich surface ($z = 0.005$ m) is shown in Fig. 6.3. The nonuniform temperature affects the response of the plate in two ways producing a thermally-induced bending moment and changing the stiffness tensor throughout the thickness.

The effect of a uniform temperature on the stiffness of the 0.01 m thick FGM plate considered in the previous examples is illustrated in Fig. 6.4 where

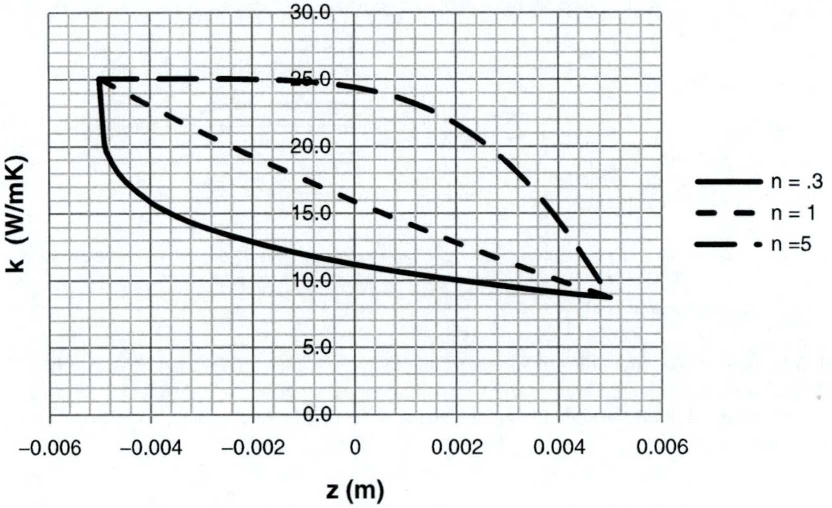

Fig. 6.2 Thermal conductivity of a 0.01 m thick Si_3N_4 FGM plate as a function of position and power n at 1,000 K

Fig. 6.3 Temperature distribution in a 0.01 m thick Si_3N_4 FGM plate ($n = 3$) subject to heat flux $q = 50\ kW/m^2$ at the surface $z = 0.005$ m

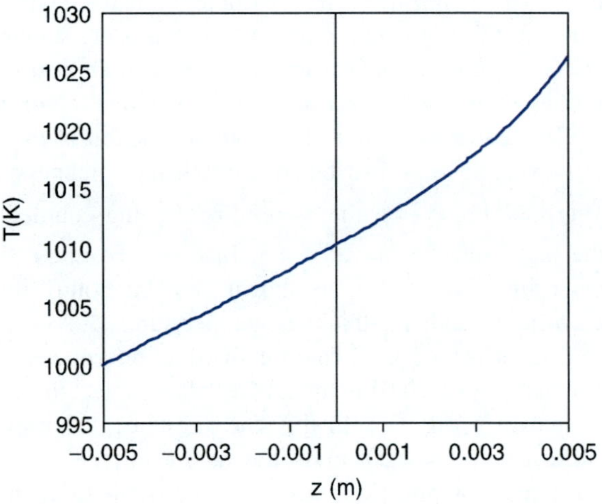

the homogenization technique for the temperature-affected stiffness of ceramic and steel phases was employed to determine variations of the extensional (A_{11}), coupling (B_{11}) and bending (D_{11}) normalized stiffnesses with temperature. The stiffnesses were normalized with respect to the corresponding values at 300 K. While the extensional and bending stiffnesses decrease with temperature, the coupling stiffness exhibits a small decrease at temperatures below 500 K followed with an increase at higher temperature values. Significant stiffness variations at elevated temperatures reflected in Fig. 6.4 emphasize the necessity to account for

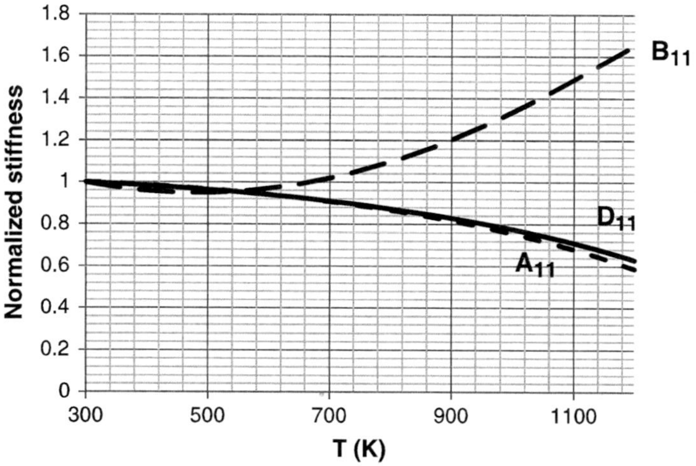

Fig. 6.4 Normalized stiffness terms A_{11}, B_{11}, and D11 for a Si_3N_4 FGM plate with $V_2(z) = \left(\frac{z}{h} + \frac{1}{2}\right)^3$ as a function of uniform temperature

all effects, i.e. explicit thermally-induced stress couples and stress resultants as well as variations of material properties, in the analysis of plates operating in high-temperature environments.

6.3 Thermal Bending and Buckling of Rectangular Isotropic Plates

In this and subsequent paragraphs it is assumed that the temperature distribution throughout the plate has already been established from the solution of the heat transfer problem. Accordingly, the analysis is limited to the structural response.

Consider a thin rectangular isotropic plate subjected to a known thermal field. The analysis is conducted by the plane stress assumption. Accordingly, the strain-displacement relations are given by (1.27) or by its geometrically linear version. The constitutive law (2.9a) has to be generalized to account for thermal expansion. This law is reproduced here in the explicit form:

$$\sigma_x = \frac{E}{1-\nu^2}\left[\varepsilon_x + \nu\varepsilon_y - (1+\nu)\,\alpha T\right]$$
$$\sigma_y = \frac{E}{1-\nu^2}\left[\varepsilon_y + \nu\varepsilon_x - (1+\nu)\,\alpha T\right]$$
$$\tau_{xy} = G\gamma_{xy} \tag{6.19}$$

where temperature is counted from the stress-free reference state (residual stresses are not considered). In isotropic materials temperature does not cause shear strains and stresses, as is reflected in (6.19).

The integration of the stresses given by (6.19) and the moments of these stresses about the middle plane through the thickness of the plate yields stress resultants and stress couples that represent a generalization of the corresponding expressions in (1.58):

$$
\begin{aligned}
N_x &= \frac{Eh}{1-\nu^2}\left\{\frac{\partial u_0}{\partial x}+\frac{1}{2}\left(\frac{\partial w}{\partial x}\right)^2+\nu\left[\frac{\partial v_0}{\partial y}+\frac{1}{2}\left(\frac{\partial w}{\partial y}\right)^2\right]\right\}-\frac{N_T}{1-\nu}\\
N_y &= \frac{Eh}{1-\nu^2}\left\{\frac{\partial v_0}{\partial y}+\frac{1}{2}\left(\frac{\partial w}{\partial y}\right)^2+\nu\left[\frac{\partial u_0}{\partial x}+\frac{1}{2}\left(\frac{\partial w}{\partial x}\right)^2\right]\right\}-\frac{N_T}{1-\nu}\\
N_{xy} &= \frac{Eh}{2(1+\nu)}\left(\frac{\partial u_0}{\partial y}+\frac{\partial v_0}{\partial x}+\frac{\partial w}{\partial x}\frac{\partial w}{\partial y}\right)\\
M_x &= -D\left(\frac{\partial^2 w}{\partial x^2}+\nu\frac{\partial^2 w}{\partial y^2}\right)-\frac{M_T}{1-\nu}\\
M_y &= -D\left(\frac{\partial^2 w}{\partial y^2}+\nu\frac{\partial^2 w}{\partial x^2}\right)-\frac{M_T}{1-\nu}\\
M_{xy} &= -D(1-\nu)\frac{\partial^2 w}{\partial x\partial y}
\end{aligned}
\tag{6.20}
$$

where all material properties are affected by temperature. "Thermal contributions" to the stress resultants and stress couples are

$$
\{N_T,\ M_T\} = \int_{-\frac{h}{2}}^{\frac{h}{2}} \alpha(T)E(T)T(z)\{1,\ z\}\,dz \tag{6.21}
$$

If the effect of temperature on properties is negligible, Eq. 6.21 becomes

$$
\{N_T,\ M_T\} = \alpha E\int_{-\frac{h}{2}}^{\frac{h}{2}} T(z)\{1,\ z\}\,dz \tag{6.22}
$$

The stresses can be evaluated in terms of stress resultants and stress couples. In particular, if the effect of temperature on material properties is neglected, the stresses are given by (Boley and Weiner 1960):

$$\sigma_x = \frac{1}{1-\nu}\left\{-\alpha E T + \frac{1}{h}\left[(1-\nu)N_x + N_T\right] + \frac{12z}{h^3}\left[(1-\nu)M_x + M_T\right]\right\}$$

$$\sigma_y = \frac{1}{1-\nu}\left\{-\alpha E T + \frac{1}{h}\left[(1-\nu)N_y + N_T\right] + \frac{12z}{h^3}\left[(1-\nu)M_y + M_T\right]\right\}$$

$$\tau_{xy} = \frac{N_{xy}}{t} + \frac{12z}{t^3}M_{xy} \tag{6.23}$$

Based on the analysis of Eqs. 6.22 and 6.23 it can be concluded that if temperature varies through the thickness, in addition to in-plane stress resultants and membrane stresses, it contributes to bending stress couples and to bending stresses in the plate.

Equations of motion in terms of stress resultants and stress couples are not explicitly affected by the presence of temperature. Accordingly, they correspond to equations (1.84) where the stress resultants and couples are given by (6.20). The first two equations (1.84) are identically satisfied through the introduction of the stress function given in (1.95). The compatibility equation (1.97) in conjunction with the strain-stress relationships available from (6.19) and with the stress function defined in (1.95) is analyzed jointly with the third equation (1.84) yielding the stress function and deflection.

The following observations are now made to gain an insight into the problem.

1. The compatibility equation (1.97) cannot be reduced to the form convenient for the analysis (similar to Eq. 1.99) if the material properties are affected by temperature that varies with in-plane coordinates. However, the compatibility equation can be reduced to a convenient form if temperature is independent of in-plane coordinates or if material properties are unaffected by temperature. In such cases, the counterpart of (1.99) is

$$\frac{1}{E}\nabla^4\varphi = -\nabla^2 N_T + \frac{\partial^2 w}{\partial x^2}\frac{\partial^2 w}{\partial y^2} - \left(\frac{\partial^2 w}{\partial x \partial y}\right)^2 \tag{6.24}$$

where $\nabla^2(...) = \frac{\partial^2(...)}{\partial x^2} + \frac{\partial^2(...)}{\partial y^2}$.

2. The third equation of equilibrium (1.84) cannot be presented in a convenient form resembling equation (2.1), unless material properties are independent of temperature or temperature is independent of in-plane coordinates. However, if at least one of these conditions is satisfied, the substitution of the stress couples given by (6.20) and using the stress function defined by (1.95) yields

$$D\nabla^4 w = p(x,y) + \frac{\partial^2\varphi}{\partial y^2}\frac{\partial^2 w}{\partial x^2} - 2\frac{\partial^2\varphi}{\partial x \partial y}\frac{\partial^2 w}{\partial x \partial y} + \frac{\partial^2\varphi}{\partial y^2}\frac{\partial^2 w}{\partial y^2} - \frac{1}{1-\nu}\nabla^2 M_T \tag{6.25}$$

Thermomechanical (structural) boundary conditions for the plate may involve kinematic conditions in terms of displacements and slopes that are not affected by temperature. However, static conditions in terms of stress resultants and stress couples are explicitly affected by thermal terms as is easily observed from (6.20). For example, the conditions of simple support along the edges x = constant and y = constant become

$$
\begin{aligned}
x = const: \quad & w = 0, \quad M_x = -D\left(\frac{\partial^2 w}{\partial x^2} + \nu\frac{\partial^2 w}{\partial y^2}\right) - \frac{M_T}{1-\nu} \\
& \qquad = -D\frac{\partial^2 w}{\partial x^2} - \frac{M_T}{1-\nu} = 0 \\
y = const: \quad & w = 0, \quad M_y = -D\left(\frac{\partial^2 w}{\partial y^2} + \nu\frac{\partial^2 w}{\partial x^2}\right) - \frac{M_T}{1-\nu} \\
& \qquad = -D\frac{\partial^2 w}{\partial y^2} - \frac{M_T}{1-\nu} = 0
\end{aligned} \tag{6.26}
$$

respectively.

Analyzing boundary conditions (6.26) or other static boundary conditions along the plate edges, additional observations are available:

3. Static boundary conditions in thermal problems are nonhomogeneous since they contain thermally-induced terms. Accordingly, such conditions imply an unavoidable thermally-induced bending in cases where $M_T \neq 0$. Consider for example a simply supported plate subject to a variable through the thickness temperature that is independent of in-plane coordinates. This seems to imply a possible trivial solution for deflections available from Eqs. 6.24 and 6.25. However, in reality thermal bending takes place due to the presence of the "loading" term in boundary conditions (6.26).
4. If the plate subject to a variable through the thickness but independent of in-plane coordinates temperature is clamped around all edges, it does not experience bending. This is because temperature does not affect the kinematic boundary conditions corresponding to clamping, while the system of Eqs. 6.24 and 6.25 has a trivial solution for deflections.
5. Irrespective of out-of-plane boundary conditions involving deflections, rotations (slopes of the middle surface), bending stress couples and transverse shear stress resultants, if the in-plane expansion of the plate is limited, thermally-induced compression is applied to the plate. This compression exaggerates bending deformations for all boundary conditions, except for those corresponding to clamping. In the latter case, a plate subject to temperature varying in the thickness direction does not experience bending and in-plane compression may result in thermal buckling.

Example 6.1: Thermal Bending of a Simply Supported Plate: Geometrically Linear Problem. Consider a simply supported isotropic plate subject to an arbitrary through-the-thickness temperature that is uniform over the plate surface, i.e. $T = T(z)$. In addition to thermally-induced stress couples, in-plane thermally-induced stress resultants are present if in-plane edge displacements are constrained.

Case 1: In-plane unconstrained edges. We start with the case where in-plane displacements are not constrained, so that the plate is free to expand in the x- and y-directions. Accordingly, thermally-induced stress resultants are equal to zero and the linear formulation enables us to reduce the governing equations to the equation of equilibrium available from (6.25):

$$D\nabla^4 w = -\frac{1}{1-\nu}\nabla^2 M_T \tag{a}$$

The edges are simply supported, so that the solution of (a) must satisfy boundary conditions (6.26).

The solution is available by representing both thermally-induced bending stress couples as well as deflections in double Fourier series:

$$M_T = \sum_{m=1}^{M}\sum_{n=1}^{N} M_{mn} \sin\alpha_m x \sin\beta_n y$$

$$w = \sum_{m=1}^{M}\sum_{n=1}^{N} W_{mn} \sin\alpha_m x \sin\beta_n y \tag{b}$$

where the amplitude of harmonics in the former series is

$$M_{mn} = \frac{4}{ab}\int_0^b\int_0^a M_T \sin\alpha_m x \sin\beta_n y dx dy \tag{c}$$

Boundary conditions (6.26) at the edges $x = 0,\ a$ and $y = 0,\ b$ are identically satisfied by series (b). The substitution of (b) into (a) yields the solution for the amplitudes in the series for deflections:

$$W_{mn} = \frac{M_{mn}}{(1-\nu)\,D\,(\alpha^2+\beta^2)} \tag{d}$$

The stresses can be evaluated from Eqs. 6.20 and 6.23 or from (6.19).

Case 2: In-plane constrained edges. The problem is more complicated if the edges of the plate prevent its expansion. For example, if the in-plane boundary conditions are

$$x = 0,\ \ x = a:\ \ u = 0,\ \ N_{xy} = 0$$

$$y = 0,\ \ y = b:\ \ v = 0,\ \ N_{xy} = 0 \tag{e}$$

it is necessary to integrate equation (6.24). In the linear problem where the temperature is uniformly distributed over the plate surface (this represents the "simplest" case for the analysis), the solution can be represented by

$$\varphi = \frac{N_T}{2h}\left(x^2 + y^2\right) \tag{f}$$

While this solution satisfies the condition of zero shear stress resultants along the plate edges, the condition for zero in-plane displacements cannot be satisfied. This is easily observed if the mid-plane strains are obtained in terms of the stress function according to (1.98). The integration of these strains results in the expressions for in-plane displacements that do not satisfy kinematic conditions (e).

In the particular case of boundary conditions (e) the solution for a simply supported plate could be obtained by substituting the linear version of the constitutive relations (6.20) into equations of equilibrium (static version of (1.84)). The resulting three linear partial differential equations could be solved by representing displacements u_0, v_0 and w as well as temperature in double Fourier series satisfying boundary conditions. While the implementation of this procedure can serve as an exercise, a similar approach is impossible in case of other boundary conditions.

The present example illustrates that the stress and deflection analysis of plates in the presence of temperature requires using numerical methods in all but several benchmark problems. This is obviously the case if the response is nonlinear and the governing equations are formulated either in the form of (6.24) and (6.25) or in terms of displacements utilizing equations of equilibrium (1.84) and nonlinear expressions for stress resultants and couples (6.20).

Example 6.2: Thermal Buckling of a Simply Supported Plate Consider a plate that is simply supported along all boundaries, prevented from in-plane expansion by adjacent structures and supports, and subject to an elevated uniform temperature. Obviously, such temperature does not produce thermal bending of the plate, while generating in-plane stress resultants determined by (6.21) or (6.22). All in-plane displacements of the edges being prevented, we can assume $u_0 = v_0 = 0$ throughout the plate.

Thermal buckling implies the existence of an out-of-plane deflection that satisfies equations of equilibrium and boundary conditions. In the present problem, geometric nonlinearity is discounted, i.e. we analyze the buckling temperature, rather than the postbuckling behavior of the plate. Accordingly, temperature being uniform throughout the plate surface, $N_T = const$ and the compatibility equation (6.24) is identically satisfied by choosing the stress function in the form (f). The equation of equilibrium (6.25) becomes

$$D\nabla^4 w = N_T\left(\frac{\partial^2 w}{\partial x^2} + \frac{\partial^2 w}{\partial y^2}\right) \tag{g}$$

The boundary conditions corresponding to simply supported edges are available from (6.26). Thermal bending being absent, these conditions revert to those for a plate without thermal loading, i.e. (2.2).

Assuming that the plate buckles into the shape characterized by the equation (h) in Example 2.7 reproduced here for convenience

$$w = W_{mn} \sin \alpha_m x \sin \beta_n y \tag{h}$$

we satisfy the boundary conditions. The buckling temperature can be obtained from the solution of the equilibrium equation (g) that upon the substitution of (h) yields

$$N_T = D\left(\alpha_m^2 + \beta_n^2\right) \tag{i}$$

In the case of a uniform temperature the thermal stress resultant is

$$N_T = \frac{Eh\alpha}{1-\nu} T \tag{j}$$

Substituting (j) into (i) we obtain the buckling temperature

$$T = \min_{m,n} \left\{ \frac{\pi^2 \left[m^2 \left(\frac{b}{a}\right)^2 + n^2\right]}{12\,(1+\nu)\,\alpha \left(\frac{b}{h}\right)^2} \right\} \tag{k}$$

It was tacitly assumed in (j) that material properties are unaffected by temperature. As is shown below, such assumption is justified since plates buckle at low temperatures. Furthermore, as is evident from (k), a uniformly heated simply supported plate buckles into the shape with one half-wave in both x- and y-directions, i.e. m = n = 1.

Representative examples were considered using (k) in the monograph of Jones (2006) for a square 6061-T6 aluminum plate with the coefficient of thermal expansion and the Poisson ratio equal to $\alpha = 23.6 \times 10^{-6} \left(\frac{1}{{}^\circ C}\right)$ and $\nu = 0.32$, respectively. At the side-to-thickness ratio $\frac{b}{h} = 100$ the plate would buckle at $T = 5.3^\circ C$, while at $\frac{b}{h} = 50$ the buckling temperature was $T = 21.1^\circ C$. Such absolute values of temperature corresponding to buckling are extremely small and require further discussion. As is obvious form this and other examples easily generated using (h), if the plate buckling resulted in failure, practically every plate used in engineering would fail as a result of normal temperature variations during its lifetime. Jones (2006) suggested that thin plates have imperfections resulting in bending, rather than buckling, due to an elevated temperature. While this explanation points to one of possible reasons for a safe behavior of plates under applied temperature, it is necessary to indicate that initial imperfections are less likely in thicker plates. However, as follows from the above results, even a relatively thick plate with the side-to-thickness ratio equal to 50 that is unlikely to have noticeable initial imperfections experiences buckling as a result of a very small temperature. Metallic plates with a smaller side-to-thickness ratio are seldom encountered in engineering.

A possible explanation to the fact that predicted thermal buckling does not cause a catastrophic failure is in the postbuckling behavior of the plate that is stable (similar to a stable postbuckling response of mechanically compressed plates depicted in Fig. 2.20). As the plate deforms in the postbuckling regime, its middle surface stretches. Plates prevented from expansion under an elevated temperature are subject to two opposing trends: (a) thermally-induced in-plane stresses due to the constraint superimposed on the expansion of the plate; (b) in-plane stresses developed in the deformed plate as a result of its stretching. While the compressive stress resultants are related to the former stresses, once out-of-plane postbuckling deformations develop, the latter stresses reduce the magnitude of compression in the plate. Accordingly, as temperature increases, out-of-plane deflections can be stopped and even reversed. This phenomenon is supported by experimental evidence as is discussed on the example of plates buckling due to fire in Sect. 6.5.

At the closing of this paragraph we emphasize the conclusion that the integration of the equations of equilibrium as well as the satisfaction of boundary conditions are complicated in the presence of temperature, except for several simple cases, such as those presented in the examples above. This difficulty may be overcome by using the Rayleigh-Ritz method that enables us to satisfy kinematic boundary conditions while violating the static ones (in thermal problems the latter conditions are often nonhomogeneous). An alternative to using the Rayleigh-Ritz method is a numerical procedure, such as the finite element or finite differences methods. The other observation refers to different in-plane boundary conditions that can affect the response of thermally loaded plates both quantitatively and qualitatively. These conditions include:

(a) Edges preventing in-plane expansion of the heated plate.
(b) Edges free to move in-plane so that thermal expansion is not prevented. Note that while the expansion of the plate may be unconstrained, tangential displacements along the edges may still be limited due to the axial stiffness of the structures supporting the edges.
(c) Edges that partially limit the expansion of the plate. For example, joints of perpendicular stringers supporting the plate at the corners are constrained by virtue of the axial stiffness of stringers. However, deformations of the stringers in the plane of the plate at the mid-length of the edge, i.e. between the joints in the direction perpendicular to the stringer axis are possible.

6.4 Thermal Bending and Buckling Problems for Rectangular Composite and Sandwich Plates

The present paragraph includes the formulation of thermal problems for shear deformable and thin composite plates. Representative examples are shown for bending and buckling of thin plates.

The stress-strain-temperature relationships for an orthotropic lamina oriented at an angle to the principal axes of the laminate are written here by assumptions of the first-order shear deformation theory neglecting the history of thermal loading (see the explanation in regards to Eq. 1.104). Adding thermal contributions to Eqs. 5.7 and 5.57 we obtain for the k-th generally orthotropic layer:

$$\begin{Bmatrix} \sigma_x \\ \sigma_y \\ \tau_{xy} \\ \tau_{yz} \\ \tau_{xz} \end{Bmatrix}_k = \begin{bmatrix} \bar{Q}_{11}(T) & \bar{Q}_{12}(T) & \bar{Q}_{16}(T) & 0 & 0 \\ \bar{Q}_{12}(T) & \bar{Q}_{22}(T) & \bar{Q}_{26}(T) & 0 & 0 \\ \bar{Q}_{16}(T) & \bar{Q}_{26}(T) & \bar{Q}_{66}(T) & 0 & 0 \\ 0 & 0 & 0 & \bar{Q}_{44}(T) & \bar{Q}_{45}(T) \\ 0 & 0 & 0 & \bar{Q}_{45}(T) & \bar{Q}_{55}(T) \end{bmatrix}_k \begin{Bmatrix} \varepsilon_x^0 + z\kappa_x - \alpha_x(T)T \\ \varepsilon_y^0 + z\kappa_y - \alpha_y(T)T \\ \gamma_{xy}^0 + z\kappa_{xy} - \alpha_{xy}(T)T \\ \gamma_{yz} \\ \gamma_{xz} \end{Bmatrix}_k \tag{6.27}$$

Transformed reduced stiffnesses $\bar{Q}_{ij}(T)$ are defined by Eqs. 5.5 and 5.58 in terms of reduced stiffnesses that depend on the material properties affected by temperature. The coefficients of thermal expansion for a generally orthotropic layer are the following functions of the coefficients $\alpha_1(T),\ \alpha_2(T)$ in the principal coordinate system of this layer:

$$\begin{Bmatrix} \alpha_x(T) \\ \alpha_y(T) \\ \frac{\alpha_{xy}(T)}{2} \end{Bmatrix}_k = \begin{bmatrix} \cos^2\theta & \sin^2\theta & -2\cos\theta\sin\theta \\ \sin^2\theta & \cos^2\theta & 2\cos\theta\sin\theta \\ \cos\theta\sin\theta & -\cos\theta\sin\theta & \cos^2\theta - \sin^2\theta \end{bmatrix}_k \begin{Bmatrix} \alpha_1(T) \\ \alpha_2(T) \\ 0 \end{Bmatrix}_k \tag{6.28}$$

The stress resultants and couples are obtained by the extension of equations (5.8) and (5.57), accounting for the effect of temperature:

$$\begin{Bmatrix} N_x \\ N_y \\ N_{xy} \\ M_x \\ M_y \\ M_{xy} \end{Bmatrix} = \begin{bmatrix} A_{11} & A_{12} & A_{16} & B_{11} & B_{12} & B_{16} \\ A_{12} & A_{22} & A_{26} & B_{12} & B_{22} & B_{26} \\ A_{16} & A_{26} & A_{66} & B_{16} & B_{26} & B_{66} \\ B_{11} & B_{12} & B_{16} & D_{11} & D_{12} & D_{16} \\ B_{12} & B_{22} & B_{26} & D_{12} & D_{22} & D_{26} \\ B_{16} & B_{26} & B_{66} & D_{16} & D_{26} & D_{66} \end{bmatrix} \begin{Bmatrix} \varepsilon_x^0 \\ \varepsilon_y^0 \\ \gamma_{xy}^0 \\ \kappa_x \\ \kappa_y \\ \kappa_{xy} \end{Bmatrix} - \begin{Bmatrix} N_x^T \\ N_y^T \\ N_{xy}^T \\ M_x^T \\ M_y^T \\ M_{xy}^T \end{Bmatrix} \tag{6.29}$$

$$\begin{Bmatrix} Q_y \\ Q_x \end{Bmatrix} = \begin{bmatrix} A_{44} & A_{45} \\ A_{45} & A_{55} \end{bmatrix} \begin{Bmatrix} \gamma_{yz} \\ \gamma_{xz} \end{Bmatrix}$$

where all elements of the matrix of stiffness coefficients may be affected by local temperature $\left(A_{ij} = A_{ij}(T),\ B_{ij} = B_{ij}(T),\ D_{ij} = D_{ij}(T),\right)$, $A_{45} = \int\limits_{z} \bar{Q}_{45} dz$ and "thermal" terms are

$$\begin{Bmatrix} N_x^T, & M_x^T \\ N_y^T, & M_y^T \\ N_{xy}^T, & M_{xy}^T \end{Bmatrix} = \int\limits_{-\frac{h}{2}}^{\frac{h}{2}} \begin{bmatrix} Q_{11}(T) & Q_{12}(T) & Q_{16}(T) \\ Q_{12}(T) & Q_{22}(T) & Q_{26}(T) \\ Q_{16}(T) & Q_{26}(T) & Q_{66}(T) \end{bmatrix} \begin{Bmatrix} \alpha_x(T) \\ \alpha_y(T) \\ \alpha_{xy}(T) \end{Bmatrix} \{1, z\}\, T dz \tag{6.30}$$

The substitution of the stress resultants and stress couples given by (6.29) into equations of equilibrium (static version of Eq. 5.62) yields the version of such equations in the case where thermal loading is present. Limiting the analysis to cross-ply symmetrically laminated plates and multi-layer angle-ply symmetrically laminated plates where $A_{16} = A_{26} = A_{45} = D_{16} = D_{26} = 0$ and $[B] = 0$ we obtained two systems of uncoupled equations in Chap. 5. However, in the presence of temperature varying through the thickness, some of the stiffness terms that were eliminated in Eqs. 5.64 and 5.65 are not equal to zero:

1. If temperature varies thorough the thickness in a symmetrically laminated cross-ply plate, $A_{16} = A_{26} = A_{45} = D_{16} = D_{26} = 0$ since the corresponding reduced stiffness in each layer is zero, i.e. $Q_{16} = Q_{26} = Q_{45} = 0$. Additionally, $B_{16} = B_{26} = 0$. However, other elements of the matrix of coupling stiffness, i.e. $B_{11},\ B_{12},\ B_{22},\ B_{66}$ are not equal to zero since the properties of each couple of layers symmetric about the middle plane but subject to different temperatures are not equal to each other. These coupling stiffness terms can be taken equal to zero only in the case where the effect of temperature on the material properties is disregarded.
2. In multi-layered symmetrically laminated angle-ply plates none of the stiffness terms can be assumed equal to zero as long as the effect of temperature on the material properties is included into consideration and temperature varies through the thickness.

The transverse shear stress resultants given by (6.29) are not explicitly affected by temperature, although the implicit effect is present through the temperature-affected transformed stiffness terms $\bar{Q}_{ij}$.

It is now possible to formulate equations of equilibrium accounting for both the effect of temperature on material properties as well as explicit thermal terms introduced in (6.30). The system of five equations incorporating all stiffness terms is very long and defies an analytical solution (e.g., Tauchert 1985). A number of analytical solutions have been developed for particular cases.

It is possible to arrive at a number of interesting conclusions relevant to both thick (shear deformable) and thin plates analyzing equations (6.29) and (6.30):

1. If temperature varies through the thickness of the plate, thermally-induced contributions to stress couples are present, i.e. $M_x^T \neq 0,\ M_y^T \neq 0,\ M_{xy}^T \neq 0$. This means that thermal bending occurs for all boundary conditions, except for

the case where all edges are clamped and temperature does not vary over the plate surface (as explained in the previous paragraph, kinematic boundary conditions corresponding to clamping are not affected by temperature). The reason for thermal bending is that it is impossible to satisfy conditions of zero bending stress couples along the edges in the presence of thermally-induced contributions.

2. Following the previous conclusion, it is evident that thermal buckling may occur only in clamped plates subject to a uniform over the surface and varying through the thickness temperature (of course, a uniform temperature is a particular case of this thermal profile).

Consider now the case where temperature varies over the surface of the plate. In such case, thermally-induced contributions to stress resultants and stress couples given by (6.30) are functions of the x- and y-coordinates. In the case of a shear-deformable plate characterized by the first-order theory, equations of equilibrium obtained as an extension of (5.62) are

$$
\begin{aligned}
&N'_{x,x} + N'_{xy,y} = N^T_{x,x} + N^T_{xy,y} \\
&N'_{xy,x} + N'_{y,y} = N^T_{xy,x} + N^T_{y,y} \\
&M'_{x,x} + M'_{xy,y} - Q_x = M^T_{x,x} + M^T_{xy,y} \\
&M'_{xy,x} + M'_{y,y} - Q_y = M^T_{xy,x} + M^T_{y,y} \\
&Q_{x,x} + Q_{y,y} + N'_x w_{,xx} + 2N'_{xy} w_{,xy} + N'_y w_{,yy} \\
&\quad + \underline{\left(N'_{x,x} + N'_{xy,y}\right) w_{,x}} + \underline{\underline{\left(N'_{xy,x} + N'_{y,y}\right) w_{,y}}} = -p + \\
&N^T_x w_{,xx} + 2N^T_{xy} w_{,xy} + N^T_y w_{,yy} \underline{\left(N^T_{x,x} + N^T_{xy,y}\right) w_{,x}} \\
&\quad + \underline{\underline{\left(N^T_{xy,x} + N^T_{y,y}\right) w_{,y}}}
\end{aligned}
\tag{6.31}
$$

where terms with the prime are the elements of stress resultants and stress couples obtained from (6.29) without accounting for the vector of explicit thermal terms in the right side of these equations. The singly and doubly underlined terms in the last equation (6.31) are cancelled out as a result of the first and second equations (6.31), respectively. If thermal terms in the right side of (6.31) are dependent on the in-plane coordinates, the system of equilibrium equations is non-homogeneous, even in the absence of transverse pressure. This leads to the following conclusion (valid for both shear-deformable and thin plates):

3. If the plate is subject to a variable over the surface thermal load, it experiences thermal bending. Thermal buckling does not occur in such case, even if temperature is uniform through the thickness and the plate is symmetrically laminated.

In the case of thin plates, Eqs. 6.27–6.30 remain valid, except for transverse shear stresses and strains in (6.27) and for transverse shear stress resultants in (6.29) that are equal to zero. Equations of equilibrium (6.31) are simplified according to the thin plate theory. Substituting the constitutive relations (6.29) into the equations of equilibrium (6.31) and retaining only linear terms in the strain-displacement relationships (1.28), (1.29) yields

$$\begin{aligned}
&A_{11}\frac{\partial^2 u_0}{\partial x^2}+2A_{16}\frac{\partial^2 u_0}{\partial x\partial y}+A_{66}\frac{\partial^2 u_0}{\partial y^2}+A_{16}\frac{\partial^2 v_0}{\partial x^2}+(A_{12}+A_{66})\frac{\partial^2 v_0}{\partial x\partial y}+A_{26}\frac{\partial^2 v_0}{\partial y^2}\\
&-B_{11}\frac{\partial^3 w}{\partial x^3}-3B_{16}\frac{\partial^3 w}{\partial x^2\partial y}-(B_{12}+2B_{66})\frac{\partial^3 w}{\partial x\partial y^2}-B_{26}\frac{\partial^3 w}{\partial y^3}=\frac{\partial N_x^T}{\partial x}+\frac{\partial N_{xy}^T}{\partial y}\\
&A_{16}\frac{\partial^2 u_0}{\partial x^2}+(A_{12}+A_{66})\frac{\partial^2 u_0}{\partial x\partial y}+A_{26}\frac{\partial^2 u_0}{\partial y^2}+A_{66}\frac{\partial^2 v_0}{\partial x^2}+2A_{26}\frac{\partial^2 v_0}{\partial x\partial y}+A_{22}\frac{\partial^2 u_0}{\partial y^2}\\
&-B_{16}\frac{\partial^3 w}{\partial x^3}-(B_{12}+2B_{66})\frac{\partial^3 w}{\partial x^2\partial y}-3B_{26}\frac{\partial^3 w}{\partial x\partial y^2}-B_{22}\frac{\partial^3 w}{\partial y^3}=\frac{\partial N_{xy}^T}{\partial x}+\frac{\partial N_y^T}{\partial y}\\
&D_{11}\frac{\partial^4 w}{\partial x^4}+4D_{16}\frac{\partial^4 w}{\partial x^3\partial y}+2(D_{12}+2D_{66})\frac{\partial^4 w}{\partial x^2\partial y^2}+4D_{26}\frac{\partial^4 w}{\partial x\partial y^3}+D_{22}\frac{\partial^4 w}{\partial y^4}\\
&-B_{11}\frac{\partial^3 u_0}{\partial x^3}-3B_{16}\frac{\partial^3 u_0}{\partial x^2\partial y}-(B_{12}+2B_{66})\frac{\partial^3 u_0}{\partial x\partial y^2}-B_{26}\frac{\partial^3 u_0}{\partial y^3}-B_{16}\frac{\partial^3 v_0}{\partial x^3}\\
&-(B_{12}+2B_{66})\frac{\partial^3 v_0}{\partial x^2\partial y}-3B_{26}\frac{\partial^3 v_0}{\partial x\partial y^2}-B_{22}\frac{\partial^3 v_0}{\partial y^3}=-p-\frac{\partial^2 M_x^T}{\partial x^2}-2\frac{\partial^2 M_{xy}^T}{\partial x\partial y}\\
&-\frac{\partial^2 M_y^T}{\partial y^2}-N_x^T\frac{\partial^2 w}{\partial x^2}-2N_{xy}^T\frac{\partial^2 w}{\partial x\partial y}-N_y^T\frac{\partial^2 w}{\partial y^2}
\end{aligned} \tag{6.32}$$

Analyzing Eq. 6.32 we can note that temperature varying through the thickness makes the plate that was symmetrically laminated at room temperature asymmetric by adding terms B_{ij} and A_{16}, A_{26}, D_{16}, D_{26}. Thus, symmetry of a laminated plate that is almost always desirable can be lost at an elevated temperature. The above-mentioned stiffness terms associated with an elevated temperature are also responsible for coupling between the first two equations and the last equation (6.32) that is absent in a symmetrically laminated plate.

Geometrically nonlinear terms can be incorporated in (6.32) resulting the formulation that usually requires a numerical analysis. An alternative formulation for the geometrically nonlinear thermoelastic problem is based on using the stress function (e.g., Tauchert 1985).

It is sometimes convenient to use the Rayleigh-Ritz method to solve thermoelastic problems. The reason is evident since assumed expressions for displacements in this method can violate static boundary conditions. Therefore, in problems where a closed form solution satisfying both the equations of equilibrium and the boundary conditions is unavailable, the Rayleigh-Ritz method may offer a satisfactory alternative.

Large deflections are seldom encountered in thick shear deformable plates that fail at relatively small deformations due to their high stiffness. Accordingly, the expression for the potential energy is shown here for thin plates, accounting for geometrically nonlinear terms. The combination of the expression for the strain energy (1.67), the stress-strain relations (6.27) and nonlinear strains in the Cartesian coordinate system (1.27) yields the potential energy for thermally loaded plate (e.g., Tauchert 1985):

$$
\Pi = \iint_A \left\{
\begin{aligned}
&\frac{A_{11}}{2}\left[\frac{\partial u_0}{\partial x}+\frac{1}{2}\left(\frac{\partial w}{\partial x}\right)^2\right]^2 + A_{12}\left[\frac{\partial u_0}{\partial x}+\frac{1}{2}\left(\frac{\partial w}{\partial x}\right)^2\right]\left[\frac{\partial v_0}{\partial y}+\frac{1}{2}\left(\frac{\partial w}{\partial y}\right)^2\right]\\
&+\frac{A_{22}}{2}\left[\frac{\partial v_0}{\partial y}+\frac{1}{2}\left(\frac{\partial w}{\partial y}\right)^2\right]^2 + A_{16}\left[\frac{\partial u_0}{\partial x}+\frac{1}{2}\left(\frac{\partial w}{\partial x}\right)^2\right]\\
&\left(\frac{\partial u_0}{\partial y}+\frac{\partial v_0}{\partial x}+\frac{\partial w}{\partial x}\frac{\partial w}{\partial y}\right) + A_{26}\left[\frac{\partial v_0}{\partial y}+\frac{1}{2}\left(\frac{\partial w}{\partial y}\right)^2\right]\left(\frac{\partial u_0}{\partial y}+\frac{\partial v_0}{\partial x}+\frac{\partial w}{\partial x}\frac{\partial w}{\partial y}\right)\\
&+\frac{A_{66}}{2}\left(\frac{\partial u_0}{\partial y}+\frac{\partial v_0}{\partial x}+\frac{\partial w}{\partial x}\frac{\partial w}{\partial y}\right)^2 - B_{11}\left[\frac{\partial u_0}{\partial x}+\frac{1}{2}\left(\frac{\partial w}{\partial x}\right)^2\right]\frac{\partial^2 w}{\partial x^2}\\
&-B_{12}\left[\left[\frac{\partial u_0}{\partial x}+\frac{1}{2}\left(\frac{\partial w}{\partial x}\right)^2\right]\frac{\partial^2 w}{\partial y^2}+\left[\frac{\partial v_0}{\partial y}+\frac{1}{2}\left(\frac{\partial w}{\partial y}\right)^2\right]\frac{\partial^2 w}{\partial x^2}\right]\\
&-B_{22}\left[\frac{\partial v_0}{\partial y}+\frac{1}{2}\left(\frac{\partial w}{\partial y}\right)^2\right]\frac{\partial^2 w}{\partial y^2} - B_{16}\left[\left(\frac{\partial u_0}{\partial y}+\frac{\partial v_0}{\partial x}+\frac{\partial w}{\partial x}\frac{\partial w}{\partial y}\right)\frac{\partial w}{\partial x}\right.\\
&\qquad\qquad\left.+2\left[\frac{\partial u_0}{\partial x}+\frac{1}{2}\left(\frac{\partial w}{\partial x}\right)^2\right]\frac{\partial^2 w}{\partial x\partial y}\right]\\
&-B_{26}\left[\left(\frac{\partial u_0}{\partial y}+\frac{\partial v_0}{\partial x}+\frac{\partial w}{\partial x}\frac{\partial w}{\partial y}\right)\frac{\partial w}{\partial y}+2\left[\frac{\partial v_0}{\partial y}+\frac{1}{2}\left(\frac{\partial w}{\partial y}\right)^2\right]\frac{\partial^2 w}{\partial x\partial y}\right]\\
&-2B_{66}\left(\frac{\partial u_0}{\partial y}+\frac{\partial v_0}{\partial x}+\frac{\partial w}{\partial x}\frac{\partial w}{\partial y}\right)\frac{\partial^2 w}{\partial x\partial y}+\frac{D_{11}}{2}\left(\frac{\partial^2 w}{\partial x^2}\right)^2+D_{12}\frac{\partial^2 w}{\partial x^2}\frac{\partial^2 w}{\partial y^2}\\
&+\frac{D_{22}}{2}\left(\frac{\partial^2 w}{\partial y^2}\right)^2+2D_{16}\frac{\partial^2 w}{\partial x^2}\frac{\partial^2 w}{\partial x\partial y}+2D_{26}\frac{\partial^2 w}{\partial y^2}\frac{\partial^2 w}{\partial x\partial y}+2D_{66}\left(\frac{\partial^2 w}{\partial x\partial y}\right)^2\\
&-N_x^T\left[\frac{\partial u_0}{\partial x}+\frac{1}{2}\left(\frac{\partial w}{\partial x}\right)^2\right]-N_{xy}^T\left(\frac{\partial u_0}{\partial y}+\frac{\partial v_0}{\partial x}+\frac{\partial w}{\partial x}\frac{\partial w}{\partial y}\right)\\
&-N_y^T\left[\frac{\partial v_0}{\partial y}+\frac{1}{2}\left(\frac{\partial w}{\partial y}\right)^2\right]+M_x^T\frac{\partial^2 w}{\partial x^2}+2M_{xy}^T\frac{\partial^2 w}{\partial x\partial y}+M_y^T\frac{\partial^2 w}{\partial y^2}
\end{aligned}
\right\} dxdy \tag{6.33}
$$

Representing the displacements of the plate in the series

$$u_0 = \sum_m \sum_n U_{mn} X'_m(x) Y'_n(y)$$

$$v_0 = \sum_m \sum_n V_{mn} X''_m(x) Y''_n(y)$$

$$w = \sum_m \sum_n W_{mn} X'''_m(x) Y'''_n(y) \tag{6.34}$$

where the functions of coordinates, as a minimum, satisfy kinematic boundary conditions, the amplitudes of terms, i.e. U_{mn}, V_{mn} and W_{mn} are found upon the substitution of (6.34) into (6.33), and applying the requirement that the potential energy must be minimal at equilibrium, so that $\frac{\partial \Pi}{\partial U_{mn}} = \frac{\partial \Pi}{\partial V_{mn}} = \frac{\partial \Pi}{\partial W_{mn}} = 0$. The solution of the resulting system of algebraic equations yields the amplitude values that can subsequently be used in equations (6.27) to specify the stresses.

Example 6.3: Thermal Bending of a Thin Symmetrically Laminated Cross-Ply Plate Neglecting the Effect of Temperature on Material Properties (Linear Problem) Consider thermal bending of a plate subject to an arbitrary temperature distribution, but neglecting its effect on material properties. Accordingly, the plate that was symmetric prior to thermal loading remains such in the presence of temperature. In this problem the equations of equilibrium (6.32) are simplified since several stiffness terms are equal to zero. Furthermore, the thermal contribution to the shear stress resultant is also absent, i.e. $N_{xy}^T = M_{xy}^T = 0$. Accordingly, upon simplifications, equations of equilibrium (6.32) become

$$A_{11}\frac{\partial^2 u_0}{\partial x^2} + A_{66}\frac{\partial^2 u_0}{\partial y^2} + (A_{12} + A_{66})\frac{\partial^2 v_0}{\partial x \partial y} = \frac{\partial N_x^T}{\partial x}$$

$$(A_{12} + A_{66})\frac{\partial^2 u_0}{\partial x \partial y} + A_{66}\frac{\partial^2 v_0}{\partial x^2} + A_{22}\frac{\partial^2 v_0}{\partial y^2} = \frac{\partial N_y^T}{\partial y}$$

$$D_{11}\frac{\partial^4 w}{\partial x^4} + 2(D_{12} + 2D_{66})\frac{\partial^4 w}{\partial x^2 \partial y^2} + D_{22}\frac{\partial^4 w}{\partial y^4}$$

$$= -\frac{\partial^2 M_x^T}{\partial x^2} - \frac{\partial^2 M_y^T}{\partial y^2} - N_x^T\frac{\partial^2 w}{\partial x^2} - N_y^T\frac{\partial^2 w}{\partial y^2} \tag{1}$$

It is observed that equations (1) are uncoupled, i.e. the first two equations include in-plane displacements, while the last equation depends on the transverse deflection only.

Boundary conditions are also uncoupled as is evidenced from the expressions for stress resultants and stress couples for a symmetric cross-ply plate

$$\begin{Bmatrix} N_x \\ N_y \\ N_{xy} \\ M_x \\ M_y \\ M_{xy} \end{Bmatrix} = \begin{bmatrix} A_{11} & A_{12} & 0 & 0 & 0 & 0 \\ A_{12} & A_{22} & 0 & 0 & 0 & 0 \\ 0 & 0 & A_{66} & 0 & 0 & 0 \\ 0 & 0 & 0 & D_{11} & D_{12} & 0 \\ 0 & 0 & 0 & D_{12} & D_{22} & 0 \\ 0 & 0 & 0 & 0 & 0 & D_{66} \end{bmatrix} \begin{Bmatrix} \varepsilon_x^0 \\ \varepsilon_y^0 \\ \gamma_{xy}^0 \\ \kappa_x \\ \kappa_y \\ \kappa_{xy} \end{Bmatrix} - \begin{Bmatrix} N_x^T \\ N_y^T \\ 0 \\ M_x^T \\ M_y^T \\ 0 \end{Bmatrix} \tag{m}$$

where mid-plane strains and changes of curvature and twist are given by linear equations (1.28) and (1.29).

Temperature can be represented in double Fourier series, so that

$$T = \sum_m \sum_n T_{mn}(x, y, z) \sin \alpha_m x \sin \beta_n y \tag{n}$$

It is noted that the temperature distribution should satisfy the heat conduction equation and thermal boundary conditions.

The thermal contributions to stress couples and stress resultants are available in double Fourier series by substituting (n) into (6.30):

$$\begin{Bmatrix} N_x^T \\ N_y^T \\ M_x^T \\ M_y^T \end{Bmatrix} = \sum_m \sum_n \begin{Bmatrix} K_{mn} \\ L_{mn} \\ G_{mn} \\ H_{mn} \end{Bmatrix} \sin \alpha_m x \sin \beta_n y \tag{o}$$

where the coefficients K_{mn}, L_{mn}, G_{mn} and H_{mn} can be specified dependent on $T_{mn}(x, y, z)$.

If the edges of the plate are simply supported by structures that do not prevent in-plane extension or contraction, but constrain tangential displacements, the corresponding boundary conditions are (2.69):

$$x = 0,\ x = a:\quad v_0 = w = 0,\quad N_x = M_x = 0$$

$$y = 0,\ y = b:\quad u_0 = w = 0,\quad N_y = M_y = 0$$

It is easy to verify that these boundary conditions are identically satisfied if the displacements are represented in series (2.70):

$$u_0 = \sum_m \sum_n U_{mn} \cos \alpha_m x \sin \beta_n y$$

$$v_0 = \sum_m \sum_n V_{mn} \sin \alpha_m x \cos \beta_n y$$

$$w = \sum_m \sum_n W_{mn} \sin \alpha_m x \sin \beta_n y$$

The substitution of (o) and (2.70) into the first two equations of equilibrium (l) yields a system of two equations for amplitudes U_{mn} and V_{mn}:

$$\begin{bmatrix} \left(A_{11}\alpha_m^2 + A_{66}\beta_n^2\right) & \alpha_m\beta_n\left(A_{12}+A_{66}\right) \\ \alpha_m\beta_n\left(A_{12}+A_{66}\right) & \left(A_{66}\alpha_m^2 + A_{22}\beta_n^2\right) \end{bmatrix} \begin{Bmatrix} U_{mn} \\ V_{mn} \end{Bmatrix} = -\begin{Bmatrix} K_{mn}\alpha_m \\ L_{mn}\beta_n \end{Bmatrix} \tag{p}$$

The equation for W_{mn} is obtained applying the Galerkin procedure to the last equation (l). For example, the (mn)-th equation is

$$\int_0^b \int_0^a \left[\begin{aligned} & D_{11}\frac{\partial^4 w}{\partial x^4} + 2\left(D_{12}+2D_{66}\right)\frac{\partial^4 w}{\partial x^2 \partial y^2} \\ & + D_{22}\frac{\partial^4 w}{\partial y^4} + \frac{\partial^2 M_x^T}{\partial x^2} + \frac{\partial^2 M_y^T}{\partial y^2} \\ & + N_x^T \frac{\partial^2 w}{\partial x^2} + N_y^T \frac{\partial^2 w}{\partial y^2} \end{aligned} \right] \sin \alpha_m x \sin \beta_n y\, dx\, dy = 0 \tag{q}$$

Upon the substitution of deflection from the third equation (2.70) and thermal terms (o), and using the orthogonality property of sine functions, Eq. q yield the system of coupled equations

$$\begin{aligned} & \left[D_{11}\alpha_m^4 + 2\left(D_{12}+2D_{66}\right)\alpha_m^2\beta_n^2 + D_{22}\beta_n^4\right] W_{mn} \\ & \quad = G_{mn}\alpha_m^2 + H_{mn}\beta_n^2 + \sum_i \sum_j \sum_s \sum_r F_{ijsr}\left(K_{ij}\alpha_s^2 + L_{ij}\beta_r^2\right) W_{sr} \end{aligned} \tag{r}$$

where

$$F_{ijsrmn} = \frac{4}{ab}\left[\int_0^a \sin\frac{i\pi x}{a}\sin\frac{s\pi x}{a}\sin\frac{m\pi x}{a}dx\right]\left[\int_0^b \sin\frac{j\pi y}{b}\sin\frac{r\pi y}{b}\sin\frac{n\pi y}{b}dy\right]$$

Once the amplitudes are found from (p) and (r), the stresses in the plate can be determined from the first three equations (6.27).

Example 6.4: Thermal Buckling of a Thin Symmetrically Laminated Cross-Ply Plate Subject to a Uniform Temperature Consider a simply supported cross-ply plate subject to uniform temperature T. In this case thermal contributions to stress couples are absent since the changes of material properties in symmetric layers occurring at the same temperature are identical. Furthermore, the thermal in-plane shear stress resultant contribution is equal to zero, as was the case in the previous example. The nonzero in-plane thermal terms N_x^T and N_y^T are constant and equal to $N_x^T = KT$, $N_y^T = LT$ where the coefficients and are determined from the corresponding equations (m).

It is noted that while such formulation neglecting the heat flow to or from boundary structures and its effect on the thermal field distribution is typical, it has a potential drawback. If the stringers or bulkheads forming the boundary serve as heat sinks, temperature within the plate under consideration may become nonuniform. In this case the response may qualitatively be affected since a nonuniform temperature distribution results in bending, rather than buckling. Therefore, the present problem is solved by assumption that the boundaries do not affect the temperature distribution.

In the case of uniform temperature, equations of equilibrium (l) are simplified:

$$A_{11}\frac{\partial^2 u_0}{\partial x^2} + A_{66}\frac{\partial^2 u_0}{\partial y^2} + (A_{12} + A_{66})\frac{\partial^2 v_0}{\partial x \partial y} = 0$$

$$(A_{12} + A_{66})\frac{\partial^2 u_0}{\partial x \partial y} + A_{66}\frac{\partial^2 v_0}{\partial x^2} + A_{22}\frac{\partial^2 v_0}{\partial y^2} = 0$$

$$D_{11}\frac{\partial^4 w}{\partial x^4} + 2(D_{12} + 2D_{66})\frac{\partial^4 w}{\partial x^2 \partial y^2} + D_{22}\frac{\partial^4 w}{\partial y^4} = -N_x^T\frac{\partial^2 w}{\partial x^2} - N_y^T\frac{\partial^2 w}{\partial y^2} \qquad \text{(s)}$$

In-plane and transverse displacements in Eq. s and boundary conditions (2.69) are uncoupled. Prebuckling in-plane displacements can be found from the first two equations (s) and in-plane boundary conditions

$$x = 0,\ x = a:\ v_0 = 0,\quad N_x = 0$$

$$y = 0,\ y = b:\ u_0 = 0,\quad N_y = 0 \qquad \text{(t)}$$

The buckling temperature is determined from the third equation (s) assuming the buckling deflection in the form of the third equation (2.70) that satisfies the boundary conditions of simple support. The substitution of this deflection into the third equation (s) and the requirement that $W_{mn} \neq 0$ yields the buckling temperature:

$$T_{cr} = \min_{m,n}\left[\frac{K\alpha_m^2 + L\beta_n^2}{D_{11}\alpha_m^4 + 2(D_{12} + 2D_{66})\alpha_m^2\beta_n^2 + D_{22}\beta_n^4}\right] \qquad \text{(u)}$$

Note that the observation regarding a small temperature increase that is sufficient to cause buckling of plates made in the previous paragraph remains applicable in the cases of composite plates. Therefore, it is often necessary to conduct postbuckling analysis using either the Rayleigh-Ritz method (e.g., Huang and Tauchert 1988) or a numerical procedure.

6.5 Example of Thermal Problem in Applications: Composite Plates Subject to Fire

The problems of durability and real-time and residual strength and stiffness of composite structures experiencing the effect of fire are of major interest to designers. In this book we refer to these problems as an example of modern issues that require a combination of experimental, analytical and numerical efforts. Composite and sandwich plates subject to a potential danger of fire are found in aerospace and naval structures as well as in civil engineering applications.

Studies of the response of composite and sandwich plates and columns to fire reflect diverse aspects of the problem, including those related to material degradation at high temperature, heat transfer and mechanical response (e.g., Mouritz and Gibson 2006; Kardomateas et al. 2009). For example, material transformation in polymer-matrix composite structures during fire involves a degradation of properties of the matrix due to an elevated temperature. This includes both the conversion of matrix into char near the surface exposed to the heat flux as well as the reduction of the properties of the material adjacent to a colder surface. Both processes are dynamic since they reflect a gradual conversion of the matrix into char, the propagation of the charred region toward the colder surface of the plate and the change of the thermal profile across the thickness. An example of the temperature distribution throughout the thickness of an E-glass/vinyl ester plate or column exposed to fire with time is shown in Fig. 6.5. In this figure, the plate or column consisted of four $[0°/90°/45°/-45°/Random]$ sub-layers. Although the original paper refers to a column, the solution of the thermal problem is also applicable to a plate subject to a uniform heat flux over the surface (the problems of heat transfer for the plate and column are identical in this study since heat transfer from side surfaces of the column was not considered). As is shown in Fig. 6.5, temperature of the colder surface ($\frac{y}{H} = 0$) increases with time that is measured in seconds since the onset of the exposure of the surface $\frac{y}{H} = 1$ to the heat flux generated by fire.

Some of the conclusions from studies of mechanical response of composite and sandwich structures subject to fire that are applicable to plates are listed below.

1. The asymmetry of material properties, decomposition of resin near the surface exposed to fire and a nonuniform temperature distribution throughout the thickness result in thermal bending stress couples. These stress couples tend to decrease with time as a result of a gradually increasing depth of the charred region adjacent to the exposed surface where the stiffness of material is negligible.
2. If the edges of the plate are constrained against contraction, in-plane tensile stress resultants develop due to heating of the plate as a result of stretching of its middle surface caused by thermally-induced bending. These stress resultants increase with time due to progressive bending. However, upon reaching a peak value, they decrease due to an increasing depth of the charred region and a degradation of material properties in the intact part of the plate. An example is shown in Fig. 6.6 for a large aspect ratio E-glass/vinyl ester composite plate. The plate that was

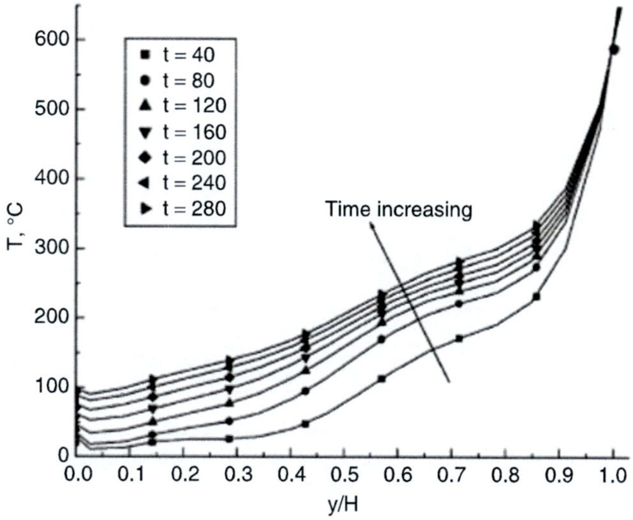

Fig. 6.5 Temperature distribution in a E-glass-vinyl ester plate or column subject to heat flux $Q = 25\,\frac{kW}{m^2}$ applied at one of the surfaces. The through-the-thickness y-coordinate is normalized by the total thickness H (From Kardomateas et al. 2009)

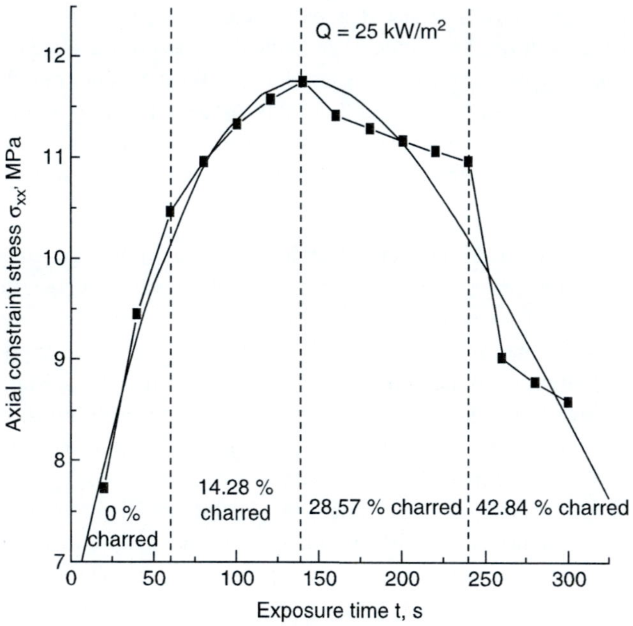

Fig. 6.6 Axial constraint stress in a large aspect ratio E-glass-vinyl ester plate subject to heat flux $Q = 25\,\frac{kW}{m^2}$ as a function of time of exposure to fire (From Kardomateas et al. 2009)

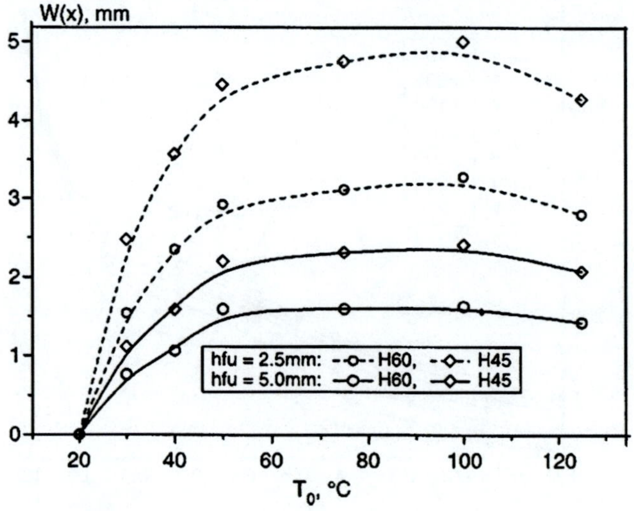

Fig. 6.7 Maximum deflections of a large aspect ratio sandwich panel as a function of the temperature of the heated surface T_0, the thickness of the facings ($h_{fu} = 2.5$ mm or 5 mm) and the properties of the Divinycell H45 or H60 core (From Birman et al. 2006)

simply supported along short edges and free along long edges was 0.15 m long, 0.025 m wide and 0.012 m thick. The increase of the depth of the charred region with time is also shown as a percentage of the thickness of the plate in this figure.

3. Deformations of plates subject to a nonuniform through thickness temperature produced by fire and an associated thermally-induced bending moment increase with time, reach a peak value and reverse their direction. This is related to a reduction of bending moment due to material decomposition and an increasing depth of the charred region.

In addition to the study concerned with the effect of fire on composite and sandwich plates, the response of such plates to an elevated surface temperature was extensively investigated (e.g., Birman et al. 2006). A typical plate experiences the reversal of deformations and stresses if the temperature of the exposed surface increases. This situation is depicted in Fig. 6.7 illustrating maximum deflections of a representative large aspect ratio sandwich plate with cross-ply facings where long edges were completely restrained against in-plane displacements as a function of temperature of the exposed surface. Cross-ply graphite-epoxy facings of the panel were 2.5 mm or 5 mm thick, while the Divinycell core considered in the example could be manufactured of two different grades H45 and H60. The panel was 101.6 mm wide, the core thickness was fixed at 20 mm and the overall panel thickness varied reflecting the thickness of the facings. The phenomenon of the reversal of deformations as the exposed surface temperature increases was also reported by Meyers and Hyer (1992) who analyzed deflections of a composite panel subject to a linearly distributed through the thickness temperature. An experimental paper of Lattimer et al. (2004) who considered deformations of sandwich panels subjected to fire also supports this observation.

The loss of strength is only one of possible failure modes for sandwich plates subjected to a combination of compressive and thermal loads (the latter loads include those generated by fire). Another possible mode of failure is wrinkling of one of the facings. This phenomenon should be analyzed accounting for temperature-induced property degradation of the facings and the core. An example of the analysis evaluating thermal wrinkling conditions in sandwich plates was conducted by Birman (2005). As follows from this analysis, solutions neglecting the effect of temperature on the properties of the facings and core produce inadequate and unconservative predictions of the wrinkling stress. Moreover, as a result of the loss of stiffness due to elevated temperatures, wrinkling may become the dominant mode of failure in sandwich panels that do not exhibit this phenomenon at room temperature.

6.6 Design Philosophy and Recommendations

Design of plates subjected to thermal loading involves several problems including heat transfer analysis, adjustment of properties accounting for the effect of temperature, and stress analysis. These problems can be coupled, e.g., heat transfer is affected by the variations of thermal conductivity due to local temperature. Furthermore, local stresses affect material properties (Dunn 1997), while being at the same time influenced by the temperature distribution, although the effect of stress on properties is usually neglected. A designer should always analyze the temperature in the structure to justify neglecting its effect on the material constants. For example, an elevation of temperature by 100°C may have a relatively mild effect on properties of steel, while its effect on properties of a glass/epoxy plate cannot be disregarded.

The effect of temperature in plates where the edges are constrained against in-plane displacements can result in either bending or buckling, dependent on the temperature distribution throughout the plate and boundary conditions. For example, an isotropic or composite plate with a through the thickness temperature variation will experience thermally-induced bending if its edges are simply supported, while the same plate buckles if the edges are clamped. Temperature that varies over the surface of the plate causes bending in all cases. Boundary structures may serve as heat sinks resulting in an outflow of heat from the plate and a nonuniform temperature over the plate surface.

A uniform temperature applied to isotropic or symmetrically laminated composite plates with edges constrained against in-plane expansion causes buckling. However, thermal buckling in such plates occurs at a very low temperature, often only several degrees over the reference stress-free value. As is well known, small temperature variations during the lifetime do not cause catastrophic failure of plates.

This is explained by a stable postbuckling behavior of plates and by tensile in-plane stresses that develop in the middle surface of a deformed plate with the edges constrained against in-plane displacements as it experiences stretching due to postbuckling deflections.

In polymeric composite plates subjected to elevated temperatures due to fire the process of transformation of matrix into char is time-dependent. Accordingly, heat transfer, property changes and structural response problems are affected by time, although the process is often sufficiently slow to be considered quasi-static. The thermal field through the thickness of the plate may vary time in problems where one surface is subject to a heat flux, while the opposite surface is in contact with ambient environment. Then, temperature increases on the colder surface of the plate with time, unless it is affected by active cooling (of course, the same trend in temperature history and distribution is present in metallic plates).

Note that effects of temperature in composite material plates occur at different scales. A microscale problem involves local stresses due to a mismatch in the coefficients of thermal expansion of fibers and matrix. These local stresses are in self-equilibrium, but they may cause debonding of fibers from matrix or local damage to the matrix. In composites operating at an elevated temperature, micromechanical thermal local stresses may actually reduce residual thermal stresses that develop due to a difference between the operational and processing temperatures. Micromechanical thermal stresses do not explicitly affect the response of the composite plate, unless they cause local damage affecting its behavior. The macromechanical thermal problem is solved upon the homogenization of the plate material, i.e. using composite thermal conductivities, coefficients of thermal expansion, etc. Macromechanical thermal stresses are not in self-equilibrium as is evidenced by the necessity to refer to boundary conditions reflecting the reaction of adjacent structures to thermally-induced deformations of the plate.

A designer of plate structures that experience effects of temperature can incorporate all above-mentioned effects in the framework of a numerical or analytical solution. While thermal bending can usually be adequately analyzed using a geometrically linear theory as long as the deflections remain smaller than half-thickness of the plate, if the plate experiences buckling, its postbuckling behavior should account for geometrically nonlinear effects.

A typical analysis and design approach can often be subdivided into two steps, preferably accounting for the effect of temperature on material properties:

1. Solution of the heat transfer problem,;
2. Solution of the structural response problem;

While the effect of temperature on material properties is referred to in both steps, different properties affect the solutions at each of these steps, e.g., thermal conductivities affects heat transfer, while structural response is dependent on moduli of elasticity and coefficients of thermal expansion. The effect of temperature on the properties of structural materials is obtained experimentally and unfortunately, in many cases such data is limited or unavailable.

In conclusion of this paragraph it is emphasized that a constraint on in-plane expansion of the plate is often imposed by adjacent plate structures. Furthermore, the presence of such structures often forces the edges of the plate to remain straight due to symmetry (e.g., adjacent edges of two identical plates subject to a thermal load symmetric about the stringer supporting these edges remain straight).

References

Birman, V. (2005). Thermally induced bending and wrinkling in large aspect ratio sandwich panels. *Composites. Part A: Applied Science and Manufacturing, 36*, 1412–1420.

Birman, V., & Byrd, L. W. (2007). Modeling and analysis of functionally graded materials and structures. *Applied Mechanics Reviews, 60*, 195–216.

Birman, V., Kardomateas, G. A., Simitses, G. J., & Li, R. (2006). Response of a sandwich panel to fire or elevated temperature on one of the surfaces, *Composites. Part A: Applied Science and Manufacturing, 37*, 981–988.

Boley, B. A., & Weiner, J. H. (1960). *Theory of Thermal Stresses*. Malabar: Robert E. Krieger Publishing Company.

Dunn, S. A. (1997). Using nonlinearities for improved stress analysis by thermoelastic techniques. *Applied Mechanics Reviews, 50*, 499–524.

Hatta, H., & Taya, M. (1985). Effective thermal conductivity of a misoriented short fiber composite. *Journal of Applied Physics, 58*, 2478–2486.

Huang, N. N., & Tauchert, T. R. (1988). Large deformation of antisymmetric angle-ply laminates resulting from nonunifrom temperature loadings. *Journal of Thermal Stresses, 11*, 287–297.

Incropera, F. P., & DeWitt, D. P. (1996). *Introduction to heat transfer*. New York: Wiley.

Jones, R. M. (2006). *Buckling of bars, plates and shells*. Blacksburg: Bull Ridge Publishing.

Kardomateas, G. A., Simitses, G. J., & Birman, V. (2009). Structural integrity of composite columns subject to fire. *Journal of Composite Materials, 43*, 1015–1033.

Lattimer. B. Y., Outllette, J., & Sorathia, U. (2004). Large-scale Fire Resistance tests on Sandwich Structures. *Proceedings of the SAMPE Meeting*, Long Beach, pp. 16–20.

Mouritz, A. P., & Gibson, A. G. (2006). *Fire properties of polymer composite materials*. Dordrecht: Springer.

Myers, C. A., & Hyer, M. W. (1992). Thermally-induced geometrically nonlinear response of symmetrically laminated composite plates (AIAA Paper AIAA-92-2539-CP).

Nowinski, J. L. (1978). *Theory of thermoelasticity with applications*. Groningen: Sijthoff and Noordhoff.

Ozisik, M. N. (1993). *Heat conduction* (2nd ed.). New York: Wiley.

Tauchert, T. R. (1985). Thermal stresses in plates – Statical problems. In R. B. Hetnarski (Ed.), *Thermal stresses I* (pp. 23–141). Amsterdam: Elsevier.

Vel, S. S., & Batra, R. C. (2003). Three-dimensional analysis of transient thermal stresses in functionally graded plates. *International Journal of Solids and Structures, 40*, 7181–7196.

Yin, H. M., Paulino, G. H., & Buttlar, W. G. (2005). Effective thermal conductivity in two-phase functionally graded particulate composites. *Journal of Applied Physics, 98*, 063704.

Chapter 7
Examples of Advanced Applications: Plates with Piezoelectric Sensors and Actuators and Functionally Graded Plates

While the previous chapters attempt to provide a comprehensive formulation of the theory of plates applied to isotropic and composite material structures, this chapter illustrates representative areas of recent research and development. Naturally, the list of such areas is extensive, including solutions where the response of plate materials is elastic-plastic, viscoelastic or viscoplastic, plates formed of several materials and incorporating inclusions of various shapes, textile composites, shape memory alloy structures, etc. In this chapter we concentrate on two developments, i.e. plates with piezoelectric sensors and actuators and functionally graded material plates. The former area is particularly relevant to thin-walled aerospace structures whose dynamic response can be monitored and controlled by piezoelectric sensors and actuators. It is also applicable to energy harvesting and structural health monitoring systems. Functionally graded plates possess a number of potential advantages over conventional composite plate designs, including elimination of delamination tendencies present in laminated structures, reduced thermal residual stresses and improved thermal response during lifetime, possible enhancement in fatigue and fracture characteristics, etc. Since functionally graded plates are often applied in thermal applications, the discussion is centered on relevant problems.

7.1 Governing Equations for Shear Deformable and Thin Plates with Piezoelectric Layers

In this paragraph, we formulate piezothermoelastic constitutive equations for thin and shear deformable laminated plates including both generally laminated composite as well as piezoelectric layers. The latter layers can serve as either sensors or actuators. The present formulation can also be reduced to a particular case where piezoelectric elements are arranged in "patches," rather than continuous layers.

An example of a plate with piezoelectric layers is found in bimorph actuators and sensors (Fig. 7.1). In a typical bimorph design, three layers forming such actuator

V. Birman, *Plate Structures*, Solid Mechanics and Its Applications 178,

DOI 10.1007/978-94-007-1715-2_7,

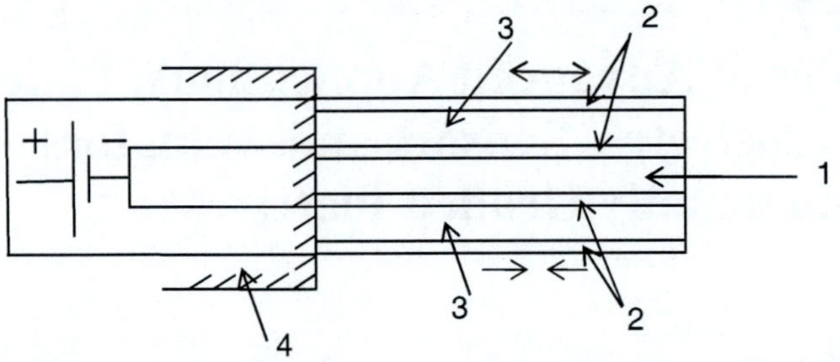

Fig. 7.1 Bimorph design. *1* = metallic substrate, *2* = electrodes, *3* = piezoelectric layers, *4* = base. Electrodes are usually so thin that their contribution to the stiffness can be neglected

include the central thin metallic substrate and two outer piezoceramic layers wired to produce a moment and zero in-plane force in response to the applied voltage. For example, in Fig. 7.1, the upper piezoelectric layer produces tensile force (horizontal arrows), while the lower layer yields an equal compressive force. The outcome is a zero net force and a bending moment equal to the product of the force by the distance between the centroids of piezoelectric layers. Although a bimorph can be modeled as a cantilevered beam structure, a more accurate analysis should represent it as a large aspect ratio plate.

The theory of a piezoelectric layer (or plate) accounting for both applied voltage as well as temperature was applied to the analysis of thin laminates by Tauchert (1992) and extended to first-order shear-deformable plates by Jonnalagadda et al. (1994). The constitutive relations for a three-dimensional piezothermoelastic body (Nye 1957) were reduced to the case of a generally orthotropic layer, relating stresses, strains and electric filed components in a three-dimensional space:

$$\begin{Bmatrix} \sigma_x \\ \sigma_y \\ \sigma_z \\ \tau_{yz} \\ \tau_{xz} \\ \tau_{xy} \end{Bmatrix}_k = \begin{bmatrix} \bar{Q}_{11} & \bar{Q}_{12} & \bar{Q}_{13} & 0 & 0 & \bar{Q}_{16} \\ & \bar{Q}_{22} & \bar{Q}_{23} & 0 & 0 & \bar{Q}_{26} \\ & & \bar{Q}_{33} & 0 & 0 & \bar{Q}_{36} \\ & & & \bar{Q}_{44} & \bar{Q}_{45} & 0 \\ & sym & & & \bar{Q}_{55} & 0 \\ & & & & & \bar{Q}_{66} \end{bmatrix}_k \begin{Bmatrix} \varepsilon_x \\ \varepsilon_y \\ \varepsilon_z \\ \gamma_{yz} \\ \gamma_{xz} \\ \gamma_{xy} \end{Bmatrix}_k$$

$$- \begin{bmatrix} 0 & 0 & \bar{e}_{31} \\ 0 & 0 & \bar{e}_{32} \\ 0 & 0 & \bar{e}_{33} \\ \bar{e}_{14} & \bar{e}_{24} & 0 \\ \bar{e}_{15} & \bar{e}_{25} & 0 \\ 0 & 0 & \bar{e}_{36} \end{bmatrix}_k \begin{Bmatrix} E_x \\ E_y \\ E_z \end{Bmatrix}_K - \begin{Bmatrix} \bar{\lambda}_1 \\ \bar{\lambda}_2 \\ \bar{\lambda}_3 \\ 0 \\ 0 \\ \bar{\lambda}_6 \end{Bmatrix}_k T_k \tag{7.1}$$

where k refers to the layer number within the laminate, z is the poling direction, transformed reduced stiffnesses $\bar{Q}_{ij}$ are defined in Chap. 5, $\bar{e}_{ij}$ are transformed piezoelectric coefficients and $\bar{\lambda}_i$ are transformed stress-temperature coefficients. Temperature T is counted from a thermally stress-free state. While the explicit effect of temperature on the stress is evident in (7.1), its implicit effect that can be quite significant is related to thermally-induced changes of the material constants (this effect will be discussed in the chapter). In general, piezoelectric properties decrease at an elevated temperature. At the Curie temperature that is specified for each piezoelectric material, it undergoes complete depolarization and the loss of the piezoelectric effect. Mechanical stresses also affect the performance of a piezoelectric material.

The above constitutive equations have provided the basis for various plate theories with piezoelectric actuators and sensors which utilize a wide range of kinematic assumptions. For the interested reader, a first review on this topic was provided by Saravanos and Heyliger (1999). Limiting the analysis to classical and first-order theories and accordingly, neglecting normal stresses in the direction perpendicular to the plane of the layer one obtains:

$$\begin{Bmatrix} \sigma_x \\ \sigma_y \\ \tau_{xy} \end{Bmatrix}_k = \begin{bmatrix} \bar{Q}_{11} & \bar{Q}_{12} & \bar{Q}_{16} \\ \bar{Q}_{12} & \bar{Q}_{22} & \bar{Q}_{26} \\ \bar{Q}_{16} & \bar{Q}_{26} & \bar{Q}_{66} \end{bmatrix}_k \begin{Bmatrix} \varepsilon_x \\ \varepsilon_y \\ \gamma_{xy} \end{Bmatrix}_k - \begin{bmatrix} 0 & 0 & \bar{e}_{31} \\ 0 & 0 & \bar{e}_{32} \\ 0 & 0 & \bar{e}_{36} \end{bmatrix}_k \begin{Bmatrix} E_x \\ E_y \\ E_z \end{Bmatrix}_k - \begin{Bmatrix} \bar{\lambda}_1 \\ \bar{\lambda}_2 \\ \bar{\lambda}_6 \end{Bmatrix}_k T_k$$

$$\begin{Bmatrix} \tau_{yz} \\ \tau_{xz} \end{Bmatrix}_k = \begin{bmatrix} \bar{Q}_{44} & \bar{Q}_{45} \\ \bar{Q}_{45} & \bar{Q}_{55} \end{bmatrix}_k \begin{Bmatrix} \gamma_{yz} \\ \gamma_{xz} \end{Bmatrix}_k - \begin{bmatrix} \bar{e}_{14} & \bar{e}_{24} & 0 \\ \bar{e}_{15} & \bar{e}_{25} & 0 \end{bmatrix}_k \begin{Bmatrix} E_x \\ E_y \\ E_z \end{Bmatrix}_k \tag{7.2}$$

Equations 7.1 or 7.2 should be complemented by three equations defining the components of the electric displacement and corresponding to the second equation (1.22) expanded to account for the effect of temperature. Electric displacements in terms of strain, electric field and temperature are (Mindlin 1974; Tzou and Bao 1995):

$$\begin{Bmatrix} D_x \\ D_y \\ D_z \end{Bmatrix}_k = \begin{bmatrix} 0 & 0 & 0 & \bar{e}_{14} & \bar{e}_{15} & 0 \\ 0 & 0 & 0 & \bar{e}_{24} & \bar{e}_{25} & 0 \\ \bar{e}_{31} & \bar{e}_{32} & \bar{e}_{33} & 0 & 0 & \bar{e}_{36} \end{bmatrix}_k \begin{Bmatrix} \varepsilon_x \\ \varepsilon_y \\ \varepsilon_z \\ \gamma_{yz} \\ \gamma_{xz} \\ \gamma_{xy} \end{Bmatrix}_k + \begin{bmatrix} \bar{\varepsilon}_{11} & \bar{\varepsilon}_{12} & 0 \\ \bar{\varepsilon}_{12} & \bar{\varepsilon}_{22} & 0 \\ 0 & 0 & \bar{\varepsilon}_{33} \end{bmatrix}_k \begin{Bmatrix} E_x \\ E_y \\ E_z \end{Bmatrix}_k + \begin{Bmatrix} 0 \\ 0 \\ \bar{p}_3 \end{Bmatrix}_k T_k \tag{7.3}$$

where $\bar{p}_3$ is the pyroelectric constant.

Transformed piezoelectric coefficients, dielectric constants and stress-temperature coefficients in Eqs. 7.1 through 7.3 are given by

$$\begin{aligned}
&\bar{e}_{31}^{(k)} = e_{31}{}^{(k)} c_k^2 + e_{32}{}^{(k)} s_k^2, \quad \bar{e}_{32}^{(k)} = e_{31}{}^{(k)} s_k^2 + e_{32}{}^{(k)} c_k^2, \quad \bar{e}_{36}^{(k)} = \left(e_{31}{}^{(k)} - e_{32}{}^{(k)}\right) c_k s_k \\
&\bar{e}_{14}^{(k)} = \bar{e}_{25}^{(k)} = \left(e_{15}{}^{(k)} - e_{24}{}^{(k)}\right) c_k s_k, \quad \bar{e}_{24}^{(k)} = e_{24}{}^{(k)} c_k^2 + e_{15}{}^{(k)} s_k^2, \quad \bar{e}_{15} = e_{15}{}^{(k)} c_k^2 + e_{24}{}^{(k)} s_k^2, \\
&\bar{e}_{33}^{(k)} = e_{33}{}^{(k)}, \\
&\bar{\varepsilon}_{11}^{(k)} = \varepsilon_{11}{}^{(k)} c_k^2 + \varepsilon_{22}{}^{(k)} s_k^2, \quad \bar{\varepsilon}_{22}^{(k)} = \varepsilon_{11}{}^{(k)} s_k^2 + \varepsilon_{22}{}^{(k)} c_k^2, \\
&\bar{\varepsilon}_{12}^{(k)} = \left(\varepsilon_{11}{}^{(k)} - \varepsilon_{22}{}^{(k)}\right) c_k s_k, \quad \bar{\varepsilon}_{33}^{(k)} = \varepsilon_{33}{}^{(k)}
\end{aligned} \tag{7.4}$$

and

$$\begin{aligned}
&\bar{\lambda}_1^{(k)} = \lambda_1{}^{(k)} c_k^2 + \lambda_2{}^{(k)} s_k^2, \quad \bar{\lambda}_2^{(k)} = \lambda_1{}^{(k)} s_k^2 + \lambda_2{}^{(k)} c_k^2, \quad \bar{\lambda}_3^{(k)} = \lambda_3{}^{(k)}, \\
&\bar{\lambda}_6^{(k)} = \left(\lambda_1{}^{(k)} - \lambda_2{}^{(k)}\right) c_k s_k
\end{aligned} \tag{7.5}$$

The functions $c_k = \cos\theta_k$ and $s_k = \sin\theta_k$ depend on the angle θ_k between the orthotropy coordinate system of the piezoelectric layer and the coordinate system of the plate. The pyroelectric constant $\bar{p}_3 = p_3$ is a property of the material of the layer. The piezoelectric and stress-temperature coefficients in (7.4) and (7.5) are material constants defined in the orthotropy axes of the material that coincide with the layer principal axes (the corresponding terms are shown in these equations without an overbar). The piezoelectric constants in the orthotropy axes are usually available from the manufacturer of the material, while the stress-temperature coefficients are determined in terms of the coefficients of thermal expansion and the elements of the stiffness tensor in the orthotropy axes as

$$\begin{Bmatrix} \lambda_1 \\ \lambda_2 \\ \lambda_3 \end{Bmatrix}_k = \begin{bmatrix} Q_{11} & Q_{12} & Q_{13} \\ & Q_{22} & Q_{23} \\ sym & & Q_{33} \end{bmatrix}_k \begin{Bmatrix} \alpha_1 \\ \alpha_2 \\ \alpha_3 \end{Bmatrix}_k \tag{7.6}$$

Note that Eqs. 7.1–7.3 are concerned with a more general case than a transversely isotropic piezoelectric layer presented in Chap. 1 (see Eqs. 1.22–1.24). This is because they refer to the situation where the principal axes of the layer do not coincide with the coordinate axes of the plate. As first explained by Lee (1990), the reason for such situation may be technological and it is related to polling and rolling of the plate necessary to produce a piezoelectric effect. While the polling direction is perpendicular to the plane of the layer, the rolling direction is parallel to its plane. The resulting stretched orthotropic layer exhibits the symmetry of an orthorhombic class mm2 crystal with respect to the laminate coordinate axes (an example of such material is a PVDF). On the other hand, piezoceramics represent an example of mm6 transversely isotropic materials. In such materials the planes yz and xz that are normal to the x and y axes, respectively, are the mirror planes, while the plane xy

perpendicular to the z-axis is the plane of twofold symmetry. In mm2 piezoelectric materials $e_{24} \neq e_{15}$, while in mm6 materials these coefficients are equal to each other. Temperature does not produce transverse stresses in the planes xz and yz since these are mirror planes for the material. Piezoelectric composite films, with embedded piezoelectric fibers, in the form of active piezoelectric composites developed by Hagood and his colleagues at MIT (e.g., Rodgers and Hagood 1998) and Macro Composite Actuators (developed at NASA Langley), provide another case where the material axes may not coincide with the structural ones.

The stress resultants and stress couples are introduced using the generalization of the first-order shear deformation theory discussed in par. 5.7 to account for the presence of the piezoelectric effect and temperature. Accordingly, the displacements of an arbitrary point of the plate are expressed in terms of the middle-plane displacements and rotations by (1.46). The strain-displacement relations including both in-plane strains and transverse shear components are given by (5.56) and (1.45), respectively. Subsequently, the integration of both the stresses given by (7.2) and the moments of these stresses with respect to the middle plane of the plate yields the stress resultants and stress couples including both piezoelectric as well as thermal contributions:

$$\begin{Bmatrix} N_x \\ N_y \\ N_{xy} \\ M_x \\ M_y \\ M_{xy} \end{Bmatrix} = \begin{bmatrix} A_{11} & A_{12} & A_{16} & B_{11} & B_{12} & B_{16} \\ A_{12} & A_{22} & A_{26} & B_{12} & B_{22} & B_{26} \\ A_{16} & A_{26} & A_{66} & B_{16} & B_{26} & B_{66} \\ B_{11} & B_{12} & B_{16} & D_{11} & D_{12} & D_{16} \\ B_{12} & B_{22} & B_{26} & D_{12} & D_{22} & D_{26} \\ B_{16} & B_{26} & B_{66} & D_{16} & D_{26} & D_{66} \end{bmatrix} \begin{Bmatrix} \varepsilon_x^0 \\ \varepsilon_y^0 \\ \gamma_{xy}^0 \\ \kappa_x \\ \kappa_y \\ \kappa_{xy} \end{Bmatrix} - \begin{Bmatrix} N_x^E \\ N_y^E \\ N_{xy}^E \\ M_x^E \\ M_y^E \\ M_{xy}^E \end{Bmatrix} - \begin{Bmatrix} N_x^T \\ N_y^T \\ N_{xy}^T \\ M_x^T \\ M_y^T \\ M_{xy}^T \end{Bmatrix} \tag{7.7}$$

$$\begin{Bmatrix} Q_y \\ Q_x \end{Bmatrix} = k \begin{bmatrix} A_{44} & A_{45} \\ A_{45} & A_{55} \end{bmatrix} \begin{Bmatrix} \gamma_{yz} \\ \gamma_{xz} \end{Bmatrix} - \begin{Bmatrix} Q_y^E \\ Q_x^E \end{Bmatrix}$$

where piezoelectric and thermal terms are identified by superscripts "E" and "T", respectively. These terms are defined by

$$\{N_x^E,\ M_x^E\} = \int_{-\frac{h}{2}}^{\frac{h}{2}} \bar{e}_{31} E_z \{1,\ z\}\, dz, \qquad \{N_y^E,\ M_y^E\} = \int_{-\frac{h}{2}}^{\frac{h}{2}} \bar{e}_{32} E_z \{1,\ z\}\, dz,$$

$$\{N_{xy}^E,\ M_{xy}^E\} = \int_{-\frac{h}{2}}^{\frac{h}{2}} \bar{e}_{36} E_z \{1,\ z\}\, dz, \qquad \begin{Bmatrix} Q_y^E \\ Q_x^E \end{Bmatrix} = \int_{-\frac{h}{2}}^{\frac{h}{2}} \begin{bmatrix} \bar{e}_{14} & \bar{e}_{24} \\ \bar{e}_{15} & \bar{e}_{25} \end{bmatrix} \begin{Bmatrix} E_x \\ E_y \end{Bmatrix} dz,$$

$$\left\{N_x^T,\ M_x^T\right\} = \int_{\frac{-h}{2}}^{\frac{h}{2}} \bar{\lambda}_1 T\{1,\ z\}\, dz, \quad \left\{N_y^T,\ M_y^T\right\} = \int_{-\frac{h}{2}}^{\frac{h}{2}} \bar{\lambda}_2 T\{1,\ z\}\, dz,$$

$$\left\{N_{xy}^T,\ M_{xy}^T\right\} = \int_{-\frac{h}{2}}^{\frac{h}{2}} \bar{\lambda}_6 T\{1,\ z\}\, dz, \tag{7.8}$$

Note that while the integration in (7.7) has to be conducted throughout the thickness of the laminated plate, the terms associated with the electric field account only for the contribution of piezoelectric layers (accordingly, in these terms the limits of integration encompass only these layers).

The equations of motion in terms of stress resultants and stress couples are not affected by the presence of piezoelectric and thermal terms that are incorporated in their expressions according to (7.7). Therefore, one can use equations (5.62) with the boundary conditions specified according to Sect. 5.7. Further details on the computational analysis of shear deformable plates with piezoelectric layers can be found in the papers of Chandrashekhara and Agarwal (1993), Jonnalagadda et al. (1994) and other relevant publications.

The simply supported laminate considered by Jonnalagadda et al. (1994) consisted of eight cross-ply graphite-epoxy T300/5208 layers $\left[0^0/90^0/0^0/90^0\right]_2$ and a double-thickness polyvinylidene fluoride (PVDF) layer at the lower surface. The latter layer served as the actuator, the applied electric field being $E_z = 30 \times 10^6\ \frac{V}{m}$. The comparison with the classical thin plate theory solution presented below relies on the results obtained by the latter theory that is also described in this paragraph.

Representative results for the nondimensional static deflection (deflection at the center divided by the thickness) for a square plate subject to a uniform pressure are shown in Fig. 7.2. As follows from this figure, the first order theory becomes significantly different from the classical theory of thin plates at the side-to-thickness ratio smaller than 25. In general, it can be shown that the trends found for shear deformable plates without piezoelectric elements remain valid in the presence of these elements. In other words, transverse shear deformability should be accounted for only if the plates are relatively "thick." However, available piezoelectric actuator materials limit the magnitude of the electric field to between 500 and 1,000 V/mm (Morgan Matroc, http://www.morganelectroceramics.com/tutorials/piezoguide11.html). Accordingly, the voltage necessary to control relatively rigid thick structures is often unrealistic. This means that shear-deformable plates cannot be effectively controlled by piezoelectric actuators employing presently available materials. However, piezoelectric sensors are not affected by such limitations, i.e. it is quite feasible to use them in shear deformable composites, including sandwich structures.

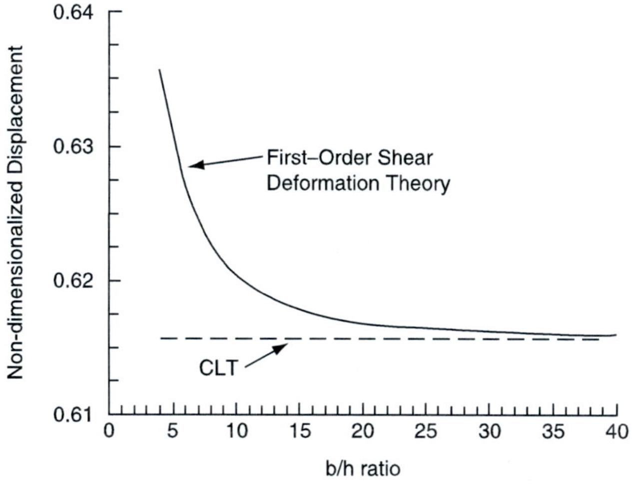

Fig. 7.2 Nondimensional deflection of a square laminate ($a = b = 300\ mm$) generated by piezoelectric actuation evaluated by the first order shear deformation and classical thin plate theories (From Jonnalagadda et al. 1994)

Considering flexible (thin) piezoelectric composites, the constitutive equations given by the first three equations (7.2) and the in-plane stress resultants and stress couples (7.7) are still applicable. Both sensor as well as actuator formulations for thin laminated piezoelectric plates were presented by Lee (1990), without accounting for the effect of temperature. Here we illustrate the corresponding formulations for vibration problems of thin laminates with piezoelectric layers. Solutions of static problems can be obtained by the corresponding simplifications of equations introduced below.

Equations of motion of a thin laminated plate with piezoelectric layers in terms of stress couples and stress resultants are identical to (1.84). In dynamic problems, the explicit effect of a constant temperature on vibrations is present only if deformations are sufficiently large to justify using the geometrically nonlinear analysis (of course, the implicit effect reflected in thermally-induced changes of the material properties is present in both geometrically linear and nonlinear formulations). If the problem is linear, the substitution of the linear versions of strain-displacement relations (1.28), (1.29) yields

$$\begin{bmatrix} L_{11} & L_{12} & L_{13} \\ L_{21} & L_{22} & L_{23} \\ L_{31} & L_{32} & L_{33} \end{bmatrix} \begin{Bmatrix} u^0 \\ v^0 \\ w \end{Bmatrix} = \begin{Bmatrix} \dfrac{\partial N_x^E}{\partial x} + \dfrac{\partial N_{xy}^E}{\partial y} \\ \dfrac{\partial N_{xy}^E}{\partial x} + \dfrac{\partial N_y^E}{\partial y} \\ -\left(\dfrac{\partial^2 M_x^E}{\partial x^2} + 2\dfrac{\partial^2 M_{xy}^E}{\partial x \partial y} + \dfrac{\partial^2 M_y^E}{\partial y^2} \right) \end{Bmatrix} + \begin{Bmatrix} -\hat{m}\ddot{u} \\ -\hat{m}\ddot{v} \\ p(x, y, t) - \hat{m}\ddot{w} \end{Bmatrix} \tag{7.9}$$

where linear differential operators in the left side are identical to those in the case of generally laminated composite plates and in-plane inertial terms are added to the first two equations. Contrary to "conventional" laminates, in a number of applications involving piezoelectric materials, such as a unimorph, the layers are often asymmetric about the middle plane of the plate. Accordingly, the corresponding operators are presented here for a general case:

$$
\begin{aligned}
L_{11}(...) &= A_{11}\frac{\partial^2(...)}{\partial x^2} + 2A_{16}\frac{\partial^2(...)}{\partial x \partial y} + A_{66}\frac{\partial^2(...)}{\partial y^2} \\
L_{12}(...) = L_{21}(...) &= A_{16}\frac{\partial^2(...)}{\partial x^2} + (A_{12}+A_{66})\frac{\partial^2(...)}{\partial x \partial y} + A_{26}\frac{\partial^2(...)}{\partial y^2} \\
L_{13}(...) = L_{31}(...) &= -B_{11}\frac{\partial^3(...)}{\partial x^3} - 3B_{16}\frac{\partial^3(...)}{\partial x^2 \partial y} - (B_{12}+2B_{66})\frac{\partial^3(...)}{\partial x \partial y^2} \\
&\quad - B_{26}\frac{\partial^3(...)}{\partial y^3} \\
L_{22}(...) &= A_{66}\frac{\partial^2(...)}{\partial x^2} + 2A_{26}\frac{\partial^2(...)}{\partial x \partial y} + A_{22}\frac{\partial^2(...)}{\partial y^2} \\
L_{23}(...) = L_{32}(...) &= -B_{16}\frac{\partial^3(...)}{\partial x^3} - (B_{12}+2B_{66})\frac{\partial^3(...)}{\partial x^2 \partial y} - 3B_{26}\frac{\partial^3(...)}{\partial x \partial y^2} \\
&\quad - B_{22}\frac{\partial^3(...)}{\partial y^3} \\
L_{33}(...) &= D_{11}\frac{\partial^4(...)}{\partial x^4} + 4D_{16}\frac{\partial^4(...)}{\partial x^3 \partial y} + 2(D_{12}+2D_{66})\frac{\partial^4(...)}{\partial x^2 \partial y^2} + 4D_{26}\frac{\partial^4(...)}{\partial x \partial y^2} \\
&\quad + D_{22}\frac{\partial^4(...)}{\partial y^4}
\end{aligned} \tag{7.10}
$$

A nonlinear version of equations (7.10) is available (e.g., Reddy 2004). Note that the contribution of in-plane inertial terms $\hat{m}\ddot{u}$ and $\hat{m}\ddot{v}$ in the right side of the first two equations (7.9) are often negligible compared with the transverse inertial term $\hat{m}\ddot{w}$. Accordingly, these terms are omitted in many solutions.

Boundary conditions are identical to those considered for thin plates, without accounting for piezoelectric and thermal effects (e.g., (1.88)). However, if these conditions are formulated in terms of stress resultants and stress couples, they include piezoelectric and thermal terms according to the constitutive relations (7.7). In a linear dynamic problem, static thermal terms do not explicitly affect the boundary conditions, i.e. static thermoelastic and dynamic piezoelastic boundary conditions are uncoupled.

In the problems where piezoelectric elements operate in the actuator mode, the voltage is usually applied in the z-direction, so that $E_x = E_y = 0,\ E_z = \frac{V}{h_k}$ where V is the voltage and h_k is the thickness of the actuator (k-th layer). If a piezoelectric layer operates as a sensor in the closed circuit, the applied electric field is equal to zero and the charge accumulated on the electrode bonded to the surface A_k of the piezoelectric layer is given by (Lee 1990):

$$\bar{q} = \int_{A_k} \left(e_{31}^{(k)} \varepsilon_x^{(k)} + e_{32}^{(k)} \varepsilon_y^{(k)} + e_{36}^{(k)} \gamma_{xy}^{(k)} \right) dA_k \tag{7.11}$$

Note that piezoelectric sensors are usually employed to measure dynamic processes since a constant charge produced by a constant mechanical load would yield a decreasing signal as a result of the losses in the circuit. The electric charge$\bar{q}$ is related to the current passing through the electrode by $i = \frac{\partial \bar{q}}{\partial t}$.

Although piezoelectric sensors take advantage of the direct piezoelectric effect which mandates generation of electric charge in response to an applied strain field, in many practical applications voltage generated in the sensor is the preferred output. This requires us to operate a sensor in the open circuit so that the total charge flux over the area of the electrode can be assumed equal to zero. This implies that electric charge is accumulated on the sensor electrodes, and voltage is generated in the sensor due to capacitor effect, which can be determined from (Birman 1996)

$$V = -\frac{1}{A_k} \int_{A_k} \int_{h_k} E_z^{(k)} dz dA_k \tag{7.12}$$

Note that the electric field in the sensor can be found from the third equation (7.3) where the electric displacement is equal to zero.

7.2 Thin Plates with Piezoelectric Sensors and Actuators

In this paragraph we present the solution of the problem of free and forced vibrations of a thin cross-ply laminated simply supported plate including transversely isotropic piezoelectric layers that serve as sensors and/or actuators. The piezoelectric and other layers are not necessarily symmetric about the middle plane.

The behavior of an asymmetric cross-ply laminated plate was analyzed by Tauchert (1992) and Lam and Ng (1999). In this case $A_{i6} = B_{i6} = D_{i6} = 0$ $(i = 1, 2)$ since $Q_{16} = Q_{26} = \bar{Q}_{16} = \bar{Q}_{26} = 0$ for all composite and piezoelectric layers. Furthermore, in-plane shear piezoelectric stress resultant and couple are $N_{xy}^E = M_{xy}^E = 0$ since $e_{36} = \bar{e}_{36} = 0$ for all piezoelectric layers.

The exact solution can be obtained for the boundary conditions given in (2.69) where the stress resultants and stress couples incorporate piezoelectric contributions according to (7.7). Such conditions correspond to the case where the plate is supported by boundary structures that are "infinitely rigid" in their plane but provide negligible rotational or out-of-plane stiffness. The absence of the in-plane stress resultant perpendicular to the boundary implies that there no adjacent plates and the bending stiffness of the boundary structure in the plane of the plate can be disregarded.

The solution of equations of motion (7.9) is obtained using double Fourier series (4.38) that satisfy all kinematic boundary conditions. These series are reproduced here in the modified form, accounting for the fact that the motion may not be periodic in time:

$$
\begin{aligned}
u &= \sum_m \sum_n U_{mn}(t) \cos \alpha_m x \sin \beta_n y \\
v &= \sum_m \sum_n V_{mn}(t) \sin \alpha_m x \cos \beta_n y \\
w &= \sum_m \sum_n W_{mn}(t) \sin \alpha_m x \sin \beta_n y
\end{aligned} \tag{7.13}
$$

The static boundary conditions as well as the equations of motion can be satisfied by representing the load and non-zero applied piezoelectric components in terms of double Fourier series:

$$
\begin{aligned}
p\,(x, y, t) &= \sum_m \sum_n p_{mn}(t) \sin \alpha_m x \sin \beta_n y \\
\left\{N_x^E, N_y^E, M_x^E, M_y^E\right\} &= \sum_m \sum_n \left\{N_x^{(mn)}(t), N_y^{(mn)}(t), M_x^{(mn)}(t), M_y^{(mn)}(t)\right\} \\
&\quad \times \sin \alpha_m x \sin \beta_n y
\end{aligned} \tag{7.14}
$$

The substitution of series (7.13) and (7.14) into (7.9) yields uncoupled systems of ordinary differential equations. For the mn-harmonics, the corresponding system is

$$
\begin{bmatrix} \hat{m} & 0 & 0 \\ 0 & \hat{m} & 0 \\ 0 & 0 & \hat{m} \end{bmatrix}
\begin{Bmatrix} \ddot{U}_{mn} \\ \ddot{V}_{mn} \\ \ddot{W}_{mn} \end{Bmatrix}
+ \begin{bmatrix} k_{11} & k_{12} & k_{13} \\ k_{12} & k_{22} & k_{23} \\ k_{13} & k_{23} & k_{33} \end{bmatrix}_{mn}
\begin{Bmatrix} U_{mn} \\ V_{mn} \\ W_{mn} \end{Bmatrix}
= \begin{Bmatrix} 0 \\ 0 \\ p_{mn}(t) \end{Bmatrix}
+ \begin{Bmatrix} \alpha_{mn} N_x^{(mn)}(t) \\ \beta_{mn} N_y^{(mn)}(t) \\ \alpha_{mn}^2 M_x^{(mn)}(t) + \beta_{mn}^2 M_y^{(mn)}(t) \end{Bmatrix} \tag{7.15}
$$

where $k_{ij}^{(mn)} = S_{ijmn}\,(r, k)$ given by Eqs. 2.72.

Equations 7.15 represent a system of ordinary differential equations, time being the independent variable. The solution of these equations can be conducted by standard methods of the theory of vibrations.

If the plate experiences forced vibrations and piezoelectric layers act as sensors operating in a closed circuit, the last vector in the right side of (7.15) is equal to zero. Then the deflections and strains are determined from the solution of accordingly simplified equations of motion that depend on piezoelectric layers only to the extent of the contribution of these layers to the stiffness. Subsequently, using the strains found from the solution in conjunction with equation (7.11), we can determine the signal from the sensors. If the piezoelectric layers are employed as sensors in the open circuit, the electric displacement is equal to zero and the electric field in the direction perpendicular to the surface of the plate can be determined similar to the approach shown in Sect. 7.4. Note that in this case piezoelectric terms in the right side of system (7.15) are not equal to zero. If the contribution of these terms is accounted for (sometimes, it can be neglected), the solution may require an iterative approach. Such approach would involve finding the strains by assumption of zero electric field, specifying the field in the sensors based on these strains, updating the strains to account for the effect of this electric field, etc. A detailed analysis along the lines outlined here is outside the scope of the book.

If piezoelectric layers are employed as actuators, the corresponding terms in the last vector in the right side of (7.15) can be found as functions of the applied voltage. The effect of the applied voltage on the motion of the plate is subsequently determined from the solution of (7.15). On the other hand, piezoelectric sensors and actuators are often used in the same structure forming a feedback system where the voltage supplied to the actuators is a function of the signal generated in the sensors. For example,

$$V(t) = G_q q(t) + G_i i(t) \tag{7.16}$$

represents a constant gain feedback in a closed-loop control system where G_q and G_i are gains. A constant-gain negative-velocity feedback corresponding to a negative value of G_i produces an effect similar to viscous damping.

While using piezoelectric layers as a part of the laminated plate is feasible, an alternative solution is based on using piezoelectric patches that are also referred to as "distributed" sensors and actuators (Fig. 7.3). Such problems have been addressed by a number of investigators, the pioneering work being published by Chandrahsekhara and Agarwal (1993) and Tzou and Fu (1994a,b).

It can sometimes be assumed that the thickness of piezoelectric patches is small as compared to the thickness of the plate enabling us to neglect the contribution of the patches to the bending stiffness. In symmetric configurations, in-plane and transverse displacements are uncoupled and the equation of motion obtained from the third equation (7.9) becomes

$$\hat{m}\ddot{w} + L_{33}(w) = p(t) + \frac{\partial^2 M_x^E}{\partial x^2} + \frac{\partial^2 M_y^E}{\partial y^2} \tag{7.17}$$

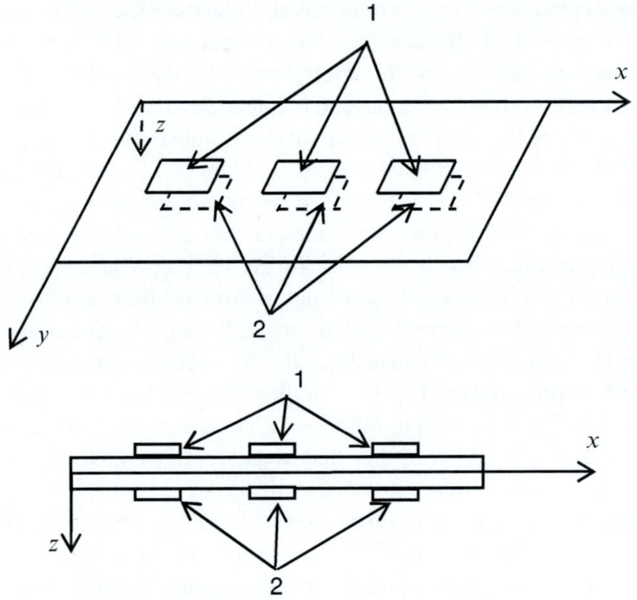

Fig. 7.3 A plate with distributed sensors and actuators (piezoelectric patches). *1* = distributed actuators, *2* = distributed sensors

Tzou and Fu (1994) represented control moments in a modal expansion form, i.e.

$$M_x^E = \sum_m \sum_n \eta_{mn}(t) z_a d_{31} E_a G V$$

$$M_y^E = \sum_m \sum_n \eta_{mn}(t) z_a d_{32} E_a G V \tag{7.18}$$

where η_{mn} is the modal participation factor, z_a is a distance from the middle plane of the plate to the middle plane of the actuator, d_{3i} is the piezoelectric strain constant representing the strain in the i-th direction generated per unit electric field applied in direction 3 (z-direction), E_a is the modulus of elasticity of the actuator and G is a gain. The feedback gain considered in this paper included displacement and velocity feedback controls. This technique can be adopted to incorporate a number of arbitrary located and shaped actuators. The analysis of the response of the plate described by (7.17) and (7.18) can be conducted using double Fourier series as shown in the paper of Tzou and Fu (1994).

Chandrashekhara and Agarwal (1993) considered a reduction in the amplitude of the transient response of a clamped symmetrically laminated cross-ply AS/3601-6 graphite-epoxy plate with collocated patch sensors and actuators. The square plate had a side length of 254 mm and the length-to-thickness ratio equal to 100 (the results for plates with such geometry generated by the first order and classical plate theories are practically identical). Piezoceramic patches (PZT G1195) were attached at the top and bottom surfaces of the plate. The plate was subject to a constant uniformly distributed pressure equal to 25 kPa and lasting for $1.6 \times 10^{-3}\,s$.

As follows from the paper, a negative-velocity feedback gain results in a much quicker damping out of the response. In particular, using the feedback characterized by a ratio of the voltage in actuators to current in the sensors equal to −250V/A, the amplitudes of vibrations were reduced to a prescribed level in about half time compared to that in identical plates without active control.

7.3 Active Control of Composite Plates Using Piezoelectric "Stiffeners-Actuators"

This paragraph illustrates the application of smart stiffeners (stringers) to active control of structures. This approach was considered for composite plates (Birman 1993), sandwich plates (Birman 1994), and spherical caps (Birman et al. 1999). Two types of smart stiffeners have been considered, i.e. piezoelectric and shape memory alloy stiffeners. In this book we analyze the first class, i.e. piezoelectric stringers operating in the active mode.

The concept of smart stiffeners originated from the ideas employed in reinforced structures where stiffeners provide an enhanced stiffness and strength with a relatively small additional weight. In smart stiffeners the added stiffness is less important than active control forces and/or moments applied by the stiffeners to the structure. The motivation for the use of active stiffeners instead of continuous piezoelectric layers is related to a potential convenience of installation, operation and inspection as well as a possible strategic placement of stiffeners over the surface of the plate maximizing the efficiency of active control applying it at the critical locations. In addition, placing piezoelectric elements at a larger distance from the plate middle surface than that available in the case of piezoelectric layers and patches provides a higher active bending moment due to a larger eccentricity.

Consider a composite plate reinforced by a system of stiffeners (Fig. 7.4). The stiffeners can be fully piezoelectric or they can be constructed from composite and piezoelectric components. In the latter case, the function of composite layers is to increase the eccentricity of the piezoelectric component, rather than the contribution to the plate stiffness. A very stiff composite component would actually be counterproductive for active control increasing local stiffness and neutralizing the beneficial effect of piezoelectrically-generated control forces. Accordingly, as shown in Fig. 7.4, it is desirable to use a light web joining the plate and a piezoelectric layer. The piezoelectric layer should be located at a maximum distance from the middle plane of the plate to increase the eccentricity of the piezoelectric force and maximize the control moment.

Equations of motion of the plate in terms of stress resultants and stress couples are (1.84) where the effect of in-plane inertia is neglected. The analysis resembles

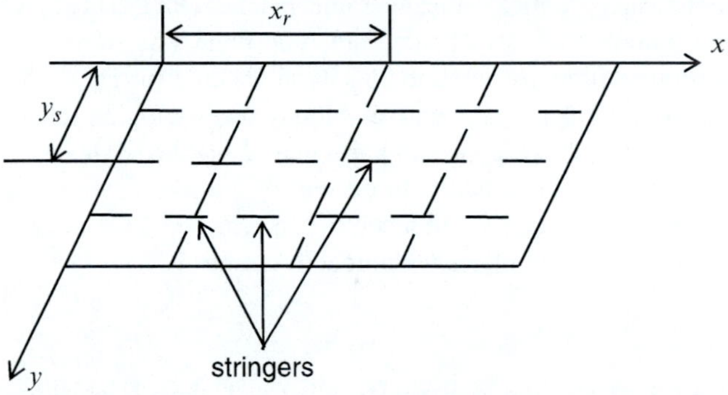

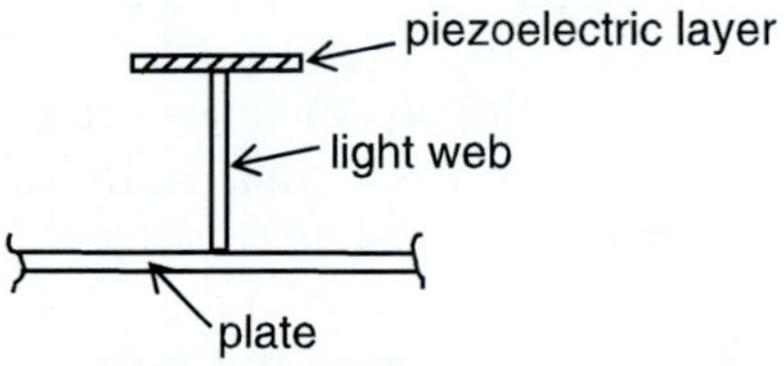

Fig. 7.4 Plate with a system of stiffeners. The stiffeners include a piezoelectric layer and a light composite web as is shown in the cross section

that of composite plates reinforced by stringers in Sect. 5.6, with additional peculiarities related to active piezoelectric control exercised through the stiffeners. The stiffeners having an open profile, their torsional stiffness is neglected. The plate reinforced by stiffeners capable of an efficient active control is thin, so that transverse shear deformability can be omitted and the analysis conducted by the classical plate theory. The stress resultants and stress couples reflect the contribution of the stiffeners that can be evaluated in the manner similar to that the previous chapters.

If the stiffeners are heterogeneous, combining piezoelectric and composite components, the mass of the plate per unit area is

$$\hat{m} = m_p + \sum_r \delta\left(x - x_r\right) \int_y \int_z \rho_r\left(y, z\right) dzdy + \sum_s \delta\left(y - y_s\right) \int_x \int_z \rho_s\left(x, z\right) dzdx \tag{7.19}$$

where m_p is the mass of the unreinforced plate (skin), $\rho_r\left(y, z\right)$ and $\rho_s\left(x, z\right)$ are local mass densities of the stiffeners oriented along the y- and x-axes, respectively, and the integration is conducted over the cross section of the corresponding stiffener. The integration is shown both over the width and the depth coordinates to emphasize that the width of piezoelectric and composite components may differ as is shown in Fig. 7.4.

The stresses in the piezoelectric layers (actuators) of the stiffeners oriented along the x- and y-axes account for the piezoelectric contribution due to the voltage component in the z (thickness) direction:

$$\sigma_x^{(a)} = E_a \left(u^0{,}_x + \frac{1}{2} w{,}_x^2 - z_{sa} w{,}_{xx} - \frac{d_{31} V}{h_{sa}} \right)$$

$$\sigma_y^{(a)} = E_a \left(v^0{,}_y + \frac{1}{2} w{,}_y^2 - z_{ra} w{,}_{yy} - \frac{d_{31} V}{h_{ra}} \right) \quad (7.20)$$

where z_{sa} and z_{ra} are distances from the plate middle plane to the centroids of the corresponding piezoelectric layers, and h_{sa} and h_{ra} are the thicknesses of these layers. Note that the strains in (7.20) include geometrically nonlinear contributions. In the composite component of the stiffener the last terms in the right side of the corresponding equation (7.20) should be omitted and the modulus of elasticity and the distance from the middle plane modified accordingly.

The stress resultants and stress couples can now be evaluated accounting for two contributions, i.e. the contribution of the skin of the plate without stiffeners and the contribution associated with the stiffeners. In particular, if the skin is symmetric cross laminated or if it is composed of multiple symmetric generally orthotropic layers, the corresponding stress resultants and stress couples are given by (5.10). In the general lamination case, the corresponding constitutive equations for the skin are (5.8). The terms contributed by the stiffeners are obtained by integrating the stresses given by (7.20) and the moments of these stresses with respect to the plate middle plane over the cross sections of the corresponding stiffeners:

$$N_x^{(s)} = \sum_s \delta(y - y_s) \left[Q_s \left(u^0{,}_x + \frac{1}{2} w{,}_x^2 \right) - F_s w{,}_{xx} - E_a A_{sa} \frac{d_{31} V}{h_{sa}} \right]$$

$$M_x^{(s)} = \sum_s \delta(y - y_s) \left[F_s \left(u^0{,}_x + \frac{1}{2} w{,}_x^2 \right) - I_s w{,}_{xx} - E_a F_{sa} \frac{d_{31} V}{h_{sa}} \right]$$

$$N_y^{(s)} = \sum_r \delta(x - x_r) \left[Q_r \left(v^0{,}_y + \frac{1}{2} w{,}_y^2 \right) - F_r w{,}_{yy} - E_a A_{ra} \frac{d_{31} V}{h_{ra}} \right]$$

$$M_y^{(s)} = \sum_r \delta(x - x_r) \left[F_r \left(v^0{,}_y + \frac{1}{2} w{,}_y^2 \right) - I_r w{,}_{yy} - E_a F_{ra} \frac{d_{31} V}{h_{ra}} \right] \quad (7.21)$$

where

$$\{ Q_s, F_s, I_s \} = \int_{A_s} E(z) \{ 1, z, z^2 \} dA_s$$

$$\{ Q_r, F_r, I_r \} = \int_{A_r} E(z) \{ 1, z, z^2 \} dA_r$$

$$\{A_{sa},\ F_{sa}\} = \int\limits_{A_{sa}} \{1,\ z\}\, dA_{sa}$$

$$\{A_{ra},\ F_{ra}\} = \int\limits_{A_{ra}} \{1,\ z\}\, dA_{ra} \tag{7.22}$$

In the first two equations (7.22), the integration is conducted over the cross sectional area of stiffeners oriented in the x- and y-directions, respectively (A_s *and* A_r). The modulus of elasticity in these equations refers to the modulus of composite or piezoelectric components located at a distance z from the plate middle plane. The last two equations (7.22) account for the piezoelectric contribution. Accordingly, the integration is conducted only over the cross sectional areas of the piezoelectric layers (A_{sa} *and* A_{ra}). As follows from (7.21), it is possible to account for a variable spacing of the stiffeners, i.e. $x_r(x),\ \ y_s(y)$, so that their effect can be maximized by applying active control at the strategically selected parts of the plate.

Similarly to the analysis of isotropic plates discussed in Sect. 2.9, in the case where stiffeners are identical and closely spaced, the Dirac functions can be replaced with the inverse of the spacing of the stiffeners according to (2.64), enabling the application of the smeared stiffeners technique.

The stress resultants and couples given by (7.21) are added to their counterparts contributed by the skin of the plate. The total stress resultants and couples are substituted into the equations of motion (1.84) yielding the system of equations:

$$\begin{bmatrix} L_{11} & L_{12} & L_{13} \\ L_{21} & L_{22} & L_{23} \\ L_{31} & L_{32} & L_{33} \end{bmatrix} \begin{Bmatrix} u^0 \\ v^0 \\ w \end{Bmatrix} + \begin{bmatrix} 0 & 0 & K_{13}(w) \\ 0 & 0 & K_{23}(w) \\ K_{31}(u,\ w) & K_{32}(v,\ w) & K_{33}(w) \end{bmatrix} + \begin{Bmatrix} 0 \\ 0 \\ N_3(w) \end{Bmatrix}$$

$$= \begin{Bmatrix} 0 \\ 0 \\ \hat{m}\ddot{w} \end{Bmatrix} + \begin{Bmatrix} P_1 \\ P_2 \\ P_3 \end{Bmatrix} V + \begin{Bmatrix} 0 \\ 0 \\ P(w,\ V) \end{Bmatrix} \tag{7.23}$$

where linear differential operators L_{ij} are identical with the corresponding operators in reinforced plates, including the contributions associated with the stiffness of the stiffeners. These operators are presented here for a general case where the constitutive relations for the skin are given by (5.8):

$$L_{11} = \bar{A}_{11}\frac{\partial^2}{\partial x^2} + 2A_{16}\frac{\partial^2}{\partial x \partial y} + A_{66}\frac{\partial^2}{\partial y^2}$$

$$L_{12} = L_{21} = A_{16}\frac{\partial^2}{\partial x^2} + (A_{12} + A_{66})\frac{\partial^2}{\partial x \partial y} + A_{26}\frac{\partial^2}{\partial y^2}$$

$$L_{13} = -L_{31} = -\left[\bar{B}_{11}\frac{\partial^3}{\partial x^3} + (B_{12} + 2B_{66})\frac{\partial^3}{\partial x \partial y^2} + 3B_{16}\frac{\partial^3}{\partial x^2 \partial y} + B_{26}\frac{\partial^3}{\partial y^3}\right]$$

$$L_{22} = A_{66}\frac{\partial^2}{\partial x^2} + 2A_{26}\frac{\partial^2}{\partial x \partial y} + \bar{A}_{22}\frac{\partial^2}{\partial y^2}$$

$$L_{23} = -L_{32} = -\left[B_{16}\frac{\partial^3}{\partial x^3} + (B_{12} + 2B_{66})\frac{\partial^3}{\partial x^2 \partial y} + 3B_{26}\frac{\partial^3}{\partial x \partial y^2} + \bar{B}_{22}\frac{\partial^3}{\partial y^3}\right]$$

$$L_{33} = -\left[\bar{D}_{11}\frac{\partial^4}{\partial x^4} + 4D_{16}\frac{\partial^4}{\partial x^3 \partial y} + 2(D_{12} + 2D_{66})\frac{\partial^4}{\partial x^2 \partial y^2} + 4D_{26}\frac{\partial^4}{\partial x \partial y^3} + \bar{D}_{22}\frac{\partial^4}{\partial y^4}\right] \quad (7.24)$$

In isotropic, cross-ply laminated and multilayered angle-ply symmetrically laminated plates these operators are simplified since $A_{16} = A_{26} = B_{16} = B_{26} = D_{16} = D_{26} = 0$. In the absence of the stiffeners these operators are reduced to the case of thin laminated plate.

The extensional, coupling and bending stiffness terms with an overbar in (7.24) include the contribution of the stiffeners:

$$\bar{A}_{11} = A_{11} + \sum_s \delta(y - y_s) Q_s, \quad \bar{A}_{22} = A_{22} + \sum_r \delta(x - x_r) Q_r$$

$$\bar{B}_{11} = B_{11} + \sum_s \delta(y - y_s) F_s, \quad \bar{B}_{22} = B_{22} + \sum_r \delta(x - x_r) F_r$$

$$\bar{D}_{11} = D_{11} + \sum_s \delta(y - y_s) I_s, \quad \bar{D}_{22} = D_{22} + \sum_r \delta(x - x_r) I_r \quad (7.25)$$

Nonlinear functions and operators K_{ij}, N_3 in the left side of (7.23) are presented in the paper of Birman (1993).

Differential operators in the right side of (7.23) are

$$P_1 = \sum_s \delta(y - y_s) R_1 \frac{\partial}{\partial x}$$

$$P_2 = \sum_r \delta(x - x_r) R_2 \frac{\partial}{\partial y}$$

$$P_3 = \sum_s \delta(y - y_s) R_1 z_{sa}\frac{\partial^2}{\partial x^2} + \sum_r \delta(x - x_r) R_2 z_{ra}\frac{\partial^2}{\partial y^2}$$

$$P(w, V) = \sum_s \delta(y - y_s) R_1 \frac{\partial(Vw_{,x})}{\partial x} + \sum_r \delta(x - x_r) R_2 \frac{\partial(Vw_{,y})}{\partial y} \quad (7.26)$$

In the above equations,

$$R_1 = E_a A_{sa}\frac{d_{31}}{h_{sa}}, \qquad R_2 = E_a A_{ra}\frac{d_{31}}{h_{ra}} \quad (7.27)$$

The solution of several representative dynamic problems using active control by piezoelectric stiffeners was discussed by Birman (1993).

As is obvious from (7.25), in the presence of stiffeners, the plate that is symmetrically laminated has nonzero coupling stiffness $\bar{B}_{11}$, $\bar{B}_{22}$ attributed to the contribution of the stiffeners. The analysis of equations (7.23) in the general case where the plate reinforced by piezoelectric stiffeners is asymmetric can be conducted analytically in the case where the problem is linear and the simple support is specified according to (2.69). Then the solution could be sought as an extension of the approach in the previous paragraph. The coupling stiffness disappears if identical stiffeners are bonded to the opposite surfaces of the plate preserving a symmetric design. If the stiffeners are symmetric about the middle plane linearized equations of motion (7.23) are reduced to a single equation for transverse vibrations:

$$\hat{m}\ddot{w} + \bar{D}_{11}\frac{\partial^4 w}{\partial x^4} + 2\left(D_{12} + 2D_{66}\right)\frac{\partial^4 w}{\partial x^2 \partial y^2} + \bar{D}_{22}\frac{\partial^4 w}{\partial y^4} + P_3(V) + P\left(w, V\right) = p\left(x, y, t\right) \tag{7.28}$$

Equation 7.28 is applicable to the analysis of forced motion as well as free vibrations (in the latter case, the driving load $p\left(x, y, t\right) = 0$). The analysis can be conducted by standard methods of the theory of vibrations of plates.

If identical stiffeners bonded to the opposite plate surfaces are driven by out-of-phase voltage, the active control can be very effective. In such case, the force applied to the plate is equal to zero, while the active moment is equal to a product of the force generated in each stiffener and the distance between the centroids of the piezoelectric components of the stiffeners bonded to the opposite surfaces of the plate. This situation resembles the operation of a bimorph shown in Fig. 7.1.

7.4 Effect of Temperature on Measurements Obtained from Piezoelectric Sensors

Temperature affects the response of composite plates incorporating piezoelectric elements through the following effects:

(a) Direct influence of temperature on the properties of the piezoelectric material (sensor or actuator);
(b) Effect of temperature on the properties of other materials constituting the plate (for example, reduction of elastic properties of the polymer matrix in a composite material, as the temperature increases towards the glass transition point);
(c) Thermally induced stresses;
(d) Thermally induced electric charge (pyroelectric effect).

In this section we concentrate on the effect of temperature on the properties of piezoelectric sensors and the related issue of the interpretation of readings from such sensors. Birman (1996) considered this problem for open-circuit sensors incorporated in thin composite plates. In such formulation the sensor is in the state of plane stress, i.e. the stresses acting perpendicular to its plane z = constant are equal to zero. Furthermore, the sensor is transversely isotropic in the plane z = constant and polarized in the thickness z-direction. It is also assumed that the problem is both geometrically and physically linear. The sensor being a relatively small patch bonded to the plate, its effect on the overall plate stiffness is usually negligible.

In some situations it may be necessary to consider the influence of the converse piezoelectric effect, i.e. the impact of the electric field generated in the sensor on the response of the structure. Additionally, the influence of the bonding layer could become essential, particularly in the vicinity to the edges of a piezoelectric patch (sensor) bonded to the composite substrate due to a local concentration of three-dimensional stresses that could cause debonding of the sensor. These complicating factors are not included in the present analysis.

The governing equations (7.2) and (7.3) for a transversely isotropic piezoelectric sensor in the state of plane stress reduce to (Birman 1996)

$$\begin{Bmatrix} \sigma_x \\ \sigma_y \end{Bmatrix} = \begin{bmatrix} Q_{11} & Q_{12} \\ Q_{12} & Q_{11} \end{bmatrix} \begin{Bmatrix} \varepsilon_x - \alpha_1 T - d_{31} E_z \\ \varepsilon_y - \alpha_1 T - d_{31} E_z \end{Bmatrix}$$

$$D_z = d_{31}\left(Q_{11} + Q_{12}\right)\left[\varepsilon_x + \varepsilon_y - 2\left(\alpha_1 T + d_{31} E_z\right)\right] + \varepsilon_{33} E_z + p_3 T \quad (7.29)$$

where reduced stiffnesses, the coefficient of thermal expansion and the piezoelectric coefficient correspond to the in-plane isotropic sensor material, i.e. $Q_{11} = Q_{22},\ \alpha_1 = \alpha_2,\ d_{31} = d_{32}$.

In open-circuit sensors the electric displacement $D_z = 0$ and the electric field generated in a sensor can be obtained from the last equation (7.29) as a function of strain and temperature. Furthermore, if temperature is quasi-static, it does not explicitly affect linear dynamic deformations, i.e. its influence is limited to the effect on the material properties. Accordingly, the component of the electric field in the polarization direction is

$$E_z = \frac{d_{31}\left(Q_{11} + Q_{12}\right)\left(\varepsilon_x + \varepsilon_y\right)}{2 d_{31}^2 \left(Q_{11} + Q_{12}\right) - \varepsilon_{33}} \quad (7.30)$$

The integration of the electric field according to (7.12) yields the voltage generated in the sensor. This voltage can be presented in a nondimensional form as a ratio of the voltage at the current temperature to that at the reference temperature (T_{ref}):

$$R = \frac{V(T)}{V\left(T_{ref}\right)} = k_1 k_2 \quad (7.31)$$

where

$$k_1 = \frac{f_1(T)}{f_1(T_{ref})}, \qquad k_2 = \frac{f_2(T)}{f_2(T_{ref})},$$
$$f_1 = \frac{d_{31}(Q_{11}+Q_{12})}{2d_{31}^2(Q_{11}+Q_{12}) - \varepsilon_{33}}, \; f_2 = \frac{1}{A_{sen}} \int\limits_{A_{sen}} \int\limits_{h_s} (\varepsilon_x + \varepsilon_y)\, dz dA_{sen} \tag{7.32}$$

where A_{sen} and h_s are the surface area and the thickness of the sensor, respectively.

Obviously, coefficients k_1 and k_2 account for the effects of temperature on the properties of the sensor and the composite substrate, respectively. As indicated above, we assume that the sensor does not noticeably contribute to the stiffness of the plate; otherwise, the coefficient $k_2$would also reflect the effect of temperature on the stiffnesses of both the composite substrate as well as the sensor. The product of coefficients k_1 and $k_2$represents a factor by which the voltage in the sensor is affected by temperature.

It remains to specify the response of the plate to dynamic loading. If this loading is represented by a uniformly distributed harmonic pressure, and temperature is uniform, the equation of flexural vibrations of a plate that can be symmetrically cross-ply laminated, multilayered symmetric angle-ply laminated or isotropic is obtained by adding thermal terms to the last equation (7.9):

$$D_{11}\frac{\partial^4 w}{\partial x^4} + 2(D_{12}+2D_{66})\frac{\partial^4 w}{\partial x^2 \partial y^2} + D_{22}\frac{\partial^4 w}{\partial y^4} + N_x^T \frac{\partial^2 w}{\partial x^2} + N_y^T \frac{\partial^2 w}{\partial y^2} = p - \hat{m}\ddot{w} \tag{7.33}$$

where in a transversely isotropic sensor $N_x^T = N_y^T = N_T$ is found from (7.8).

The solution of the problem of the response of a simply supported plate is sought by representing the pressure and deflections in double Fourier series as it was done in Chap. 4. The substitution of (4.2) and (4.11) into the equation of motion (7.33) yields the amplitudes of harmonics in the series representing deflections. Subsequently, the strains in the sensor are found as

$$\{\varepsilon_x, \varepsilon_y\} = -z_s \left\{ \frac{\partial^2 w}{\partial x^2}, \frac{\partial^2 w}{\partial y^2} \right\} \tag{7.34}$$

wherez_s is a distance from the middle plane of the sensor to the plate middle plane.

The sum of strains in the sensors that is needed to evaluate f_2 according to (7.32) is

$$\varepsilon_x + \varepsilon_y = \sum_m \sum_n z_s \left(\alpha_m^2 + \beta_n^2\right) S_{mn}^{-1} p_{mn} \sin\alpha_m x \sin\beta_n y \sin\omega t$$
$$S_{mn} = \alpha_m^4 D_{11}(T) + 2\alpha_m^2 \beta_n^2 [D_{12}(T) + 2D_{66}(T)] + \beta_n^4 D_{22}(T) - \left(\alpha_m^2 + \beta_n^2\right) N_T - \hat{m}\omega^2 \tag{7.35}$$

where the plate is assumed to undergo periodic vibrations with frequency ω.

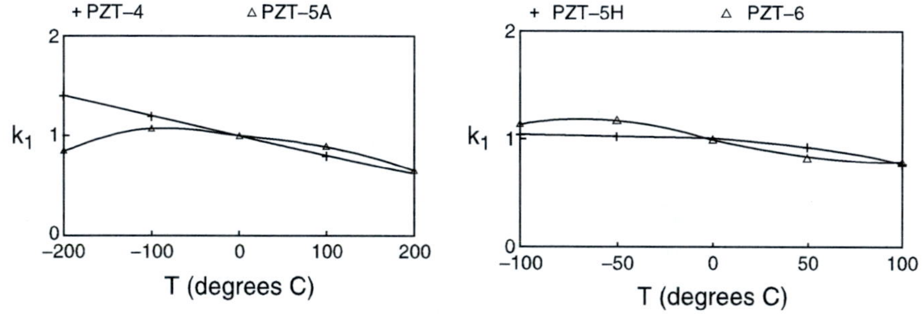

Fig. 7.5 Effect of temperature on the component of the sensor voltage dependent on the properties of piezoelectric sensors (From Birman 1996)

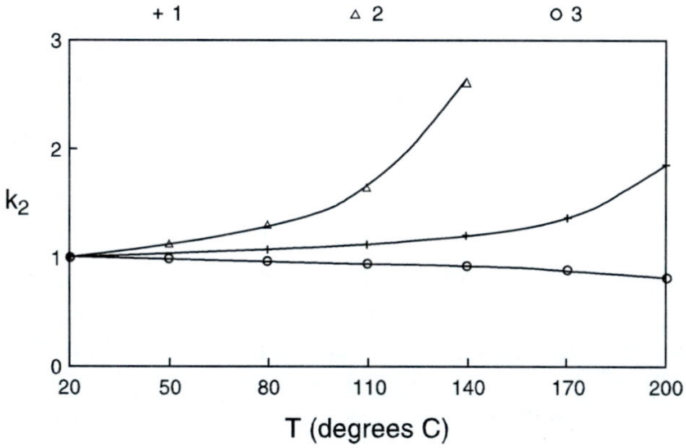

Fig. 7.6 Effect of temperature on the component of the sensor voltage dependent on the strains in large aspect ratio specially-orthotropic Kevlar-epoxy plates. Curves 1, 2 and 3 correspond to the driving frequencies equal to zero (static case), 25 and 50 rad/s, respectively. (From Birman 1996)

The effect of temperature on the coefficient k_1 associated with the effect of thermally-induced changes in piezoelectric sensor properties was evaluated for several classes of piezoceramics produced by Morgan Matroc, i.e. PZT-4, PZT-5A, PZT-5H and PZT-8 (Birman 1996). As follows from Fig. 7.5, the coefficient k_1 decreases due to an elevated static temperature, although the trend may be reversed at low temperatures.

The effect of temperature on the properties of the plate can further complicate measurements from piezoelectric sensors as is reflected in the coefficient k_2. A somewhat extreme case is shown in Fig. 7.6 for a large aspect ratio specially orthotropic Kevlar-epoxy plate where the fibers are oriented along long edges. We call this case "extreme" since such lamination is highly unlikely in practical applications where the stiffness has to be maintained in the direction of both long and short edges (accordingly, a cross-ply or angle-ply plate is more likely to be used in design). A specially orthotropic plate is shown here since it better elucidates the effect of a thermally-induced matrix degradation (the effect of temperature on the fibers is insignificant in the temperature range shown in Fig. 7.6). A difference in the response to an elevated temperature is dependent on the driving frequency as can be observed in Fig. 7.6. This is related to the relationship between the driving and fundamental frequencies. An increase in the deflections and strains in the static and low-frequency cases (curves 1 and 2) is attributed to degradation in the stiffness at a higher temperature. This degradation both increases the plate flexibility as well as results in a smaller fundamental frequency that approached the driving frequency. An increasing temperature could result in a resonance as is evident from curve 2. On the other hand, if the driving frequency exceeds the fundamental frequency of the plate at a room temperature, a higher temperature results in a larger gap between the fundamental and driving frequencies and smaller amplitudes (curve 3). As follows from Figs. 7.5 and 7.6, using piezoelectric sensors in plate structures exposed to temperature variations can result in inaccurate readings if thermal effects both on the sensor as well as on the plate are disregarded.

While the solution in the paper of Birman (1996) applied a decomposition of the sensory voltage in the coefficients reflecting the effect of temperature on the properties of sensors and that associated with the response and property variations of the composite substrate, Lee and Saravanos (1997, 2000) formulated the problem at a more general level. The motivation emerges from other practical situations requiring for example to consider the influence of the converse piezoelectric effect, i.e. the impact of the electric field generated in the sensor on the response of the structure. In other cases, the influence of the bonding layer could become essential, particularly in the vicinity to the edges of a piezoelectric patch (sensor) bonded to the composite substrate due to a local concentration of three-dimensional stresses that could cause debonding of the sensor. In the former paper a layer-wise formulation was employed to account for mechanical, thermal and electric coupling as well as temperature-dependence of the composite and piezoelectric materials. A finite element model was derived based on this generalized formulation and applied to the evaluation of stresses, displacements, and voltages in piezoelectric sensors bonded to a composite plate. An example of voltage obtained in PZT-5A sensors on the surface of a rectangular symmetrically laminated angle-ply AS4/3501-6 carbon-epoxy plate is shown in Fig. 7.7 where "Constant" and "Temperature-dependent" curves correspond to the cases where the properties of the sensors and plate are affected and unaffected by local temperature, respectively. The plate was simply supported along two opposite edges and free along the other couple of edges. The sensors were bonded to the surface subject to a higher temperature. As is shown in Fig. 7.8, the effect of temperature-dependent properties results in a nonlinear relationship between the voltage and thermal gradient compared to the case of a temperature-insensitive plate material where this relationship is linear.

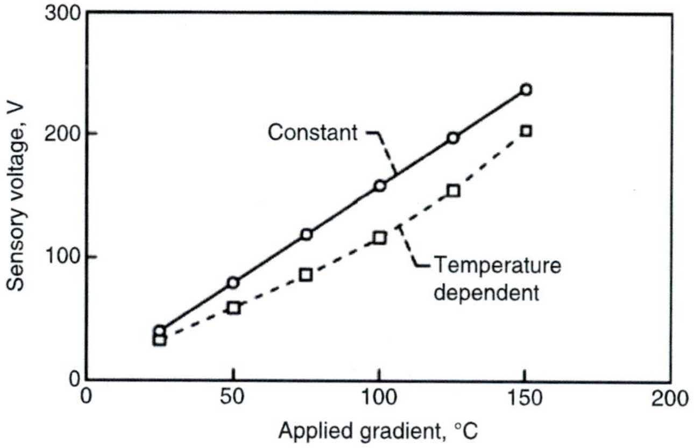

Fig. 7.7 Effect of temperature-dependence of material properties on voltage in piezoelectric sensors bonded to a composite plate as a function of thermal gradient through the plate thickness. (From Lee and Saravanos 1997)

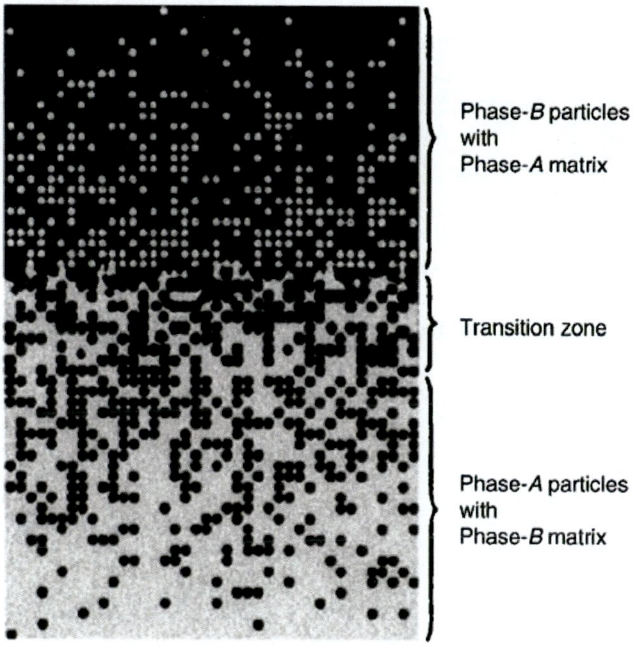

Fig. 7.8 Section of a particulate FGM plate where the volume fraction of two constituent phases varies in the thickness (*vertical*) direction. (From Yin et al. 2004)

7.5 Concept of Functionally Graded Material (FGM) Plates

Functionally graded materials (FGM) are composite materials formed of two or more constituent phases with a continuously variable composition. These materials are composite by nature, but their properties reflect a distribution of the constituent phases that is a monotonic, rather than piece-wise function of coordinates. Accordingly, FGMs possess a number of advantages over laminated composites, including a potential reduction of in-plane and transverse through-the-thickness stresses, absent or severely reduced delamination trend, an improved residual stress distribution, enhanced thermal properties, higher fracture toughness, and reduced stress intensity factors. A number of reviews dealing with various aspects of FGM have been published (e.g., Suresh and Mortensen 1998; Miyamoto et al. 1999; Birman and Byrd 2007).

Functionally graded plates are often understood as particulate-material systems. However, alternative configurations, including functional grading of fibers, using several variable fiber systems, and combinations of fibers and particles embedded within the matrix have also been discussed in literature.

Functionally graded plates are often employed in applications that involve thermal effects. One of typical high-temperature applications is found in ceramic-metal plates where ceramic material is mostly concentrated in the vicinity to the surface exposed to thermal load, while metal dominates at the opposite surface of the plate. A schematic illustration of a functionally graded plate where the concentration of two constituent phases varies through the thickness is presented in Fig. 7.8.

An example of a natural biological multifunctional grading is found in the tendon-to-bone insertion site where the orientation of collagen fibers as well as the content of mineral varies from tendon to bone (Fig. 7.9). Such double grading reduces the stress concentration at the site of junction of two materials (tendon and bone) that possess a huge difference in the stiffness ($E_{bone} \approx 20\ GPa,\quad E_{tendon} \approx 0.4\,GPa$). The analysis of the effect of grading on the properties and stress concentration at the tendon-to-bone insertion site can be found in the papers of Genin et al. (2009) and Liu et al. (2011).

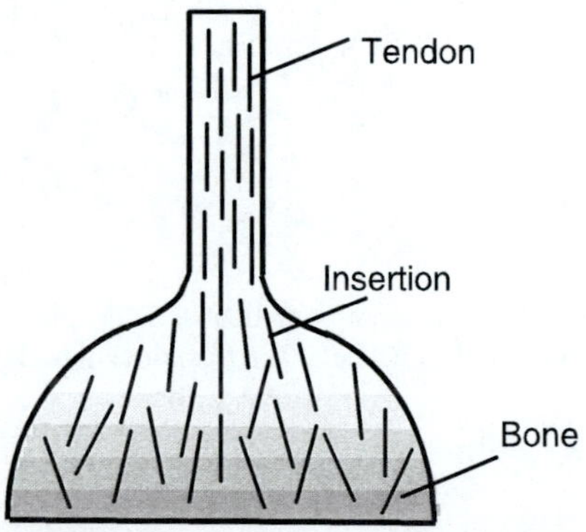

Fig. 7.9 The tendon-to-bone insertion site (schematic illustration). The orthotropic tendon attaches to the bone through the narrow insertion region that is characterized by gradients in mineral volume fraction (represented by *shading*) and collagen fiber orientation (represented by *lines*)

The analysis of functionally graded plates should address a number of issues, each of them deserving close attention:

1. Micromechanics should be employed to evaluate the properties of the material, dependent on the concentration of constituent phases. For example, the Mori-Tanaka micromechanics is suitable for the evaluation of properties of a FGM plate with arbitrary-shape inclusions embedded within the matrix. However, the accuracy of this theory is retained only for the inclusion volume fraction limited to about 35%. Other micromechanical (homogenization) methods include the self-consistent method, the finite element method, the method of cells and its modifications, etc.
2. In thermomechanical problems, the issue of heat conduction and temperature distribution throughout the plate has to be addressed prior to the solution of the mechanical static and/or dynamic response problems. The solution is complicated since thermal properties of the FGM plate that should be determined from a micromechanical model in terms of the properties of the constituent materials depend on local temperature. In particular, high temperature ceramic-metal FGM plates often operate in the range of temperatures where it is impossible to disregard their effect on the properties of either of constituent materials. Accordingly, the effect of temperature on the properties of such FGM plates should be incorporated in the relevant analysis, the composite material properties being evaluated as functions of temperature using a micromechanical theory since the effect of temperature on each constituent material is different. The outcome of the thermal analysis includes a distribution of temperature in the plate and its effect on the local composite properties (such as stiffness, strength and conductivity).

   While the heat conduction problem should be addressed accounting for the effect of temperature on the properties of constitutive materials, these properties are in turn affected by local temperature. An analytical solution can be sought if relationships between the properties and temperature are available in the form of empirical equations (e.g., Touloukian 1967). Alternatively, an iterative solution could be employed, evaluating the heat transfer problem without accounting for thermal effects on the properties in the first iteration. Subsequently, the distribution of temperature obtained in the first iteration can be used to update local properties, solve the heat transfer problem in the second iteration, etc.
3. The problems of static or dynamic response of the plate to external loading include a multitude of possible issues, such as the strength analysis, stability, dynamic response, natural frequencies, fatigue and fracture. The methods of analysis employed in the theory of plates may often be applicable to the analysis of FGM plates concerned with their macromechanical response. Although FGM plates can be pseudo-isotropic at the local micromechanical scale, the variation of the material content through the thickness usually makes them asymmetric with respect to the middle plane. Accordingly, the analysis should rely on the theory of composite plates, including nonzero elements of the matrix of coupling stiffness$[B]$. A noticeable exception to asymmetry of FGM plates is found

in symmetric sandwich FGM structures where identical functional grading is employed between each facing and the core to eliminate or reduce delamination tendencies.

In addition to the micromechanical, thermal and structural analyses, the issue of manufacturability of FGM plates should be closely monitored by engineers. It is sometimes impossible or impractical to guarantee a prescribed variation of the material properties if it is governed by a complicated analytical expression (Birman and Byrd 2007).

In the next two paragraphs we illustrate the formulation of the heat conduction and thermomechanical problems. The micromechanical formulation necessary to evaluate the properties of FGM in terms of the properties of the constituent materials is outside the scope of the book (references to some of micromechanical theories suitable for the analysis of FGM plates and examples of their application can be found in the papers of Genin and Birman (2009) and Pindera et al. (2009)).

7.6 Thermal Problems of FGM Plates

In the solutions of heat transfer problems in FGM plates authors often concentrate on one-dimensional heat conduction, i.e. temperature is assumed to vary in the thickness direction only. However, such one-dimensional approach may become invalid in the vicinity to the edges if supporting structures serve as heat sinks. Also, in case where the properties of the plate vary in-plane as well as through the thickness, thermal analysis has to be three-dimensional. In such cases it is necessary to generalize the Fourier heat conduction equation (6.2) for a linear three-dimensional problem where conductivities vary in all directions and internal heat generation does not take place (Boley and Weiner 1960):

$$\frac{\partial}{\partial x}\left[k_x(x,y,z,T)\frac{\partial T}{\partial x}\right]+\frac{\partial}{\partial y}\left[k_y(x,y,z,T)\frac{\partial T}{\partial y}\right]+\frac{\partial}{\partial z}\left[k_z(x,y,z,T)\frac{\partial T}{\partial z}\right]$$
$$=\rho(x,y,z)\,c(x,y,z,T)\frac{\partial T}{\partial t} \tag{7.36}$$

where the mass density of the material is assumed independent of temperature. Note that all material properties, except for mass density, can be affected by the local temperature $T(x,y,z,t)$.

The solution of (7.36) should be subject to thermal boundary conditions that are similar to those outlined in Chap. 6 (see Sect. 6.1). Possible conditions on the surfaces of the plate include a prescribed temperature or heat flux, perfect insulation or convection boundary conditions.

If the boundaries of the plate are supported by structures that provide a perfect thermal contact, the in-plane thermal boundary conditions enforce the continuity of temperature and heat flux. For example, such conditions along the boundary $x = \bar{x}$ are

$$T(\bar{x}, y, z, t) = T_b(\bar{x}, y, z, t)$$
$$k_x \frac{\partial T(\bar{x}, y, z, t)}{\partial x} = k_{xb} \frac{\partial T_b(\bar{x}, y, z, t)}{\partial x} \tag{7.37}$$

where the subscript "b" refers to the boundary structure.

The solution of the heat transfer problem in the general case can be obtained by numerical methods. If the properties vary in the z-direction only (as is the case in the majority of functionally graded materials), the solution is available by the perturbation method and other numerical methods (e.g., Noda (1999)). An example of the analytical solution for a particular distribution of thermal conductivity through the thickness is demonstrated in Sect. 6.2.

Examples of the solution of the heat transfer problem resulting in a temperature distribution throughout the thickness of the FGM plate are discussed below. Praveen and Reddy (1998) considered thermal stresses in a ceramic-metal FGM plate subject to constant temperatures on the opposite metal-rich and ceramic-rich surfaces. All properties of the plate, including thermal conductivity, were assumed independent of temperature and varying through the thickness according to the rule of mixtures:

$$P(z) = (P_c - P_m)\left(\frac{2z + h}{2h}\right)^n + P_m \tag{7.38}$$

where P is the property, subscripts "c" and "m" refer to the ceramic and metal phases, $-\frac{h}{2} \leq z \leq \frac{h}{2}$, and n is a non-negative power, $n = 0$ corresponding to a ceramic plate, while increasing values of n reflect a higher metal content. Note that a relationship between the power n and the volume fraction distribution of ceramic and metal phases should be specified from micromechanics.

The mathematical formulation of the static heat transfer problem is available from the heat conduction equation (7.36) and prescribed thermal boundary conditions on the surfaces:

$$\frac{d}{dz}\left[k_z(z)\frac{dT}{dz}\right] = 0,$$
$$T\left(z = \frac{h}{2}\right) = T_c, \quad T\left(z = -\frac{h}{2}\right) = T_m \tag{7.39}$$

where $k_z(z)$ is given according to (7.38).

The distribution of temperature throughout the thickness of an aluminum-zirconia plate is shown for various values of power n and the temperatures of the ceramic and metal surfaces equal to $300^\circ C$ and $20^\circ C$, respectively in Fig. 7.10. It can be observed that temperature in fully ceramic or fully metallic plates varies

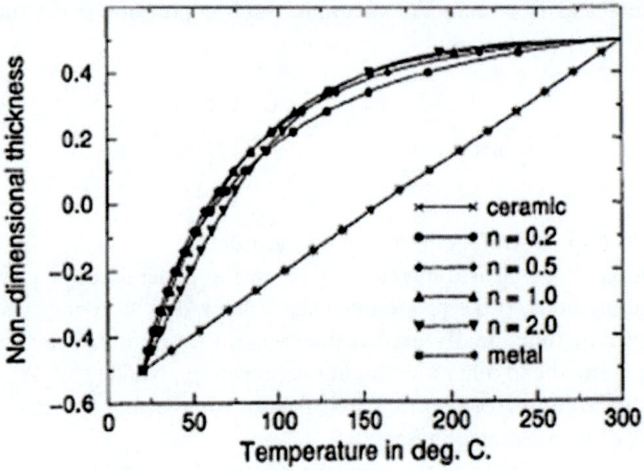

Fig. 7.10 Distribution of temperature throughout the thickness of an aluminum-zirconia FGM plate. The thickness coordinate is normalized with respect to the plate thickness. (From Praveen and Reddy 1998)

linearly throughout the thickness (the corresponding lines coincide). Such linear variation is due to neglecting temperature-dependence of the thermal conductivity. If this simplifying assumption was discarded, the corresponding $T_i(z),\ i = c, m$ relationships would be nonlinear. As reflected in Fig. 7.10, it is possible to significantly modify the thermal profile through functional grading, reducing temperature in the interior of the plate.

One-dimensional transient heat transfer problems in FGM plates were studied by Carslaw and Jaeger (1959), Jin and Paulino (2001) and Byrd and Birman (2010). Consider, for example, the situation where the surfaces of a FGM strip (or plate) were suddenly cooled down from the initial zero reference temperature to prescribed temperatures, i.e. the initial and boundary conditions were

$$T\,(z, t = 0) = 0,\ \ T\,(z = 0,\, t \geq 0) = -T_0,\ \ T\,(z = b,\, t \geq 0) = -T_b \qquad (7.40)$$

where $z = 0$ and $z = b$ are the surface coordinates. Neglecting the effect of temperature on the material properties, the dynamic heat conduction equation is obtained from the one-dimensional version of (7.36) or from (6.2) as

$$\frac{\partial}{\partial z}\left[k_z(z)\frac{\partial T(t)}{\partial z}\right] = \rho(z)c(z)\frac{\partial T(t)}{\partial t} \qquad (7.41)$$

The solution of this problem was obtained by applying a discrete model, i.e. dividing the plate into a number of layers and assuming that the properties within each layer are constant (Jin and Paulino 2001). Using such approach the heat conduction equation within each homogeneous layer is reduced to a standard one-dimensional dynamic heat conduction problem:

$$\frac{\partial^2 T(t)}{\partial z^2} = \frac{1}{\kappa_n}\frac{\partial T(t)}{\partial t}, \; z_n < z < z_{n+1}, \; n = 0, \; 1, \; ..., \; N \tag{7.42}$$

where (N + 1) is the total number of layers and $\kappa_n = \frac{k_{zn}}{\rho_n c_c}$ is the thermal diffusivity of the layer (the thermal conductivity, mass density and specific heat in this formula refer to the n-th layer).

The solution relies on the interface temperatures, so that the initial and boundary conditions for the n-th layer ($z_n \leq z \leq z_{n+1}$) can be written as

$$T\,(z_n \leq z \leq z_{n+1}, t = 0) = 0, \; T\,(z_n, t > 0) = T_n(t), \; T\,(z_{n+1}, t > 0) = T_{n+1}(t) \tag{7.43}$$

where T_n and T_{n+1} are interface temperature.

These conditions superimpose the continuity for temperature at the layer interfaces at an arbitrary time instant. Additionally, the heat flux continuity is enforced at the layer interfaces resulting in N equations used to determine N unknown interfacial temperatures $T_n(t), \; n = 1, \; 2, \; ..., \; N$:

$$\left[k_{z(n-1)}\frac{\partial T_{n-1}(t)}{\partial z}\right]_{z \to z_n} = \left[k_{zn}\frac{\partial T_n(t)}{\partial z}\right]_{z \to z_n} \tag{7.44}$$

where the left and right sides refer to the heat flux in the $(n-1)$-th and n-th layers, respectively.

Numerical solutions of the present and other relevant heat transfer problems in FGM plates are outside the scope of this book. Thus we limit the discussion to a representative mathematical formulation of the problem outlined above.

In conclusion, it is possible to find analytical solutions of thermal problems for specific cases of FGM plates. However, in the general case, particularly if it is necessary to account for the effect of temperature on the properties of the constituent materials, the solution of the heat transfer problem should be numerical. Note that the effect of temperature on the properties of composite materials with the volume fraction of constituent phases varying through the thickness of the plate is probably impossible to predict since such information requires a large number of experiments. Accordingly, it may be necessary to employ the knowledge of temperature-dependent properties of each constituent material and apply a micromechanical theory to evaluate the effect of temperature on the composite material. The problem becomes even more complicated if the properties of a FGM plate vary both through the thickness as well as in its plane. The three-dimensional thermal problem for such structure should be solved numerically.

7.7 Thermomechanical Problems of FGM Plates

Once the problem of temperature distribution throughout the FGM plate is solved, the analysis can be carried out either using equations of the theory of elasticity or by one of the plate theories accounting for a variable property distribution throughout the thickness. In this paragraph we illustrate both methods. In the following examples it is assumed that the problems of micromechanics and heat transfer have already been solved, so that it is possible to concentrate on the response of the plate to external loading.

Example 7.1: Three-Dimensional Thermoelastic Analysis of FGM Plates (Vel and Batra 2003) The approach based on thermoelasticity utilizes a through-the-thickness power representation of the properties, displacements, strains and stresses combined with in-plane trigonometric series for displacements, strains and stresses. The solution that is applicable only to a certain class of boundary conditions is exact and can be used as a benchmark. Although the thermal problem in the paper was solved for a dynamic process, this process was assumed sufficiently slow to neglect inertial terms when studying the mechanical response.

The constitutive equations governing the quasi-static behavior of a quasi-isotropic linear elastic material are obtained by a generalization of the corresponding equations in Chap. 1:

$$\sigma_{ij} = \lambda(T)\varepsilon_{kk}\delta_{ij} + 2G(T)\varepsilon_{ij} - \beta(T)\delta_{ij}T \tag{7.45}$$

where the Lame constants $\lambda(T)$ and $G(T)$ and the stress-temperature modulus $\beta(T)$ are functions of temperature and δ_{ij} is the Kronecker delta. The solution illustrated below was obtained by assumption that material properties are independent of temperature, though they vary with the thickness z-coordinate reflecting the grading of the plate material. The strains ε_{ij} in the geometrically linear problem are given by equations (1.6).

The solution can be obtained for a plate with the edges that are unrestrained in the in-plane direction perpendicular to the boundary but infinitely stiff in the boundary plane. Accordingly,

$$\begin{aligned} x = 0, \quad x = a: \quad & \sigma_x = w = v = 0 \\ y = 0, \quad y = b: \quad & \sigma_y = w = u = 0 \end{aligned} \tag{7.46}$$

Additionally, the surfaces of the plate are free of external stresses, so that

$$z = \pm\tfrac{h}{2}: \quad \sigma_{zz} = \tau_{xz} = \tau_{yz} = 0 \tag{7.47}$$

Material properties of the FGM plate are represented in the Taylor series about the middle surface, so that every property is given by

$$P = \sum_{n=0}^{\infty} \bar{P}_n z^n \tag{7.48}$$

Note that varying the coefficients $\bar{P}_n$ of series (7.48) it is possible to represent arbitrary variations of material properties (e.g., the shear modulus and thermal conductivity can vary through the thickness according to different laws, etc.).

The solution of the heat transfer problem was obtained by Vel and Batra (2003) for the case where the boundaries were maintained at a constant zero temperature that was also the initial temperature in the plate. Thermal boundary conditions on the plate surfaces were chosen in the form that allowed modeling a prescribed temperature or heat flux as well as an exposure to the ambient temperature. Then the solution of the transient heat transfer problem yielded a distribution of temperature in the form

$$T\,(x, y, z, t) = \left(\sum_{r=0}^{\infty} T_r(t) z^r\right) \sin \alpha_m x \sin \beta_k y \tag{7.49}$$

where r, m and k are positive integers and the coefficients $T_r(t)$ are as shown in the paper. The problem being linear, the solution can be obtained by superposition of results obtained for various combinations of m and k.

The solution that satisfies the boundary conditions was obtained by representing displacements by the series

$$u = \left(\sum_{r=0}^{\infty} U_r z^r\right) \cos \alpha_m x \sin \beta_k y$$
$$v = \left(\sum_{r=0}^{\infty} V_r z^r\right) \sin \alpha_m x \cos \beta_k y$$
$$w = \left(\sum_{r=0}^{\infty} W_r z^r\right) \sin \alpha_m x \sin \beta_k y \tag{7.50}$$

Displacements given by (7.50) can be substituted into (1.6), the resulting strains and temperature given by (7.49) being subsequently employed in the constitutive relations (7.45). Material properties being expressed in terms of series (7.48), the constitutive relations substituted into the static version of equations of equilibrium (1.9), $\frac{\partial \sigma_{ij}}{\partial x_j} = 0$, yield coupled recurrence algebraic equations for every non-negative integer power n in series (7.48). These equations presented in the paper of Vel and Batra (2003) were used to express U_r, V_r, W_r $(r \geq 2)$ in terms of six unknown coefficients U_r, V_r, W_r $(r = 0,\ 1)$. These six coefficients were subsequently evaluated from six boundary conditions (7.47) yielding displacements and stresses throughout the plate.

Example 7.2: Thermoelastic Analysis Based on the Theory of Plates Once the thermal problem is solved and the FGM properties and temperature distribution have been determined, the problems of stress analysis, buckling, postbuckling and dynamic response, can be addressed using a suitable theory of composite plates. For example, if the plate is analyzed by the first-order shear deformation theory, one can use equations of motion (5.62), while free vibrations of thin FGM plates can be analyzed by (1.84). The in-plane stress-strain relationships are obtained from a generalized version of (5.7) accounting for the presence of temperature:

$$\begin{Bmatrix} \sigma_x \\ \sigma_y \\ \tau_{xy} \end{Bmatrix} = \begin{bmatrix} \bar{Q}_{11}(T) & \bar{Q}_{12}(T) & \bar{Q}_{16}(T) \\ \bar{Q}_{12}(T) & \bar{Q}_{22}(T) & \bar{Q}_{26}(T) \\ \bar{Q}_{16}(T) & \bar{Q}_{26}(T) & \bar{Q}_{66}(T) \end{bmatrix} \begin{Bmatrix} \varepsilon_x^0 + z\kappa_x - \alpha_x T \\ \varepsilon_y^0 + z\kappa_y - \alpha_y T \\ \gamma_{xy}^0 + z\kappa_{xy} - \alpha_{xy} T \end{Bmatrix} \tag{7.51}$$

where the reference to the k-th layer is eliminated in understanding that a FGM plate consists of a single layer with continuously varying properties.

In numerous applications a FGM plate consists of two phases, such as metal and ceramic, and while the properties of the plate vary through its thickness, the material can be assumed isotropic at each point. In such case bending-twisting coupling is absent, i.e. $Q_{16} = Q_{26} = 0$ and the constitutive relations become

$$\begin{Bmatrix} N_x \\ N_y \\ N_{xy} \\ M_x \\ M_y \\ M_{xy} \end{Bmatrix} = \begin{bmatrix} A_{11} & A_{12} & 0 & B_{11} & B_{12} & 0 \\ A_{12} & A_{22} & 0 & B_{12} & B_{22} & 0 \\ 0 & 0 & A_{66} & 0 & 0 & B_{66} \\ B_{11} & B_{12} & 0 & D_{11} & D_{12} & 0 \\ B_{12} & B_{22} & 0 & D_{12} & D_{22} & 0 \\ 0 & 0 & B_{66} & 0 & 0 & D_{66} \end{bmatrix} \begin{Bmatrix} \varepsilon_x^0 \\ \varepsilon_y^0 \\ \gamma_{xy}^0 \\ \kappa_x \\ \kappa_y \\ \kappa_{xy} \end{Bmatrix} - \begin{Bmatrix} N_x^T \\ N_y^T \\ N_{xy}^T \\ M_x^T \\ M_y^T \\ M_{xy}^T \end{Bmatrix}$$

$$\begin{Bmatrix} Q_y \\ Q_x \end{Bmatrix} = \begin{bmatrix} A_{44} & 0 \\ 0 & A_{55} \end{bmatrix} \begin{Bmatrix} \gamma_{yz} \\ \gamma_{xz} \end{Bmatrix} \tag{7.52}$$

Equations 7.52 can be applied to the analysis of shear deformable (first order theory) or thin FGM plates using the appropriate strain-displacement relationships. Geometrically nonlinear effects can be accommodated as well.

The properties of ceramic-metal plates considered by Praveen and Reddy (1998) were evaluated in terms of properties of metal and ceramic by Eq. 7.38. The finite element analysis was conducted for a square $0.2m \times 0.2m \times 0.01m$ plate subject to a uniform pressure. The nondimensional load parameter was defined as

$$P = \frac{p_0 a^4}{E_m h^4},$$

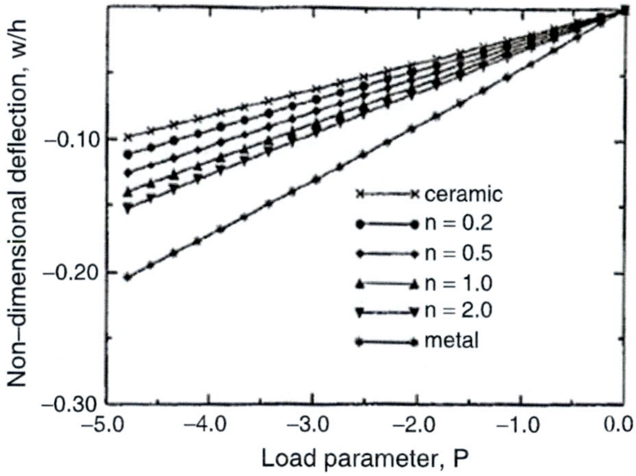

Fig. 7.11 Nondimensional deflections at the center of a square FGM aluminum-zirconia plate under a uniform pressure and varying through the thickness temperature. (From Praveen and Reddy 1998)

In addition to pressure, the plate was subject to temperature of 300°C at the ceramic surface, while the metallic surface was maintained at 20°C.

Variations of the nondimensional deflection of aluminum-zirconia plates are shown in Fig. 7.11 as a function of the nondimensional load (a temperature gradient is also present as explained above). Larger deflections in metallic and predominantly metallic plates are due to a lower stiffness compared to that in predominantly ceramic plates. The results shown in Figs. 7.11 illustrate that it is possible to manipulate the response of FGM plates by an appropriate grading of the material. Among other conclusions from this paper, the linear analysis was shown to predict higher values of deflections, the difference between the linear and nonlinear solutions increasing with a higher overall content of metal.

The paper by Na and Kim (2006) represents another example of a three-dimensional finite element analysis of ceramic-metal FGM plates where material properties varied through the thickness according to a power law. Geometric nonlinearity was incorporated into the analysis through the Green-Lagrange strain-displacement relationships. Temperature distributions were assumed, rather than evaluated from the heat transfer analysis. Thus, the useful results are limited to a uniform temperature distribution that does not depend on the solution of the heat transfer problem. The plate was clamped along all edges and in-plane edge displacements were prevented. As was shown in this paper, the magnitude of bending stresses was significantly affected by grading of the plate confirming the observation regarding a considerable potential for the modification of the response available in FGM plates.

As a practical matter, it should be noted that while the stresses can be reduced by an appropriate material grading, the effect of such grading on the strength should also be considered. This is because the beneficial effect of reduced stresses in the plate may be neutralized by a reduced strength. Thus the comprehensive solution should involve a micromechanical strength analysis. Furthermore, the weight of the plate affected by grading should be taken into consideration since adding a heavier material to the plate may contradict design requirements. Accordingly, a design of FGM plates is a comprehensive process involving coupled micromechanical, heat transfer (in case thermal effects are present) and macromechanical aspects.

7.8 Conclusions and Recommendations

New materials and technologies have changed the face of engineering, including design and development of such structures as plates. Today's engineer has at his or her disposal the spectrum of possible materials, tailoring tools and design and computational capacities that were unimaginable several decades ago. However, the fundamental concepts and theoretical foundations have not been altered by the latest developments.

The introduction of new materials and structural concepts into practical engineering have emphasized the necessity to address the problem of design in its entirety, including all aspects, such as micromechanics, heat propagation, and macromechanics, rather than trying to address separate elements of the problem. This has been illustrated in our discussion on functionally graded materials subject to thermomechanical loading where we observe coupling of several theoretical formulations and problems that are dependent on each other:

- Micromechanics, including stiffness, local strength, and thermal properties;
- Heat transfer;
- Effect of temperature on local properties of individual material phases and on properties of FGM;
- Macromechanical analysis, including stress analysis, buckling, dynamic, fatigue and fracture problems.

Another consequence of the introduction of new materials is related to an interdisciplinary nature of the relevant problem. For example, the analysis of a plate with piezoelectric sensors and actuators involves both mechanical as well as electric aspects. If such plate is subject to thermal effects, the influence of temperature on the properties of sensors and actuators as well as its effect on the material of the plate have to be addressed, combined with the solution of the heat transfer problem.

Multiscale problems are also typical in modern engineering. While we do not discuss the subject of nanotubes in this book, adding single-wall or multi-wall nanotubes to the material substrate can drastically increase the strength and stiffness of the structure. For example, Qian et al. (2000) reported a nearly 25% increase in the strength and 36–42% increase in stiffness of composites reinforced by nanotubes constituting only 1% of the weight of the material. The analysis of the response of nanotube-reinforced structures should be conducted at the nanoscale that is necessary to characterize the response of nanotubes, microscale concerned with the evaluation of the properties of the nanotube-reinforced material, and macroscale where we address the issues of structural response.

In this chapter, we considered plates incorporating piezoelectric components that can serve as both sensors and actuators providing active control to the plate and functionally graded plates. In particular, piezoelectric layers or patches are used to control vibrations of plates. They can also be applicable in energy harvesting applications where forced vibrations produce electric energy that can be either stored or immediately used. The application of piezoelectric elements in plates is restricted by material limitations. In particular, voltage limitations are imposed to avoid a depolarization of the material by an electric field with polarity that is opposite to the poling voltage. The limits of electric field that can be applied to a piezoelectric element are typically of an order of 1,000 V/mm, although recently developed piezoelectrics may be operated at a higher electric field. Temperature limits on piezoelectric sensors have been discussed in Sect. 7.4. As a result of the degradation of piezoelectric properties in the vicinity to the Curie temperature, culminating in an irreversible loss of these properties at this temperature, piezoelectric sensors and actuators should be operated with caution in cases where the structure is exposed to thermal loads. Aging of piezoelectric elements is another consideration that should be considered by a designer. This phenomenon is related to a degradation of piezoelectric properties of the material with time that may reduce the sensitivity of piezoelectric sensors and the efficiency of piezoelectric actuators. Finally, it should be noted that high mechanical stresses can cause a depolarization. While we do not attempt to quantify all these limitations on the use of piezoelectric materials in plate structures, the issues raised above should be addressed by a designer.

The choice of a particular piezoelectric material should account for two different classes of these materials, i.e. piezoceramics (PZT) and piezoelectric polymers (PVDF), that possess drastically different properties. For example, PZT can withstand maximum stresses that are about 25 times higher than their counterparts in PVDF. It is worth emphasizing that piezoelectric materials can operate at a very high frequency reaching 200 MHz, providing an extraordinary quick response that is often valuable in applications.

Functionally graded plates incorporate two or more material phases in a composite structure with varying properties. The composition of the constituent materials and the corresponding property variations are usually in the thickness direction. However, it is also possible to use variable in-plane material distribution, resulting in variable stiffness and strength in the plane of the plate. In cases where circular inclusions or particles of an arbitrary shape that are randomly oriented in three dimensions are embedded within the substrate matrix, the properties of the plate at each point are isotropic. However, using a preferential orientation of non-circular particles, it is possible to manufacture non-isotropic particulate FGM plates.

Functional grading provides a designer with an attractive tailoring and optimization tool, enabling him to monitor and control the response of the structure. While designing a FGM plate, the engineer should be concerned with the issue of its manufacturability. A distribution of material throughout a small thickness of the plate governed by a complicated analytical law is often either impossible or expensive to reproduce in real life. Thus, technological aspects should prominently feature in such designs.

In conclusion of this chapter, it is necessary to emphasize that new material and structural concepts can alter "traditional" engineering disciplines, including design of plates. The implementation of these concepts usually requires extensive analytical, experimental and numerical efforts. However, potential benefits associated with an increased safety and reliability, a broader range of potential applications and allowable loading conditions, lighter weight and lower cost during lifetime of structure often justify the necessary effort.

References

Birman, V. (1993). Active control of composite plates using piezoelectric stiffeners. *International Journal of Mechanical Sciences, 35*, 387–396.

Birman, V. (1994). Analytical models of sandwich plates with piezoelectric strip-stiffeners. *International Journal of Mechanical Sciences, 36*, 567–578.

Birman, V. (1996). Thermal effects on measurements of dynamic processes in structures using piezoelectric sensors. *Smart Materials and Structures, 5*, 379–385.

Birman, V., & Byrd, L. W. (2007). Modeling and analysis of functionally graded materials and structures. *Applied Mechanics Reviews, 60*, 195–216.

Birman, V., Knowles, G. J., & Murray, J. J. (1999). Application of piezoelectric actuators to active control of composite spherical caps. *Smart Materials and Structures, 8*(2), 218–222.

Boley, B. A., & Weiner, J. H. (1960). *Theory of thermal stresses*. New York: John Wiley and Sons.

Byrd, L. W., & Birman, V. (2010). Transient heat conduction in a one dimensional functionally graded material. *Heat Transfer Engineering, 31*, 212–221.

Carslaw, H. S., & Jaeger, J. G. C. (1959). *Conduction of heat in solids*. Oxford: Clarendon.

Chandrashekhara, K., & Agarwal, A. N. (1993). Active vibration control of laminated composite plates using piezoelectric devices: A finite element method. *Journal of Intelligent Material Systems and Structures, 4*, 496–508.

Genin, G. M., & Birman, V. (2009). Micromechanics and structural response of functionally graded particulate-matrix, fiber-reinforced composites. *International Journal of Solids and Structures, 46*, 2136–2150.

Genin, G. M., Kent, A., Birman, V., Wopenka, B., Pasteris, J. D., Marquez, P. J., et al. (2009). Functional grading of mineral and collagen in the attachment of tendon to bone. *Biophysical Journal, 97*, 976–985.

Jin, Z.-H., & Paulino, G. H. (2001). Transient thermal stress analysis of an edge crack in a functionally graded material. *International Journal of Fracture, 107*, 73–98.

Jonnalagadda, K. D., Blanford, G. E., & Tauchert, T. R. (1994). Piezothermoelastic composite plate analysis using first-order shear deformation theory. *Computers and Structures, 51*(1), 79–89.

Lam, K. Y., & Ng, T. Y. (1999). Active control of composite plates with integrated piezoelectric sensors and actuators under various dynamic loading conditions. *Smart Materials and Structures, 8*, 223–237.

Lee, C. K. (1990). Theory of laminated piezoelectric plates for the design of distributed sensors/actuators/Part I: Governing equations and reciprocal relationships. *The Journal of the Acoustical Society of America, 87*, 1144–1158.

Lee, H-J., & Saravanos, D. A. (1997). The effect of temperature dependent material nonlinearities on the response of piezoelectric composite plates (NASA TM-97-206216). *Journal of Intelligent Material Systems and Structures, 9*(7), 503–508.

Lee, H.-J., & Saravanos, D. A. (2000). Mixed multi-field finite element formulation for thermopiezoelectric composite shells. *International Journal of Solids and Structures, 37*, 4949–4967.

Liu, Y., Birman, V., Chen, C., Thomopoulos, S., & Genin, G. M. (2011). Mechanisms of bimaterial attachment at the interface of tendon to bone. *Journal of Engineering Materials and Technology, 133*(1), 011006.

Mindlin, R. D. (1974). Equations of high frequency vibrations of thermopiezoelectric crystal plates. *International Journal of Solids and Structures, 10*, 625–632.

Miyamoto, Y., Kaysser, W. A., Rabin, B. H., Kawasaki, A., & Ford, R. G. (1999). *Functionally graded materials: Design processing and applications*. Dordrecht: Kluwer.

Na, K.-S., & Kim, J.-H. (2006). Nonlinear bending response of functionally graded plates under thermal loads. *Journal of Thermal Stresses, 29*, 245–261.

Noda, N. (1999). Thermal stresses in functionally graded materials. *Journal of Thermal Stresses, 22*, 477–512.

Nye, J. F. (1957). *Physical properties of crystals*. Oxford: Oxford University Press.

Pindera, M.-J., Khatam, H., Drago, A., & Basal, Y. (2009). Micromechanics of spatially uniform heterogeneous media. *Composites. Part B: Engineering, 40*, 349–378.

Praveen, G. N., & Reddy, J. N. (1998). Nonlinear transient thermoelastic analysis of functionally graded ceramic-metal plates. *International Journal of Solids and Structures, 35*, 4457–4476.

Qian, D., Dickey, R., Andrews, R., & Rantell, T. (2000). Load transfer and deformation mechanisms in carbon nanotube-polystyrene composites. *Applied Physics Letters, 76*, 2868–2870.

Reddy, J. N. (2004). *Mechanics of laminated composite plates and shells: Theory and analysis* (2nd ed.). Boca Raton: CRC Press.

Rodgers, J. A., Hagood, N. W. (1998). Design, manufacture and testing of an integral twist-actuated rotor blade. In Y. Murotsu, C.A. Rogers, P. Santini, H. Okubo (Eds.), *Proceedings of the Eight International Conference on Adaptive Structures and Technologies* (pp. 63–72). Lancaster, PA: Technomic.

Saravanos, D. A., & Heyliger, P. R. (1999). Mechanics and computational models for laminated piezoelectric beams plates, and shells. *Applied Mechanics Reviews, 52*, 305–320.

Suresh, S., & Mortensen, A. (1998). *Fundamentals of functionally graded materials*. London: IOM Communications.

Tauchert, T. R. (1992). Piezothermoelastic behavior of a laminated plate. *Journal of Thermal Stresses, 15*, 25–37.

Touloukian, Y. S. (1967). *Thermophysical Properties of High Temperature Solid Materials*. New York: Macmillan.

Tzou, H. S., & Bao, Y. (1995). A theory of anisotropic piezothermoelastic shell laminates with sensor/actuator applications. *Journal of Sound and Vibration, 184*, 453–473.

Tzou, H. S., & Fu, H. Q. (1994). A study of segmentation of distributed piezoelectric sensors and actuators, Part I: Theoretical analysis. Part II: Parametric study and active vibration control. *Journal of Sound and Vibration, 172*, Part I, 247–259, Part II, 261–275.

Vel, S. S., & Batra, R. C. (2003). Three-dimensional analysis of transient thermal stresses in functionally graded plates. *International Journal of Solids and Structures, 40*, 7181–7196.

Yin, H. M., Sun, L. Z., & Paulino, G. H. (2004). Micromechanics-based elastic model for functionally graded materials with particle interactions. *Acta Materialia, 52*, 3535–3543.

Index

V. Birman, *Plate Structures*, Solid Mechanics and Its Applications 178,
DOI 10.1007/978-94-007-1715-2, © Springer Science+Business Media B.V. 2011

T

V